AF380342

STATISTICS FOR AGRICULTURAL SCIENCES

THIRD EDITION

STATISTICS FOR AGRICULTURAL SCIENCES

THIRD EDITION

Dr. G. Nageswara Rao

Ph.D

Professor and University Head (Retd.),
Department of Statistics and Mathematics,
Acharya N.G. Ranga Agricultural University,
Rajendranagar,
Hyderabad-500 030

BSP BS Publications

A unit of **BSP Books Pvt., Ltd.**
4-4-309/316, Giriraj Lane,
Sultan Bazar, Hyderabad - 500 095

Statistics for Agricultural Sciences, *Third Edition by G. Nageswara Rao*

Published by :

BSP **BS Publications**
A unit of **BSP Books Pvt., Ltd.**
4-4-309/316, Giriraj Lane, Sultan Bazar,
Hyderabad – 500 095.
Phone : 040 – 23445688, 23445600
e-mail : info@bspbooks.net
www.bspbooks.net

ISBN: 978-93-88305-30-3

To my wife

Shrimati Gadiraju Gruhalakshmi

Preface to Third Edition

I am happy that the second edition of the book was well received by students and staff for whom it was intended. I felt that there is need to include SPSS data analysis Procedure for statistical methods and experimental designs in view of the latest trend of using statistical packages in data analysis. Accordingly SPSS data analysis procedure is included in this edition taking simple example in each case for most popular statistical methods and experimental designs.

A new chapter on "Econometrics" is added to cater to the needs of M.Sc. and Ph.D. students of Agricultural Economics and Business management and Business Administration.

I have taken care to cover all the courses in Statistics taught to B.Sc., M.Sc. and Ph.D students in Agriculture, Veterinary and Home Science Faculties in Agricultural Universities under I.C.A.R system in this edition. I have thoroughly revised and updated the material presented in this edition of the book. This book is also useful to all the P.G. and Ph.D. students of Biology and Bio-technology students. This book also serves as useful reference book to staff members for guiding M.Sc., Ph.D. students in statistical analysis of research data for their thesis work and also for their personal research projects.

I am thankful to M/S Nikhil Shah and Anil Shah of B.S. Publications, Hyderabad for taking initiative in bringing the third edition of this book.

Hyderabad   **G. Nageswara Rao**

Preface to Second Edition

When I was following the present day syllabi of various universities I found some units such as Multivariate Statistical Methods etc., are added compared to my first edition. I was considering to bring second edition for a long time to update the Statistical Theory to match the requirements of the courses.

In this edition a new Chapter on Multivariate Statistical Methods is added which is useful to Postgraduate and Ph.D. students in Agriculture, Veterinary and Home sciences. Good number of examples are added, besides practical applications, after each chapter of Statistical Methods, which are useful to undergraduate and postgraduate students.

This book also covers courses such as 'Data Analysis for Managers taught to Agricultural Business Management PG students. Quantitative methods and Analysis Techniques taught to B.Plan and MURP students in JNTU and also courses on statistics taught to Biotechnology PG students in Agricultural and other traditional universities.

This book is also intended to serve as an effective reference book to Research Workers in Agriculture, Veterinary and Home Science faculties.

I am thankful to M/S Nikhil Shah and Anil Shah of BS Publications, Hyderabad for taking initiative in bringing the second edition of this book.

G. Nageswara Rao

Acknowledgements

With reverence I owe my gratitude to my parents Late Sri G.V. Krishna Rao and Late Srimati G. Venkata Ramanamma in shaping my career with their affection, love and encouragement.

I am grateful to the Literary Executor of the Sri Ronald A. Fisher, F.R.S., and Dr. Frank Yates, F.R.S. and to Longman Group Ltd., London, for giving me permission to reprint Tables I, II, III, V, VI, VII from their book Statistical Tables for Biological, Agricultural and Medical Research (6th Edition 1974) and also to authors Drs Swed Frieda, S. and Eisenhart, C., Finney, D. G., Mann, H.B. and Whitney, D. R. and Smirnov, N. and publishers of Annals of Mathematical Statistics and Biometrika for kindly permitting me to reprint tables from their papers.

I express my grateful thanks to all the authors and publishers whose books I frequently consulted and referred in this book and especially to Cochran (1953) *Sampling Techniques;* Cochran and Cox (1957) *Experimental Designs;* Croxton and Cowden (1966) *Applied General Statistics;* Goulden (1939) *Methods of Statistical Analysis;* Panse and Sukhatme (1967) Statistical Methods for Agricultural Workers; Scheffe (1967) *The Analysis Of Variance;* Siegel (1956) *Non-Parametric Statistics for the behavioural Sciences;* Snedecor and Cochran (1968) *Statistical Methods;* Sukhatme (1953) *Sampling Theory of Surveys with Applications.* D. F. Morrison (1967) *Multivariate statistical methods.* J. Johnston (1971) *Econometric methods.* Umarji R.R.(1962) *Probability and mathematical statistics.*

Hyderabad   **G. Nageswara Rao**

Contents

PART I
STATISTICAL METHODS

Contents   (xv)

Part II
EXPERIMENTAL DESIGNS

PART III
SAMPLE SURVEYS, ECONOMIC AND NON-PARAMETRIC STATISTICS

18. ECONOMIC STATISTICS

19. NON-PARAMETRIC STATISTICS

PART IV
MULTIVARIATE STATISTICAL METHODS

20. MULTIVARIATE STATISTICS METHODS

PART V
ECONOMETRICS

21. TWO-VARIABLE LINEAR MODEL

PART I
STATISTICAL METHODS

CHAPTER 1

INTRODUCTION

In recent days we hear talking about 'Statistics' from a common person to highly qualified person. It only shows how 'Statistics' has been intimately connected with wide range of activities in daily life.

Statistics can be used either as plural or singular. When it is used as plural, it is a systematic presentation of facts and figures. It is in this context that majority of people use the word 'Statistics'. They only meant mere facts and figures. These figures may be with regard to production of foodgrains in different years, area under cereal crops in different years, per capita income in a particular state at different times, etc., and these are generally published in Trade Journals, Economics and Statistics Bulletins. Newspapers, etc. When statistics is used as singular, it is a science which deals with collection, classification, tabulation, analysis and interpretation of data.

Statistics as a science is of recent origin. The word 'Statistics' has been derived from a Latin word which means 'State' which in turn means 'politically organised people' i.e., government. Since governments used to collect the relevant data on births and deaths, defence personnel, financial status of the peoples, import and export, etc. Statistics was identified with Government. Recently, it pervades all branches of sciences, social sciences and even in Humanities like English literature. For example, in English literature the style of a particular poet or an author can be assessed with the help of statistical tools.

In the opinion of Fisher 'Statistics' has got three important functions to play (*i*) Study of statistical populations (*ii*) study of the variation within the statistical populations (*iii*) study of the methods of reduction of data.

P.C. Mahalanobis compares 'Statistician' with a 'Doctor' where Doctor prescribes medicine according to the disease of the patient whereas statistician suggests statistical technique according to the data in hand for proper analysis and interpretation.

Bowley defined statistics as 'the science of measurements of the social organism regarded as a whole in all its manifestations.' Another definition says that it is 'quantitative data affected to a marked extent by a multiplicity of causes.' Y.t another definition says that it is a 'Science of counting' or 'Science of averages' and so on. But all these definitions are incomplete and are complementary to each other.

There are some of the limitations of 'Statistics' also when the data are not properly handled. People start disbelieving in statistics when the (1) data are not reliable (2) computing spurious relationships between variables (3) generalizing from a small sample to a population without taking care of error involved.

If one is ensured that data are reliable and is properly handled by a 'skilled statistician', the mistrust of statistics will disappear and in place of it precise and exact revelation of data will come up for reasonable conclusions.

CHAPTER 2

COLLECTION, CLASSIFICATION AND TABULATION OF DATA

2.1 COLLECTION OF DATA

The data are of two kinds: *(i)* Primary data *(ii)* Secondary data.

Primary data are based on primary source of information and the secondary data are based on secondary source of information.

2.1.1 Primary Data

Primary data are collected by the following methods: *(i)* By the investigator himself. *(ii)* By conducting a large scale survey with the help of field investigators. *(iii)* By sending questionnaires by post.

(i) The first method is limited in scope since the investigator himself cannot afford to bear the expenses of a large scale survey and also the time involved therein. Therefore, this method is of much use only in small pilot surveys like case studies. This method is being adopted by individual Investigators who submit dissertation for Master's and Doctoral degrees in rural sociology, Ag. Extension, Ag. Economics, Home Management, etc.

(ii) In this method the schedules which elicit comprehensive information will be framed by the Chief Investigator with the help of other experts based on objectives of the survey. The field investigators would be trained with the methodology and survey, mode of filling the schedules and the skill of conducting interviews with the respondents, etc. The field investigators will furnish the schedules by personal interview method and submit the schedules to the Chief Investigator for further statistical analysis. This method of collecting data requires more money and time since wide range of information covering large area is to be collected. But the findings based on the large scale survey will be more comprehensive and helpful for policy making decisions. The Decennial Census in India, National sample survey rounds conducted by Govt. of India, Cost of Cultivation schemes, PL 480 schemes, etc., are some of the examples of this method.

(iii) In the third method, the questionnaire containing different types of questions on a particular topic or topics systematically arranged in order which elicit answers of the type yes or no or multiple choice will be sent by post and will be obtained by post. This method is easy for collecting the data with minimum expenditure but the respondents must be educated enough so as to fill the questionnaires properly and send them back realizing the importance of a survey. The Council of Scientific and Industrial Research (CSIR) conducted a survey recently by adopting this method for knowing the status of scientific personnel in India.

2.1.2

The secondary data can be collected from secondary source of information like newspapers, journals and from third person where first hand knowledge is not available. Journals like Trade Statistics, Statistical Abstracts published by State Bureau of Economics and Statistics, Agricultural situation in India, import and export statistics and Daily Economic times are some of the main sources of information providing secondary data.

2.2   CLASSIFICATION OF DATA

The data can be classified into two ways: *(i)* classification according to attributes (Descriptive classification) *(ii)* classification according to measurements (Numerical classification).

2.2.1 Descriptive Classification

The classification of individuals (or subjects) according to qualitative characteristic (or characteristics) is known as descriptive classification.

(a) *Classification by Dichotomy:* The classification of individuals (or objects) according to one attribute is known as simple classification. Classification of fields according to irrigated and un-irrigated, population into employed and unemployed, students as hostellers and not-hostellers, etc., are some of the examples of simple classification.

(b) *Manifold Classification:* Classification of individuals (or objects) according to more than one attribute is known as manifold classification. For example, flowers can be classified according to colour and shape; students can be classified according to class, residence and sex, etc.

2.2.2 Numerical Classification

Classification of individuals (or objects) according to quantitative characteristics such as height, weight, income, yield, age, etc., is called as numerical classification.

EXAMPLE: 227 students are classified according to weight as follows:

TABLE 2.1

Weight (lbs)	90-100	100-110	110-120	120-130	130-140	140-150
No. of students	20	35	50	70	42	10

2.3 TABULATION OF DATA

Tabulation facilitates the presentation of large information into concise way under different titles and sub-titles so that the data in the table can further be subjected to statistical analysis. The following are the different types of tabulation:

2.3.1 Simple Tabulation

Tabulation of data according to one characteristic (or variable) is called as simple tabulation.

Tabulation of different high yielding varieties of wheat in a particular state, area under different types of soils are some of the examples.

2.3.2 Double Tabulation

Tabulation of data according to two attributes (or variables) is called double tabulation. For example, tabulation can be done according to crops under irrigated and unirrigated conditions.

2.3.3 Triple Tabulation

Tabulation of data according to three characteristics (or variables) is called triple tabulation.

For example, population tabulated according to sex, literacy and employment.

2.3.4 Manifold Tabulation

Tabulation of data according to more than three characteristics is called manifold tabulation.

EXAMPLE: Tabulated data of students in a college according to native place, class, residence and sex is given in Table 2.2.

TABLE 2.2 STUDENTS

Class	Rural				Urban			
	Male		Female		Male		Female	
	Hostellers	Day Scholars	Hostellers	Day Scholars	Hostellers	Day Scholars	Hostellers	Day Scholars
Intermediate								
Graduate								
Post graduate								

2.3.5

The following are some of the precautions to be taken in tabulation of data.

(a) The title of the table should be short and precise as far as possible and should convey the general contents of the table.

(b) The sub-titles also should be given so that whenever a part of information is required it can be readily obtained from the marginal totals of the table.

(c) The various items in a table should follow in a logical sequence. For example, the names of the states can be put in an alphabetical order, the crops can be written according to importance on the basis of consumption pattern, the age of students in an ascending order, etc.

(d) Footnotes should be given at the end of a table whenever a word or figure has to be explained more elaborately.

(e) Space should be left after every five items in each column of the table. This will not only help in understanding of the items for comparison but also contributes for the neatness of the table.

EXERCISES

1. Draw up two independent blank tables, giving rows, columns and totals in each case, summarising the details about the members of a number of families, distinguishing males from females, earners from dependants and adults from children.

2. At an examination of 600 candidates, boys outnumber girls by 16 per cent. Also those passing the examination exceed the number of those failing by 310. The number of successful boys choosing science subjects was 300 while among the girls offering arts subjects there were 25 failures. Altogether only 135 offered arts and 33 among them failed. Boys failing the examination numbered 18. Obtain all the class frequencies.

3. In an Agricultural University 1200 teachers are to be classified into 600 Agricultural, 340 Veterinary, 200 Home Science and 60 Agricultural Engineering Faculties. In each Faculty there are three cadres such as Professors, Associate Professors and Assistant Professors and in each Cadre there are three types of activity as teaching, Research and Extension. Draw the appropriate table by filling up the data.

4. Classify the population into Male and Female; Rural and Urban, Employed and unemployed, Private and Government and draw the table by filling up with hypothetical or original data.

CHAPTER 3

FREQUENCY DISTRIBUTION

3.1 FREQUENCY DISTRIBUTION

Frequency may be defined as the number of individuals (or objects) having the same measurement or enumeration count or lies in the same measurement group. Frequency distribution is the distribution of frequencies over different measurements (or measurement groups). The forming of frequency distribution is illustrated here.

EXAMPLE: Below are the heights (in inches) of 75 plants in a field of a paddy crop.

17,	8,	23,	24,	26,	13,	31,	16,	14,	35,	6,	11,
12,	11,	15,	21,	10,	4,	3,	19,	35,	36,	19,	40,
28,	17,	12,	2,	27,	31,	11,	21,	16,	34,	39,	1,
7,	12,	13,	10,	6,	21,	24,	22,	26,	28,	17,	6
5,	15,	11,	16,	28,	4,	3,	19,	27,	35,	37,	14,
2,	9,	8,	16,	13,	22,	8,	26,	13,	12,	16,	14,
27,	31,	6.									

The difference between highest and lowest heights is $40 - 1 = 39$.

Supposing that 10 groups are to be formed, the class interval for each class would be $39/10 = 3.9$. The groups (or classes) will be formed with a class interval of 4 starting from 1 continuing upto 40. The number of plants will be accounted in each class with the help of vertical line called 'tally mark' After every fourth tally mark the fifth mark is indicated by crossing the earlier four marks. This procedure is shown in the following Table 3.1.

TABLE 3.1

Class	Tally marks	Frequency
1-4	卌 ‖	7
5-8	卌 ‖‖	9
9-12	卌 卌 ‖	11
13-16	卌 卌 ‖‖	14
17-20	卌 ‖	6
21-24	卌 ‖‖	8
25-28	卌 ‖ ‖	9
29-32	‖‖	3
33-36	卌	5
37-40	‖‖	3
		75

3.1.1 Inclusive Method of Grouping

The different groups formed in Table 3.1 belong to inclusive method of grouping since both upper and lower limits are included in each class. For example, in the first group, plants having heights 1" and 4" are included in that group itself. The width of each class is called class interval. The mid-value of the class interval is called class mark.

3.1.2 Exclusive Method of Grouping

In this method the upper limit of each group is excluded in that group and included in the next higher group. The inclusive method of grouping in Table 3.1 can be converted to exclusive method of grouping by modifying the classes 1-4, 5-8, 9-12, 13-16..., to 0.5-4.5, 4.5-8.5, 8.5-12.5, 12.5-16.5. However, the class interval in each group in exclusive method increased is 4. Here the upper limit, 4.5 is excluded in the first group and included in the next higher group 4.5-8.5. In other words, plants having. heights between 0.5" to 4.4" are included in the group 0.5-4.5and having heights from 4.5" to 8.4" are included in the next group 4.5 to 8.5 and so on.

3.1.3 Discrete Variable

A variable which can take only fixed number of values is known as discrete variable. In other words, there will be a definite gap between any two values. The number of children per family, the number of petals per flower, the number of tillers per plant, etc., are discrete variables. This variable is also called as 'discontinuous variable'.

3.1.4 Discrete Distribution

The distribution of frequencies of discrete variable is called discrete distribution. The frequency distribution of plants according to number of tillers is given in Table 3.2.

TABLE 3.2

No. of tillers	0	1	2	3	4	5	6	7
No. of plants	10	25	42	65	72	18	16	3

3.1.5 Continuous Variable

A variable which can assume any value between two fixed limits is known as continuous variable. The height of plant, the weight of an animal, the income of an individual, the yield per hectare of paddy, crop etc., are continuous variables.

3.1.6 Continuous Distribution

The distribution of frequencies according to continuous variable is called continuous distribution. For example, the distribution of students according to weights, is given in Table 3.3.

TABLE 3.3

Weight (lbs)	90-100	100-110	110-120	120-130	130-140	140-150
No. of students	6	15	42	18	12	5

3.2 DIAGRAMMATIC REPRESENTATION

The representation of data with the help of a diagram is called diagrammatic representation.

(i) **Bar Diagram:** In this diagram, the height of each bar is directly proportional to the magnitude of the variable. The width of each bar and the space between bars should be same.

EXAMPLE: The yearwise data on area under irrigation in a particular state is represented by bar diagram in Fig. 3.1.

TABLE 3.4

Year	Area under irrigation (million hectares)
1970	15
1971	17
1972	18
1973	18
1974	20
1975	22

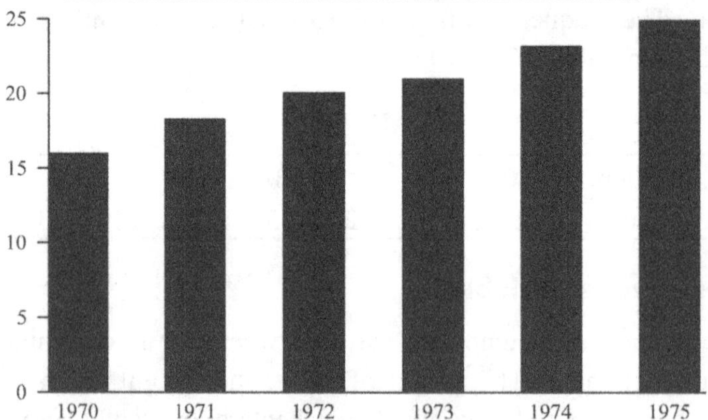

Fig. 3.1 Bar diagram.

(ii) ***Component Bar Diagram:*** In this case, the heights of the component parts of the bar are directly proportional to the magnitude of the constituent parts of the variable. Here also the width of the bars and the space between the bars should be same. This diagram would not be much of advantage if the component parts are more than three.

EXAMPLE: The area under irrigation in Table 3.4 is further sub-divided according to source of irrigation and is presented in Table 3.5.

TABLE 3.5

	Area under irrigation (in million tons)			
Year	*Canal (C)*	*Tank (T)*	*Well (W)*	*Total*
1970	7	5	3	15
1971	7	6	4	17
1972	8	6	4	18
1973	8	6	4	18
1974	8	6	6	20
1975	9	6	7	22

The component bar diagram representing the data in Table 3.5 are given in Fig. 3.2.

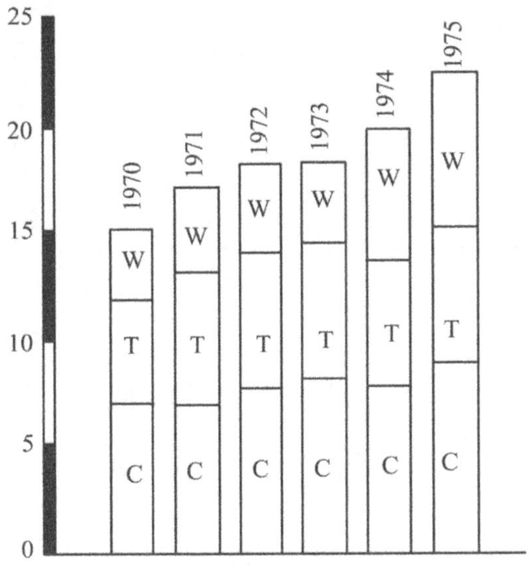

Fig. 3.2 Component bar diagram.

(iii) ***Multiple Bar Diagram:*** In this diagram, the height of each bar in a group of bars is directly proportional to the magnitude of individual item in a group of items. For example, yearwise cereal production, sex-wise literacy in different years, election yearwise, number of seats secured by different political parties in a Parliament (or State assembly) can be represented by Multiple bar diagram

EXAMPLE: The following is the data on wages for different categories of agriculture labour in different years.

TABLE 3.6 LABOUR WAGES

Year	Male (M)	Female (F)	Child (C)
1950	0.75	0.50	0.30
1960	1.50	1.00	0.75
1970	2.50	2.00	1.50
1975	4.00	3.00	2.50

The multiple bar diagram representing the data in Table 3.7 is given in Fig. 3.3.

(iv) *Pie Diagram:* This is also known as Pie-chart. It is useful when the number of component parts of the variable is more than three. Here the areas of different sectors of a circle is directly proportional to the magnitudes of the different component parts of the variable.

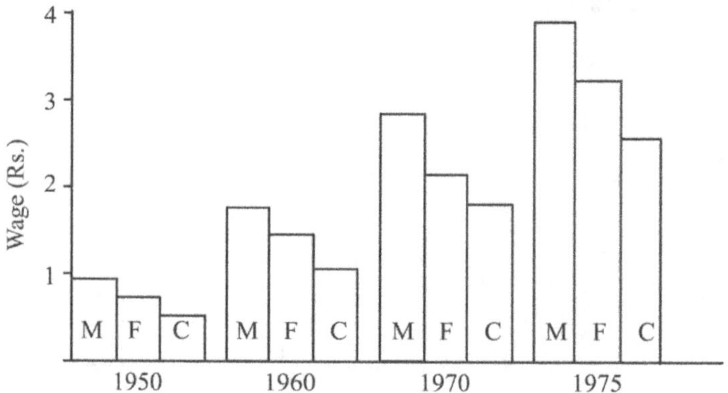

Fig. 3.3 Multiple bar diagram.

Let m_1 be the magnitude of the first component out of m, the total magnitude of the variable.

Then,
$$\theta_1 = 2\pi.\frac{m_1}{m} = 360.\frac{m_1}{m}$$

Where θ_1 is the angle of a first sector. Similarly θ_2, θ_3 ......can be obtained by multiplying 2π with $\frac{m_2}{m}, \frac{m_3}{m}$, etc.

After obtaining $\theta_1, \theta_2, \ldots\ldots\ldots$ the different sectors can be drawn on a circle each representing the individual component. The radius of the circle is proportional to the total magnitude of the variable.

EXAMPLE: Represent the expenditure of a salaried employee on different items by Pie-diagram. The details are given in Table 3.7.

The Pie-diagram representing the data in Table 3.7 is given in Fig. 3.4.

TABLE 3.7

Items	Expenditure (Rs.)	Sector angles (θ_i)
Food	120	$360 \times \dfrac{120}{350} = 123.43$
House rent	70	$360 \times \dfrac{70}{350} = 72.00$
Clothing	50	$360 \times \dfrac{50}{350} = 51.43$
Education for children	35	$360 \times \dfrac{35}{350} = 36.00$
Transport	25	$360 \times \dfrac{25}{350} = 25.71$
Miscellaneous	50	$360 \times \dfrac{50}{350} = 51.43$
	360	

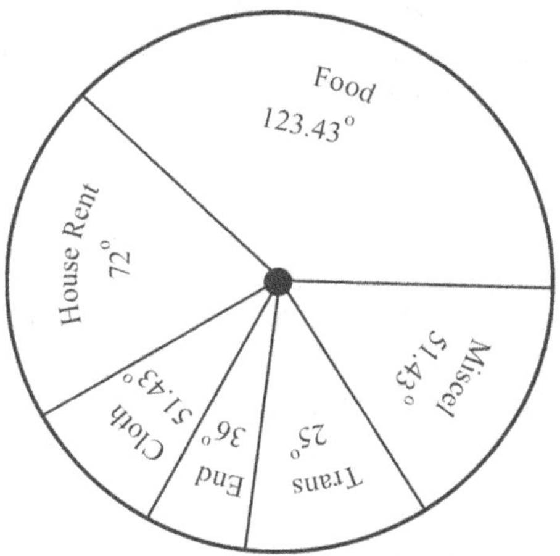

Fig. 3.4 Pie diagram.

It may be noted that the expenditure on food and house rent is accounted for a major share of the employee's salary.

Also the expenditure targets on different items in five year plans can be purposefully represented by pie diagram. If more than one employee is involved in the above example, as many circles may be drawn representing as many employees with the radius of each circle is proportional to the square root of the total salary .of the corresponding employee.

(v)   *Pictograms:* These are also called as pictorial charts. In this each variable is represented by the corresponding picture and the volume of a picture is directly proportional to the magnitude of the variable. For example, wheat production can be represented by the size of the wheat bag (or wheat ear or the number of wheat bags of the same size) according to particular scale, the size of the army by the size of the soldier (or soldiers of same size), the strength of navy by the size of battleship (or the battle ships of same size), number of tractors by the size of tractor (or the tractors of same size) according to particular scale, etc.

Advantages: A diagram is always more appealing to eye than mere numerical data. It is easy for making comparisons and contrasts when more than one diagram is involved. It is easy to understand even for a layman.

Disadvantages: The main disadvantage of this representation is that it only gives rough idea of the variable but not the exact value. Also whenever the number of items are more it is difficult to depict on the diagrams since they require more space, time and un weildy for comparison.

3.3   GRAPHIC REPRESENTATION

Just as in the case of diagrammatic representation, here different methods of graphic representation are presented.

3.3.1 Histogram

It consists of rectangles erected with bases equal to class intervals of frequency distribution and heights of rectangles are proportional to the frequencies of the respective classes in such a way that the areas of rectangles are directly proportional to the corresponding frequencies.

EXAMPLE: Represent the following frequency distribution of farms according to area in a particular village by a histogram given in Table 3.8

From Fig. 3.5 one can infer that the maximum number of farms are lying in the group (2-4) and the minimum number in the group (10-12). The total area under the histogram is equal to the total frequency.

TABLE 3.8

Area (hectares)	0-2	2-4	4-6	6-8	8-10	10-12
No. of farms	40	48	25	18	12	7

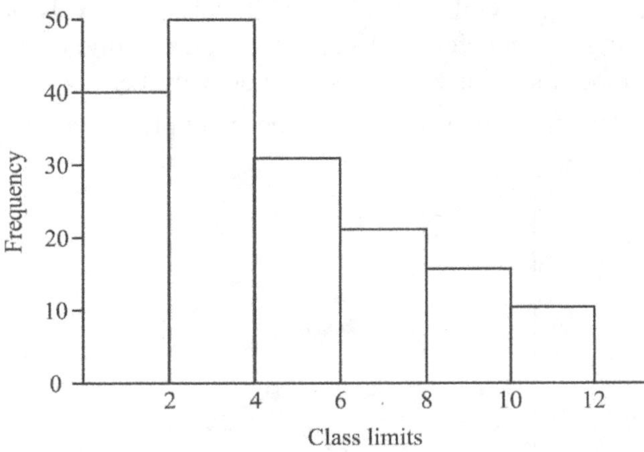

Fig. 3.5 Histogram.

3.3.2 Frequency Polygon

If the points are plotted with midvalues of the class intervals on the X-axis and the corresponding frequencies on the Y-axis, the figure obtained by joining these points with the help of a scale is known as frequency polygon.

EXAMPLE: The frequency polygon for the data given in Table 3.8 is as follows.

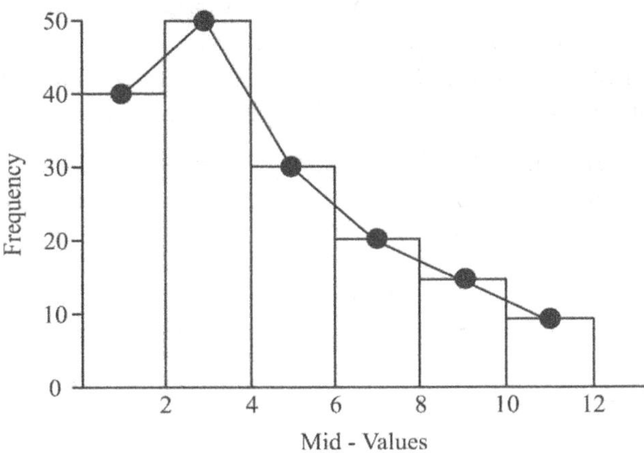

Fig. 3.6 Frequency polygon.

The frequency polygon in Fig. 3.6 is drawn with the assumption that the frequencies are concentrated at the midvalues of the corresponding classes. It may be noted that the area under histogram is equal to the area under frequency polygon.

3.3.3 Frequency Curve

If the points are plotted with midvalues of the class intervals on X-axis and the corresponding frequencies on Y-axis, the figure formed by joining these points with a smooth hand is known as frequency curve.

EXAMPLE: The frequency curve for the example given in Table 3.8 is given in Fig. 3.7.

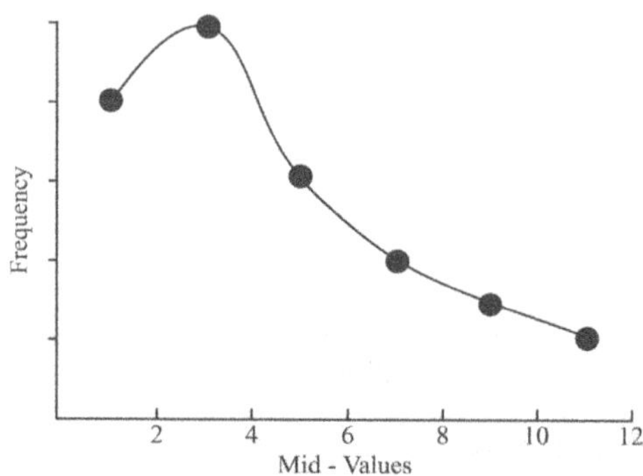

Fig. 3.7 Frequency curve.

3.3.4 Cumulative Frequency Curve (ogive)

If the points are plotted with upper limits of classes on X-axis and the, corresponding cumulative frequencies (less than) on Y-axis, the figure formed by joining these points with a smooth hand is known as cumulative frequency curve (less than). If the lower limits of classes are taken on X-axis and the corresponding cumulative frequencies (greater than) on Y-axis, the curve so obtained is called cumulative frequency curve (greater than).

EXAMPLE: Represent the distribution of rainfall on different days from July to September months in a particular locality and in a particular year by cumulative frequency curves.

TABLE 3.9

Rainfall (in cm)	No. of days	Cum.fre. (less than)	Cum.fre. (greater than)
0-3	6	6	92
3-6	9	15	86
6-9	10	25	77
9-12	25	50	67
12-15	19	69	42
15-18	15	84	23
18-21	8	92	8

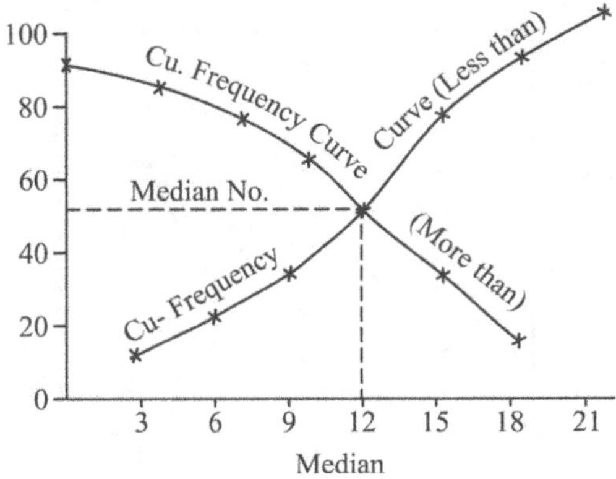

Fig. 3.8

The X-co-ordinate of the point of intersection of two cumulative frequency curves is the median value. The Y-co-ordinate will correspond to median no. i.e., $\dfrac{N+1}{2}$ where N is the total frequency. From the Fig. 3.8 first quartile and third quartile can be obtained from X-axis for the corresponding values (N+1)/4 and $\dfrac{(3N+1)}{4}$ respectively on Y-axis. The reader is advised to refer Sections 4.2 and 5.2 respectively for definitions of median and quartiles.

3.3.5 Lorenz Curve

It is the curve drawn between two variates which are expressed in percentage cumulative frequencies. This curve is useful to depict the income distribution' of individuals where cumulative percentage of individuals are taken on the X-axis and the corresponding cumulative percentage of incomes are taken on the Y-axis. This is commonly used in graphic representation of the inequality aspect of the income distribution. This is due to Italian statisticians. Gini and Lorenz. This curve can also be used for the distribution of any non-negative variate, with a continuous type of distribution as for example, for the distribution of factories by capital size, (or number of employees), etc. The equality of the income distribution is depicted as a straight line drawn with 45°connecting the two diagonal points, and which is known as 'egalitarian line'. If the income distribution is not even then the egalitarian line will take a curve shape. This curve is called 'Lorenz curve', If Lorenz curve is closer towards 'egalitarian line' there is less of inequality of income distribution. If the Lorenz curve is away from the 'egalitarian line' there is more of inequality of income distribution.

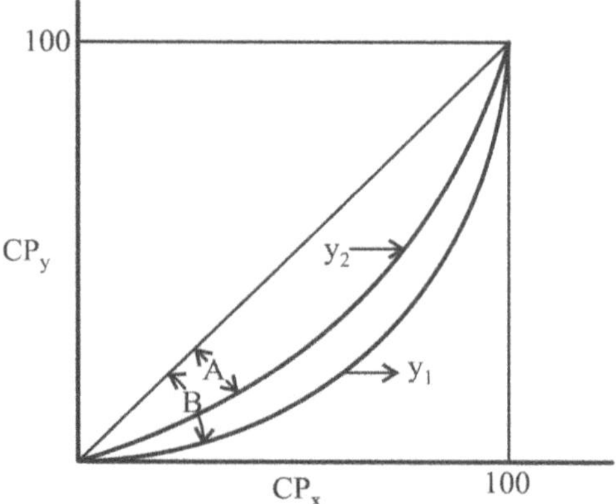

Fig. 3.9 Lorenz curve

From Fig. 3.9, it can be inferred that the distribution of income in year y_2 has tended towards equality in comparison to the year y_1.

Since A< B where A is the area between y_2 Lorenz curve and egalitarian line and B is the area between y_1 and egalitarian line as shown in. Fig. 3.9. The procedure for finding out the area A or B is given in the following sub-section of 'Fitting of Lorenz Curve'. If any curve coincides with the 'egalitarian line then the area would become zero and the Gini's concentration' ratio would be zero.

3.3.6   Fitting of Lorenz Curve

"The approximate procedure of fitting Lorenz curve as will as the method of finding out the area between Lorenz curve and egalitarian line is given here.

TABLE 3.10

Income class	No. of persons	Mid value of income value	Prop. of persons (P_x)	Prop. of income (P_y)	Cum. Prop. of persons (CP_x)	Cum. Prop. of income (CP_y)
$Y_0 - Y_1$	f_1	y_1^1	P_1	q_1	$p_1 = p_1^1$	$q_1 = q_1^1$
$Y_1 - Y_2$	f_2	y_2^1	p_2	q_2	$p_1 + p_2 = p_2^1$	$q_1 + q_2 = q_2^1$
$Y_2 - Y_3$	f_3	y_3^1	p_3	q_3	$p_1 + p_2 + p_3 = p_3^1$	
...	:	:	:	:	:	$q_1 + q_2 + q_3 = q_3^1$:
$Y_{k-1} - Y_k$	f_k	y_k^1	p_k	q_k	p_k^1	q_k^1
	Σf_i	Σyo_i	$\Sigma p_i = 1$	$\Sigma q_i = 1$		

Let Δ be the area of the trepezium between Lorenz curve and the X-axis in Fig. 3.10.

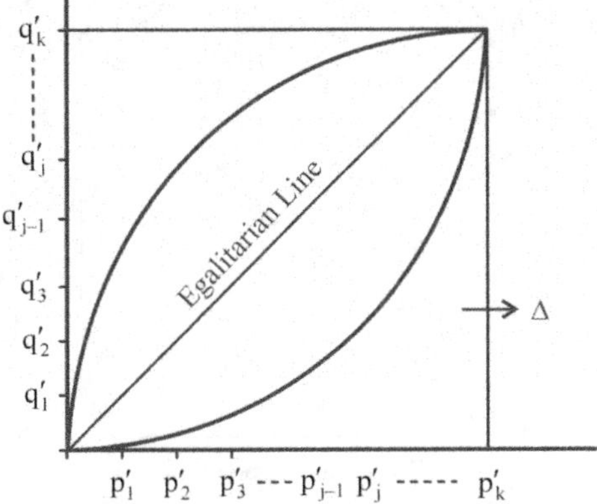

Fig. 3.10 Lorenz curve.

The area between the 'Lorenz curve' "and 'egalitarian line' can be obtained by subtracting Δ from 0.5.

$$\text{Area of trepezium, } \Delta = \sum_{i=1}^{k} \frac{\left(q_j^1 + q_{j-1}^1\right)}{2}\left(p_j^1 - p_{j-1}^1\right)$$

Area between 'Lorenz curve' and 'egalitarian line' is $L = \dfrac{1}{2} - \Delta = \dfrac{1-2\Delta}{2}$

It may be noted that the above method is an approximate one for finding out the area of trepezium.

EXAMPLE: The following is the distribution of income of different staff in an educational institution. Represent the data by Lorenz curve and also find the proportionate number of persons having income upto 20 per cent.

TABLE 3.11

Income Group	No.of Staff members (X)	Mid. value of income group (Y)	Percentage Prop.of Y(P_y)	Percentage Prop. of X(P_x)	Cum. of Prop. of Y C_py	Cum. of Prop of X C_px
Less than 200	20	100	2.06	9.39	2.06	9.39
200-400	35	300	6.19	16.43	8.25	25.82
400-600	62	500	10.31	29.11	18.56	54.93
600-800	48	700	14.43	22.54	32.99	77.47
800-1000	25	900	18.56	11.73	51.55	89.20
1000-1200	16	1100	22.68	7.51	74.23	96.71
1200 & above	7	1250	25.77	3.29	100.00	100.00
	213	4850				

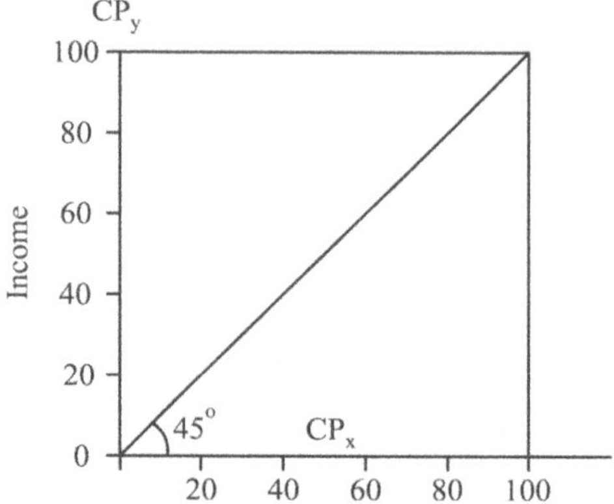

Fig. 3.11 Egalitarian line.

From Fig, 3.11, the proportion of persons having income upto 20 per cent is 60 per cent.

3.3.7 Remarks

The graphic representation generally depicts the trend when the number of observations is large. Also it provides intermediary values, though roughly.

EXERCISES

1. The following is the distribution of heights of plants of a particular crop.

Height (inches)	35-39	40-44	45-49	50-54	55-59	60-64
No. of plants	12	20	15	29	9	3

   Draw the (i) Histogram (ii) Frequency Polygon (iii) Frequency curve and (iv) ogive.

2. Draw the histogram of the following distribution of marriages classified according to the age of the bridegroom, and give your comments.

Class boundaries (age in years)	18-21	21-24	24-27	27-30	30-33
	11	61	73	57	33
Frequencies (in thousands)	33-36	36-39	39-42		
	21	14	9		

   (B.Sc. Madras, April, 1969)

3. Draw the 'Ogive' for the following frequency distribution.

Age (in years)	10-15	15-20	20-25	25-30	30-35	35-40
No. of persons	42	60	150	70	35	20

4. The following are the data regarding the area under grape cultivation in different years.

Year	1960	1961	1962	1963	1964	1965
Area (100 acres)	20	22	27	30	32	34

Represent the above data by a suitable diagram.

5. Represent the following data by a bar diagram and comment on their relationship. *(B.Sc Madras, Sept., 1969)*

Country	Birth rate	Death rate	Infant mortality
A	15.2	11.3	56
B	16.9	10.4	49
C	26.8	17.2	112
D	32.6	23.1	165

6. Represent the following data by sub-divided bar diagram drawn on the percentage basis. *(B.Sc. Madras, April, 1969)*

Heads of expenditure	State A	State B
	(in lakhs of rupees)	
Agriculture	517	578
Irrigation	648	910
Industry	186	496
Transport	566	984
Miscellaneous	148	106

7. The data given below relates to the income of workers' families in an Industrial area

Income per week	Number of families	Average income per family to the nearest rupee
Less than Rs. 25	92	16
25-35	335	24
35-45	402	36
45-55	246	44
55-65	144	52
65-75	42	65
Above Rs. 75	36	82

Draw a Lorenz curve to represent the data and determine there from, what percentage of the total income of the working classes is earned by the highly paid 25 per cent of the families.

(B.Sc. Madras, Aprtl.1967)

8. The expenditure pattern of two cultivators on one hectare farm for different items of agricultural inputs and the corresponding sector angles are given in the following table.

Item	Cultivator I		Cultivator II	
	Expenditure (Rs.)	Sector angle (degrees)	Expenditure (Rs.)	Sector angle (degrees)
Land reclamation	200	54.14	100	22.78
Hybrid seeds	300	81.20	350	79.75
Fertilisers	600	162.41	800	182.28
Tractor rent	80	21.65	120	27.34`
Electricity charges for pumping water	30	8.12	60	13.67
Labour charges	120	32.48	150	34.18
	1330	360.00	1580	360.00

CHAPTER 4

MEASURES OF LOCATION

It is always advisable to represent group of data by a single observation provided it does not loose any important information contained in the data and brings out every important information from it. This single value, which represents the group of values, is termed as a 'measure of central tendency' (or a measure of location or an 'average'). This should be a representative value or a typical member of the group. The different measures of location are 1. Arithmetic mean, 2. Median, 3. Mode, 4. Geometric Mean and 5. Harmonic mean.

4.1 ARITHMETIC MEAN

It is defined as the sum of the observations divided by its number.

Let $X_1, X_2, \ldots, X_n$ be n observations then the Arithmetic mean (A.M), $\overline{X}$ is defined as $\dfrac{X_1 + X_2 + \ldots + X_n}{n}$ which can be written as $\dfrac{1}{n}\Sigma X_i$, where 'Σ' is the summation which indicates the summing up of the observations from X_1 to X_n.

EXAMPLE: Compute the mean daily milk yield of a buffalo given the following milk yields (in kgs) for the consecutive 10 days.

$$15, 18, 16, 9, 13, 20, 16, 17, 21, 19$$

$$\overline{X} = \frac{15 + 18 + \ldots + 19}{10} = 16.4 \text{kg}$$

4.1.1 Linear Transformation Method

If the observation values are large, more in number and the deviation among themselves is small, the linear transformation method will save time in computation.

Let $d_i = X_i - A$ where A is called arbitrary mean and which is taken as round figure mid way between highest and lowest values.

$$\overline{X} = A + \overline{d}$$

$$= A + \frac{\Sigma d_i}{n}$$

For the above example, let A = 15

TABLE 4.1

Sl. No	X_1	$d_l = (X_l - A)$
1.	15	0
2.	18	3
3.	16	1
4	9	−6
5.	13	−2
6.	20	5
7.	16	1
8.	17	2
9.	21	6
10	19	6
		14

$$\overline{X} = A + \frac{\Sigma d_1}{n}$$

$$= 15 + \frac{14}{10} = 16.4 \text{kg}$$

4.1.2 Discrete Frequency Distribution

Let $f_1, f_2,..., f_n$ be n frequencies corresponding to the variate values $X_1, X_2,...,$ X_n respectively, then $\overline{X} = \dfrac{\Sigma f_i X_i}{\Sigma f_i}$

EXAMPLE: Compute the mean number of flowers per plant for the following data.

TABLE 4.2

No. of flowers (X_i)	No. of plants (f_i)	$f_i X_i$
0	5	0
1	10	10
2	12	24
3	16	48
4	8	32
5	7	35
6	2	12
Total	60	161

$$\overline{X} = \frac{161}{60} = 2.68$$

4.1.3 Grouped Frequency Distribution

(a) *Direct method.* Let f_1, f_2, ... f_n be n frequencies corresponding to the mid-values of the class intervals X_1, X_2, ... , X_n then the A.M. is given by

$$\overline{X} = \frac{\Sigma f_i\, X_i}{\Sigma f_i} = \frac{\Sigma f_i\, X_i}{N}$$

EXAMPLE: Find the mean breadth of leaf given the following distribution.

TABLE 4.3

Breadth of leaf (in cms)	No. of leaves (f_i)	Mid-value	$F_i x_i$	D_i $X_i - A$ over C	$f_i d_i$
2-4	7	3	21	-2	-14
4-6	10	5	50	-1	-10
6-8	19	7	133	0	0
8-10	15	9	135	1	15
10-12	9	11	99	2	18
12-14	3	13	39	3	9
Total			477		18

$$\overline{X} = \frac{477}{63} = 7.57 \text{ cm}$$

(b) *Linear transformation method*

$$\overline{X} = A + C\frac{\Sigma f_d\, X_i}{N} \text{ where } d_i = \frac{X_i - A}{C}, \text{ Here } A = 7$$

From Table 4.3, we have $\overline{X} = 7 + \dfrac{18 \times 2}{63} = 7.57$ C = class interval.

Whenever the class interval is same it is always convenient to take $d_i = \dfrac{X_i - A}{C}$ where C is class interval to simplify the calculations,

Consequently in the formula of $\overline{X}$ the second expression is multiplied by C.

The characteristics of a satisfactory average are listed here.

Characteristics of a satisfactory average: *(a)* It should have well defined formula, *(b)* It should be based upon all the observations, *(c)* It should be comprehensible, *(d)* It should be least affected by sampling fluctuations, *(e)* It should be easily computed, and *(f)* It should be capable of algebraic treatments.

Merits of A.M.: It possesses all the characteristics of satisfactory average which include algebraic properties such as

(i) The algebraic sum of the deviations taken from A.M. is zero, i.e., $\Sigma(X_i - \overline{X}) = 0$.

(ii) Let $\overline{X_1}$ be the mean of n_1 observations, $\overline{X_2}$ be the mean of the n_2 observations, ... , $\overline{X_k}$ be the mean of n_k observations then the mean $\overline{X}$ of $n = (n_1 + n_2 + ... + n_k)$ observations is given by

$$\overline{X} = \frac{n_1\overline{X_1} + n_2\overline{X_2} + ... + n_k\overline{X_k}}{n_1 + n_2 + ... + n_k}$$

Demerits of A.M.: *(a)* It may not be always identified with any one of the observations from which it is calculated, *(b)* It gives more weight age to extreme items whenever they are present, *(c)* It is difficult to calculate whenever the extreme classes in the continuous frequency distribution are not well defined.

4.2 MEDIAN

It is defined as that value of the variate below which half of the values lie and above which the remaining half lie when the variate values are arranged in ascending order of magnitude.

Case *(i) Variate values:*

(a) The number of observations is odd:

EXAMPLE: Find the median score of the following scores obtained by students in a particular one hour examination.

$$6, 9, 13, 4, 11, 8, 12, 9.5, 7$$

Arranging the scores in ascending order of magnitude, we have 4, 6, 7, 8, 9, 9.5, 11, 12, 13

Median number $= \dfrac{n+1}{2} = \dfrac{9+1}{2} = 5$, where n = number of observations.

The value of 5th observation is 9 which is the median value.

(b) If n is even

Let the scores be: 6, 9, 13, 4, 11, 8, 12, 9.5, 7, 8.5.

After arranging in ascending order of magnitude, we have 4, 6, 7, 8, 8.5, 9, 9.5, 11, 12, 13

Median No. $= = \dfrac{n+1}{2} = \dfrac{10+1}{2} = 5.5$

Median value: 5th value + 0.5 (6th value − 5th value)

$$= 8.5 + 0\,5\,(9 - 8.5) = 8.75$$

Case *(ii)* *Grouped frequency distribution*:

Let f_1, f_2,...f_n be n frequencies corresponding to the mid-values of the classes X_1, X_2, ... , X_n respectively then the Median is given by

$$M = 1 + \frac{\frac{N+1}{2} - m}{f} \times C$$ where 1 = lower limit of the median Class,

C = Class interval of the median class, $\frac{N+1}{2}$ = median number,

m = cum. fre. just preceding to the median class, f = frequency of the median class.

Here we assume that the groups are formed in ascending order of magnitude. Median class is that class in which the median number $\frac{N+1}{2}$ lies.

EXAMPLE: Obtain the median from the following distribution of weights of children in a particular locality.

TABLE 4.4

	Weights (Kg.)	No. of children	Cum. fre.
	0-4	3	3
	4-8	9	12
	8-12	18	30
	12-16	20	50
Median class	16-20	16	66
	20-24	7	73

$$M = 1 + \frac{\left(\frac{N+1}{2} - m \right)}{f} \times C$$

$\frac{N+1}{2} = \frac{73+1}{2} = 37$ Since 37 lies between the cumulative frequencies 30 and 50, the class (12-16) is the median class. 1 = 12, C= 4, m = 30, f = 20.

$$M = 12 + \frac{37 - 30}{20} \times 4 = 13.4$$

Merits of Median: *(a)* It can be calculated even if the extreme classes are not well defined, *(b)* It can be easily located on frequency curve, *(c)* It is useful whenever the qualitative characters are under consideration, *(d)* It is having one important algebraic property *i.e.,* the sum of the absolute values of the deviations is least when the deviations are taken from the Median.

Demerits of Median: *(a)* It is not based on all the observations, *(b)* It is not widely used in practice, and *(c)* It is not well defined.

4.3 MODE

Mode is that value of the, variate which occurs most frequently.

Case *(i) Variate values*

EXAMPLE: Find the model height (in inches) from the heights of 20 students.

60, 65, 64, 58, 69, 72, 64, 64, 65, 60, 61, 67, 64, 63, 67, 64, 68, 63, 64 and 66.

Here height 64" is repeated more number of times and hence model height = 64".

Case *(ii) Grouped distribution*

Let f_1, f_2 , ... , f_n be n frequencies corresponding to the mid-values of the classes X_1, X_2, ..., X_n respectively then the Mode (M') is given by the formula

$$M' = 1 + \frac{f - f_p}{2f - f_p - f_s} \times C \text{ where } 1 = \text{lower limit of the modal class,}$$

f = frequency of the modal class, f_p = frequency just preceding to modal class, f_s = frequency just succeeding to modal class and C = class interval of the modal class and modal class is that class in which maximum frequency exists.

EXAMPLE: Calculate the 'Mode' for the following distribution of wages in a certain factory.

TABLE 4.5

Daily wages	2-4	4-6	6-8	8-10	10-12	12-14	14-16	16-18
No. of employees	29	43	75	135	90	60	35	33

$$M' = 1 + \frac{f - f_p}{2f - f_p - f_s} \times C$$

$$1 = 8, f = 135, f_p = 75, f_s = 90$$

$$M' = 8 + \frac{135 - 75}{270 - 75 - 90} \times 2 = 9.14$$

If the maximum frequency occurs more than once in the distribution, the method of grouping is adopted for locating the modal class and then the above formula will be used for finding the modal value.

Merits: *(a)* It can be easily located on frequency curve, *(b)* It can be calculated even when extreme classes are not well defined except the modal class, *(c)* It is used mostly in business. For example, Cloth merchant would like to keep a certain quality of cloth having maximum sales, shoe-maker would like to keep a shoe of size or sizes having maximum sales, and *(d)* It can easily be calculated except in the case where the maximum frequency occurs more than once.

Demerits: *(a)* It is not based upon all the observations, *(b)* It is not having algebraic properties, and *(c)* It is not stable, since different methods of forming class intervals would lead to different modal values.

The empirical relationship between mean, median and mode is Mean – Mode = 3 (Mean – Median).

4.4   GEOMETRIC MEAN

Let $X_1, X_2, ..., X_n$ be n observations then the geometric Mean is the n-th root of their product.

Case *(i) Variate values*

The geometric mean of n observations, say, $X_1, X_2, ... , X_n$ is the n-th root of their product.

Hence $G.M, = (X_1 . X_2 ... X_n)^{1/n}$

EXAMPLE: Compute the G.M. of the following observations: 6, 8, 11, 12, 21, 13.

$$G.M. = (6 \times 8 \times 11 \times 12 \times 21 \times 13)^{1/6}$$

Log $$G.M = \frac{1}{6} (\log 6 + \log 8 + ... + \log 13)$$

$$= 1/6(0.7782 + 0.9031 + 1.0414 + 1.0792 + 1.3222) + 1.1139) = 1.0397$$

$$G.M. = \text{Anti log } (1.0397) = 10.96$$

Case *(ii) Grouped distribution*

Let $f_1, f_2,..., f_n$ be n frequencies corresponding to the mid-values of the class intervals $X_1, X_2, ..., X_n$ then the Geometric mean is given by the formula:

$$G.M. = \left(X_1^{f_1}, X_2^{f_a} ... X_n^{f_n}\right)^{1/n}$$

$$\log G.M. = 1/N(f_1 \log X_1 + f_2 \log X_2 + ... + f_n \log X_n)$$

$$G.M. = \text{Anti log } 1/N(f_1 \log X_1 + f_2 \log X_2 + \ldots + f_n \log X_n)$$

EXAMPLE: Find the G.M. for the following distribution.

TABLE 4.6

Class	Frequency (f_i)	Mid value (X_i)	Log X_i	$f_i \log X_i$
0-5	4	2.5	0.3979	1.5916
5-10	10	7.5	0.8751	8.7510
10-15	28	12.5	1.0969	30.7132
15-20	17	17.5	1.2430	21.1310
20-25	6	22.5	1.3522	8.1132
25-30	2	27.5	1.4393	2.8786
	67			73.1786

Log G.M. = 1/67 (73.1786) = 1.0922

G.M. = 12.37

Merits: *(a)* It has well defined formula, *(b)* It is based upon all the observations, *(c)* It is used in computing index numbers and also in time series analysis whenever ratios are under consideration. It is also used in finding out rate of change in population and computing compound interest, and *(d)* It possesses algebraic properties.

Demerits: *(a)* It is difficult to calculate, *(b)* *It* cannot be calculated whenever zero value is present in the observations, and *(c)* It may not be identified with any of the given observations.

4.5 HARMONIC MEAN

It is defined as the reciprocal of the arithmetic mean of the reciprocals.

Case *(i) Variate values*

Let $X_1, X_2, \ldots, X_n$ be n observations, then the Harmonic mean is given by the formula

$$H.M = \cfrac{1}{\cfrac{1}{n}\left(\cfrac{1}{X_1} + \cfrac{1}{X_2} + \ldots + \cfrac{1}{X_n}\right)} = \cfrac{n}{\left(\cfrac{1}{X_1} + \cfrac{1}{X_2} \ldots + \cfrac{1}{X_n}\right)}$$

EXAMPLE: The following are the quantities of onion (in kgs) sold per rupee in 6 markets. Find the average quantity per rupee.

$$1.5, 0.75, 0\ 5, 2.0, 2.5, 1.4$$

$$\text{H.M} = \frac{6}{\left(\dfrac{1}{1.5} + \dfrac{1}{0.75} + \ldots + \dfrac{1}{1.4}\right)} = 1.07 \text{ kg}$$

Case (ii) Grouped distribution

Let f_1, f_2,...,f_n be n frequencies corresponding to the mid-values of the classes X_1, X_2, ..., X_n respectively, then the Harmonic mean is given by the formula:

$$\text{H.M} = \frac{N}{\dfrac{f_1}{X_1} + \dfrac{f_2}{X_2} + \ldots + \dfrac{f_n}{X_n}}$$

EXAMPLE: Compute the H.M. For the following grouped distribution.

TABLE 4.7

Class	Frequency (f_i)	Mid-value (X_i)	$\dfrac{f_i}{X_i}$
3-6	6	4.5	1.33
7-10	9	8.5	1.06
11-14	14	12.5	1.12
15-18	20	16.5	1.21
19-22	10	20.5	0.49
23-26	2	24.5	0.08
	61		5.29

$$\text{H.M} = \frac{61}{5.29} = 11.53$$

Merits: *(a)* It is well defined, *(b)* It is based upon all the observations, and *(c)* It is useful in the cases like finding out average rate of work per hour, average quantity of a commodity per rupee, average distance travelled per hour, etc.

Demerits: *(a)* It is not much used in practice except in few cases mentioned above, *(b)* It is difficult to calculate, and *(c)* It gives more weightage to smaller values.

The relation between A.M, G.M & H.M is A. M. $\geq$ G. M $\geq$ H.M.

EXERCISES

1. The following table gives the yield of wheat from 10 equal plots.

Plot No.	1	2	3	4	5	6	7	8	9	10
Yield (in kg.)	60	40	50	45	60	55	65	50	65	55

If the area of each plot is 242 square yards, find the average yield per acre?   *(U.P. Board, 1963)*

2. The following is the distribution of heights of 85 plants.

Height (cms)	30-32	33-35	36-38	39-41	42-44	45-47
No. of plants	8	13	20	29	10	5

Find the Mean, Mode and Median heights of 85 plants.

3. Find the median for the following table relating to the number of grains per wheat blade.

No. of grains	20-24	24-28	28-32	32-36	36-40	40-44	44-48
No. of wheat ears	6	10	25	35	14	5	8

Locate also the 'median' from the cumulative frequency curve.

4. Find the median and mode for the following table.

No. of days absent		No. of students
More than	40	10
"	30	25
"	25	47
"	15	47
"	10	49
"	5	67
"	0	85

5. Compute the Geometric average of relative prices of the following commodities for the year 1939. (Base year 1938 - Price 100).

Commodity	Rice	Corn	Wheat	Oats	Barley	Potates	Sugar
Relative price	118	129	100	131	150	144	126
Weight	17	1385	561	408	100	194	142

Calculate also the weighted Geometric Mean using the weights.

6. If a city had a population of 2,50,000 in a given year and 3,00,000 five years later. What was the average annual per cent change ?

7. Rice is being sold at the following rates (kg. per 5 rupees) at 10 different markets.

    1.00, 0.80, 0.90, 0.70, 0.60, 1.10, 0.90, 0.75, 0.65, 0.45

    Compute the average quantity of rice per 5 rupees.

8. The following is the distribution of Fat (percentage) in100 samples collected from different milk centres in villages

Fat (%)	1-3	3-5	5-7	7-9	9-11
Samples	40	26	30	2	2

    Compute Mean, Median, Mode, G.M and H.M. of Fat content per sample.

9. The following is the distribution of body weights of 100 calves at the 1st lactation

Body weight (kg)	30-40	40-50	50-60	60-70	70-80
Calves	12	26	34	20	8

    Find Mean, Median, Mode, G.M. and H.M. of body weight of calves.

10. Compute the Arithmetic Mean yield (bags) of paddy given in the following distribution.

Yields (bags)	Less than 20	Less than 25	Less than 30
Farms	6	18	30

Yield (bags)	Less than 35	Less than 40	Less than 45
Farms	34	16	14

    Yield (bags) less than 20 less than 25 less than 30

11. The following is the distribution of Annual Income (Rs.) of families in a locality of a city.

Less than 20,000	20,000-40,000	40,000-80,000
64	45	32
80,000-1,60,000	1,60,000-3,20,000	3,20,000 and above
26	10	13

    Find A.M.?

12. The following is the distribution of grades (10-point scale) obtained by a student in a semester final examination in different subjects.

Subject	A	B	C	D	E	F	G
Credits	3	4	2	3	1	2	4
Grade	8.2	7.6	8.7	9.0	8.5	7.4	8.8

Find the Grade point average obtained by the student in that semester.

13. Compute the average rainfall in a rainy season in a city of a particular year.

Rain fall (cms)	Less than 2	2-4	4-6	6-8
Days	10	14	20	26

Rain fall (cms)	8-10	10 and above
Days	8	12

CHAPTER 5

MEASURES OF DISPERSION

Measure of Dispersion: It is a measure which can give the wide spread or scattering of observations among themselves or from a central point. The different measures of dispersion are *(i)* Range, *(ii)* Quartile deviation, *(iii)* Mean deviation, and *(iv)* Standard deviation.

5.1 RANGE

It is defined as the difference between the highest and lowest values in a series of observations.

EXAMPLE: Find the 'Range' for the following weights of 15 goals.

30, 25, 14, 42, 18, 26, 21, 11, 35, 32, 29, 23, 20, 19, 13.

Range: 49 – 11 = 31.

Range is not much used in practice since it depends upon two extreme values. Therefore presence of any extremely high and low values in the observations will affect the range considerably. However, this measure is easy to compute. This is useful when the data are of homogeneous nature. It is also used in the preparation of control charts and for the data based on daily temperatures, rainfall, etc.

5.2 QUARTILE DEVIATION: (SEMI-INTER QUARTILE RANGE)

It is given by the formula, Q.D. = $(Q_a - Q_1)/2$ where Q_1 = First quartile, Q_3 = Third quartile. First and third quartiles are also called as lower and upper quartiles respectively.

5.2.1 First Quartile

It is that value of the variate below which one-fourth of the values lie and above which the remaining three-fourth of the values lie when the values are arranged in ascending order of magnitude.

5.2.2 Third Quartile

It is that value of the variate below which three-fourth of the values lie and above which the remaining one-fourth of the values lie when the values are arranged in ascending order of magnitude.

Case *(i) Variate values*

EXAMPLE: Find the Q.D. for the following observations on number of mesta plants in 10 equisized plots.

$$13, 9, 16, 4, 8, 19, 7, 23, 21, 12.$$

Arranging the values in ascending order of magnitude, we have

$$4, 7, 8, 9, 12, 13, 16, 19, 21, 23.$$

$$\text{Q.No.} = \frac{n+1}{4} = \frac{10+1}{4} = 2.75$$

Q_1 No. = 2nd value + 0.75 (3rd value – 2nd value)

$$= 7 + 0.75 \times 1 = 7.75$$

$$Q_3 \text{No.} = \frac{3n+1}{4} = \frac{3 \times 10 + 1}{4} = 7.75$$

Q_3 No. = 7th value + 0.75 (8th value – 7th value)

$$= 16 + 0.75 \,(19 - 16) = 18.25$$

$$\text{Q.D.} = \frac{18.25 - 7.75}{2} = 5.25$$

Case *(ii) Continuous distribution*

Let f_1, f_2, ..., f_n be n frequencies corresponding to the mid-values of the classes X_1, X_2, ..., X_n respectively then the first quartile, Q_1 is given by

$$Q_1 = l_1 + \frac{\dfrac{(N+1)}{4} - m_1)}{f_1} \times C_1 \quad \text{where } l_1 = \text{ lower limit of the first quartile}$$

class, $N = \Sigma f_1$, $\dfrac{N+1}{4}$ = first quartile number, m_1, = cumulative frequency just preceding to the first quartile class, f_1 = frequency of the first quartile class, C_1 = Class interval of the first quartile class, First quartile class is that class in which the cumulative frequency $\dfrac{N+1}{4}$ exists.

Similarly the third quartile is obtained by the formula

$$Q_3 = l_3 + \frac{\left(\dfrac{3N+1}{4}\right) - m_3)}{f_3} \times C_3$$

where the symbols indicate the same as in the case of first quartile with first quartile replaced by third quartile.

EXAMPLE: Compute the Q.D. for the following distribution of marks obtained in an examination by 80 students.

TABLE 5.1

Marks	No. of students	Cum. fre
0-5	3	3
5-10	10	13
10-15	18	31
15-20	25	56
20-25	9	65
25-30	8	73
30-35	7	80

$$Q_1 \text{ No.} = \frac{N+1}{4} = \frac{80+1}{4} = 20.25$$

, $(10 - 15)$ is the first quartile class since 20.25 lies in that class.

$$Q_1 = 10 += \frac{(20.25 - 13)}{18} \times 5 = 12.01$$

$$Q_3 \text{ No.} = \frac{3N+1}{4} = \frac{3 \times 80 + 1}{4} = 60.25$$

, $(20 - 25)$ is the third quartile class since 60.25 lies in that class.

$$Q_3 = 20 + \frac{(60.25 - 56)}{9} \times 5 = 22.36$$

$$\text{Q.D.} = \frac{Q_3 - Q_1}{2} = \frac{22.36 - 12.01}{2} = 5.18$$

Unlike Range, the presences of abnormal values do not affect the Q.D. since the end values on either side do not figure in the definition.

5.3   MEAN DEVIATION

It is the mean of the absolute values of the deviations taken from some average.

5.3.1 Mean Deviation about Mean

Let $X_1, X_2, ..., X_n$ be n observations, the Mean deviation about mean is given by the formula

$$\text{M.D. about mean} = \frac{1}{n}\sum |X_i - \overline{X}| \text{ where } \overline{X} = \text{A.M.}$$

By linear transformation method, we have

$$\text{M.D. about mean} = \frac{1}{n}\sum |d_i - \overline{d}| \text{ where } di = (X_1 - A) \text{ and } \overline{d} = \sum \frac{d_i}{n}$$

EXAMPLE: Find the M.D. about mean for the following data.

4, 9, 10, 14, 7, 8, 6, 14.

TABLE 5.2

									Total
Xi	4	9	10	14	7	8	6	14	72
$\|X_i - \overline{X}\|$	5	0	1	5	2	1	3	5	22

$\overline{X} = 9$, M.D. about mean $= 22/8 = 2.75$

Case *(ii) Continuous frequency distribution*

Let $f_1, f_2, \ldots, f_n$ be n frequencies corresponding to the mid-values of the classes $X_1, X_2, \ldots, X_n$ respectively then the M.D. about mean is given by the formula

$$\text{M. D. about mean} \quad \frac{1}{N}\sum f_i \left| X_i - \overline{X} \right| \quad \text{where } \overline{X} = \text{mean.}$$

EXAMPLE: Find the M.D. about mean for the following grouped distribution.

TABLE 5.3

Yield of milk per day (in kgs)	No. of dairy animals (f_i)	Mid-value X_1	f_iX_i	X_i − X̄	f_i| X_1 − X̄|
0-2	6	1	6	5.66	33.96
2-4	10	3	30	3.66	36.60
4-6	14	5	70	1.66	23.24
6-8	18	7	126	0.34	6.12
8-10	11	9	99	2.34	25.74
10-12	7	11	77	4.34	30.38
12-14	5	13	65	6.34	31.70
	71		473		187.74

$\overline{X} = 473/71 = 6.66$

M.D. about mean $= 187.74/71 = 2.64$

M. D. can be computed by taking deviations from any other average like Median, Mode, etc. on the similar lines given above wherein Median, Mode, etc. will be substituted in the formula in place of Mean. M. D. is least when deviations are taken from the Median.

5.4   STANDARD DEVIATION

It is defined as the square root of the mean of the squares of the deviations taken from Arithmetic mean.

5.4.1   Variate Values

Let X_1, X_2, ... , X_n be n observations then standard deviation (S.D.) is given by the formula

$$\sigma = \sqrt{\frac{1}{n}\Sigma(X_i - \overline{X})^2} \quad \text{where } \overline{X} = \text{A.M.}$$

Simplifying the above formula, we have

$$\sigma = \sqrt{\frac{1}{n}\left[\Sigma(X_i^2 - \frac{(\Sigma X_i)^2}{n}\right]}, \ \sigma^2 = \text{variance}$$

By linear transformation method, we have
If $d_i = (X_i - A)$ where $A = $ Arbitrary mean

$$\sigma = \sqrt{\frac{1}{n}\left[\Sigma d_i^2 - \frac{(\Sigma d_i)^2}{n}\right]}$$

EXAMPLE: Compute the S.D. of the following data based on number of seeds germinated out of 20 in each of the ten petty dishes.

15, 13, 10, 17, 8, 12, 14, 11, 13, 15

TABLE 5.4

											Total
Xi	15	13	10	17	8	12	14	11	13	15	128
Xi²	225	169	100	189	64	144	196	121	169	225	1702

$$\sigma = \sqrt{\frac{1}{10}\left[1702 - \frac{(128)^2}{10}\right]} = 2.52$$

$$\sigma^2 = 6.35$$

5.4.2   Continuous Frequency Distribution

Let f_1, f_2, ..., f_n be n frequencies corresponding to the mid-values of the classes X_1, X_2, ..., X_n respectively, then the standard deviation is given by the formula

$$\sigma = \sqrt{\frac{1}{N}\Sigma f_i(X_1 - \overline{X})^2}$$

Simplifying, we have

$$\sigma = \sqrt{\frac{1}{N}\left[\Sigma f_i X_1^{\,2} - \frac{\left(\Sigma f_i X_1\right)^2}{N}\right]}$$

By linear transformation method, we have

$$\sigma = CX\sqrt{\frac{1}{N}\left[\Sigma f_i d^2_{\,i} - \frac{\left(\Sigma f_i d_1\right)^2}{N}\right]}$$

where $d_i = \dfrac{X_i - A}{C}$, A = Arbitrary mean and C = class interval.

EXAMPLE: Find the standard deviation and variance for the following distribution of lengths of wheat ears.

(i)   Direct method

TABLE 5.5

Length of wheat ear	No. of ears (f_i)	Mid-value(x_i)	$f_i X_i$	$f_i X^2_{\,i}$	d_i	$f_i d_i$	$f_i d_i^2$
7-9	8	8	64	512	−2	−16	32
9-11	18	10	180	1800	−1	−18	18
11-13	25	12	300	3600	0	0	0
13-15	15	14	210	2940	1	15	15
15-17	6	16	96	1536	2	12	24
	72		850	10388		−7	89

$$\sigma = \sqrt{\frac{1}{72}\left[10388 - \frac{\left(850\right)^2}{72}\right]} = 2.22$$

Variance = 4.91

$$\sigma = 2X\sqrt{\frac{1}{72}\left[89 - \frac{\left(-7\right)^2}{72}\right]} = 2.22$$

Variance = 4.91

This is most commonly used measure of dispersion as a counterpart of mean in the case of 'measures of location'. This gives minimum value when the deviations are taken from the mean.

5.5   COEFFICIENT OF VARIATION

Sometimes it is necessary to express variation of a series of data relative to an average. For example, the variation of 2 to 3 quintals per acre in the production would be significant for a local variety of paddy but not so in the case high yielding variety. Hence, coefficient of variation (C.V.) which is the percentage ratio of S.D. to Mean is calculated for each of the variety. The one which is having less coefficient of variation (C.V.) is considered more consistent variety. Since C.V. is independent of units, it is useful for comparison of any two series with different units. The C.V. is also found useful to compare two players with respect to their consistency in scoring.

$$C.V. = \frac{S.D}{mean} \times 100$$

EXAMPLE: The scores of two candidates A and B in different one-hour examinations are given below. Examine who is the more consistent scorer.

TABLE 5.6

Candidate	One-hour examination					
	I	*II*	*III*	*1V*	*V*	*VI*
A	9.0	8.0	7.5	8.5	9.0	8.0
B	5.5	9.5	6.5	8.5	10.0	8.0

Arbitrary Mean = 8.0, $\Sigma d_i = 2.00$, $\Sigma d^2_i = 2.50$

Candidate A: Mean = 8 + 2/6 = 8.33

$$S.D. = \sqrt{1/6\left[2.5 - \frac{(2)^2}{6}\right]} \qquad C.V. = \frac{0.55}{8.33} \times 100$$

$$= 0.55 \qquad\qquad = 6.60$$

Candidate B: Mean = 8 + 0/6 = 8.00, $\Sigma d_2 = 0$, $\Sigma d^2_2 = 15.00$

$$S.D. = \sqrt{1/6\left[15 - \frac{(0)^2}{6}\right]} \qquad C.V. = \frac{1.58}{8.00} \times 100$$

$$= 1.58 \qquad\qquad = 19.75$$

Therefore, candidate A is more consistent.

5.6   STATISTICAL POPULATION

An aggregate of animate or inanimate objects is called statistical population. For example, large group of data on heights, weights, etc., is known as Statistical population.

5.7 SAMPLE

To study particular character it is always not possible to study the whole lot or whole population as it requires more time and money. Therefore, we have to rely upon a part of the population for our study. This part or portion of a population is called sample. However, the sample should be as far as possible representative of the population with respect to character under consideration. This can be ensured by drawing the units from population at random so that each and every sample of equal size will be selected in the sample with equal probability.

Now the value of the mean based on sample of observations need not be equal to the population mean. The difference between the sample mean and the population mean is called 'sampling error'. Suppose if we take all possible samples of equal size, the means based on these samples follow a distribution known as 'sampling distribution'. The mean of all the means of samples of equal size is an estimate of the population mean, and the standard deviation of the means of these samples is known as 'standard error of mean'.

Since it is difficult, in general, to study all the possible samples, we have to depend on a single sample. The standard error of mean based on a single sample is given as

$$\text{S.E } (\overline{X}) = \frac{\sigma}{\sqrt{n}} = \text{Where } \sigma = \text{S.D. in the population, } \overline{X} = \text{Mean of a}$$

sample, n = size of the sample.

If σ is not known, it is estimated from a sample of observations as follows:

Case (*i*) If n is large sample (Say > 30)

$$\text{S.E. } (\overline{X} = S/\sqrt{n} \text{ where } S = S + \sqrt{1/n\ \Sigma(X_1 - \overline{X})^2}$$

Case (*ii*) If n is small sample (Say < 30)

$$\text{S.E. } (\overline{X}) = \frac{s}{\sqrt{n}} \text{ where } s = \text{'unbiased' estimate of } \sigma$$

$$s = \sqrt{\frac{1}{n-1}\Sigma(X_i - \overline{X})^2}$$

EXERCISES

1. The country's foodgrains output (in million tons) for 20 years are given as :

   75, 74, 80, 81, 85, 86, 84, 81, 90, 87, 92, 94, 95, 93, 98, 96, 94, 99, 109, 110.

Obtain the values of Range, Mean deviation about Mean and Median and Coefficient of Variation.

2. The following table gives the yield of paddy in maunds per acre based on crop cutting experiments in a certain area. During 1940-41.

Yield (Maunds Acre)	Freq.	Yield (Maunds/Acre)	Freq.
0	4	24	128
3	4	27	73
6	32	30	50
9	81	33	13
12	135	36	12
15	198	39	5
18	210	42	1
21	144		

Calculate the Arithmetic mean, Standard deviation and quartile deviation of the distribution. *(I.A.S., 1949)*

3. Find the 'mean deviation about mode' for the following grouped distribution.

Class	3-7	7-11	11-15	15-19	19-23	23-27
Freq.	5	7	13	31	18	4

4. A distribution consists of three components with frequencies 200, 250 and 300 having means 25, 20 and 15 with standard deviations 3,6 and 5 respectively. Find the 'mean' and 'standard deviation' of the combined distribution.

5. If any two series, where d_1 and d_2 represent the deviations from the same arbitrary mean, 15, the following results are given,

$n_1 = 12$ $\Sigma d_1 = 25$ $\Sigma d_1^2 = 650$

$n_2 = 20$ $\Sigma d_2 = -20$ $\Sigma d_2^2 = 480$

Compute the coefficient of variation for both the series and determine which is more consistent series.

6. Below are the scores of two cricketers in 10 innings. Find who is the more consistent scorer.

A	204	68	150	30	70	95	60	76	24	19
B	99	190	130	94	80	89	69	85	65	40

7. Compute the 'coefficient of variation' and 'standard error of mean' given the following distribution of protein (percentage) content in 100 samples of red gram collected from different farms.

Protein (%)	Less than 4	4-8	8-12	12-16	60-20
Samples	28	20	26	10	16

8. The following is the distribution of yields (quintals) per hectare in different farms in a Research Zone.

Yield (quintals)	10-16	16-22	22-28	28-34	34 and above
Farms	8	14	26	22	10

   Compute coefficient of variation for the above distribution of yields.

9. Find the standard deviation of the following distribution of leaf areas (square mm) of sun flower crop in an experimental field.

Leaf area (sq.mm)	20-30	30-40	40-50	50-60
Leaves	8	14	28	16

Leaf area (sq.mrn)	60-70	70-80	80 and above
Leaves	10	20	10

10. The following is the distribution of ear lengths (cm) of paddy crop of high yielding variety in an experimental field.

Ear length (cm)	less than 4	4-6	6-8	8-10
Ears	16	20	24	18

Ear length (cm)	10-12	12 and above
Ears	15	7

   Obtain coefficient of variation, quartile deviation and mean deviation from mean for the above data.

11. The following is the distribution of heights of maize plants (cms) in an experimental field in a reasearch station.

Height (cms)	80-100	100-120	120-140	140-160
Plants	28	16	30	17

Height (cms)	160-180
Plants	9

   Obtain 'standard deviation' and 'standard error of mean' for the aobve distribution of heights.

12. Find mean deviation from 'Median' and 'Mode' for the following distribution of rain fall (mm) in July month at an Agricultural Research station.

60,	76,	16,	18,	32,	18,	34,	46,	76,
76,	90,	92,	93,	84,	80,	54,	62,	70,
100,	85,	22,	23,	26,	48,	78,	76,	91,
56,	50,	66,	81					

13. Compute mean deviation from 'Mode' for the following distribution of yields (kg) of grapes in different grape gardens in an year.

Grape yields (kgs)	300-400	400-500	500-600
Gardens	8	16	19
Grape yields (kgs)	600-700	700-800	
Gardens	9	3	

14. The following are the minimum temperatures (celcius) recorded in a hill station of North India in the month of January.

6,	3,	2,	0,	−1,	−6,	3,	2,	10,	7,
8,	9,	3,	4,	11,	5,	3,	0,	6,	3,
−4,	−2,	1,	0,	3,	−1,	6,	2,	1,	3,
5.									

   Compute 'Range' and 'Quartile' deviation',

15. The following is the frequency distribution of number of Custard Apples per tree in a graden

Custard Apples	150	172	175	184	189
Trees	6	10	15	8	12
Custard Apples	201	210			
Trees	13	5			

   Find Mean deviation from 'Median' for the above distribution.

CHAPTER 6

MOMENTS, SKEWNESS AND KURTOSIS

6.1 MOMENTS

Let $X_1, X_2, ..., X_n$ be n observations, then 'k' - th raw moment is defined by

$$V_k = 1/n \, \Sigma(X_i - X)^k$$

where A = Arbitrary mean

The 'k'-th central moment is given by

$$\mu_k = 1/n \, \Sigma(X_i - \overline{X})^k \quad \text{where } \overline{X} = \text{A.M.}$$

In the case of a frequency distribution, the k-th raw and central moments are given by V_k and μk respectively, as

$$V_k = \frac{1}{N} \, \Sigma f(X_i - A)^k, \quad \mu_k = \frac{1}{N}\Sigma f(X_i - \overline{X})^k \qquad .....(6.1)$$

where $N = \Sigma f_1$ = Total frequency.

The relation between k-th central moment and raw moments is given by

$$\mu_k = V_k - \binom{k}{1} V_{k-1} \, V_1 + \binom{k}{2} V_{k-2} \, V_1^2 - \binom{k}{3} V_{k-3} V_1^3 + ... + (-1)^k V_1^k \quad .....(6.2)$$

where $V_{k-1}, V_{k-2}, ... V_1$, are (k–1)-th, (k–2)-th ...,

1st raw moments respectively. $\binom{k}{1}, \binom{k}{2}$ etc., are the number of

combinations taking 1 at a time, 2 at a time, etc., respectively out of k values.

If k = 1, $\mu_1 = V_1 - V_1 = 0$

If k = 2, $\mu_2 = V_2 - V_1^2$ = variance

If k = 3, $\mu_3 = V_3 - 3V_2 V_1 + 2V_1^3$

If k = 4, $\mu_4 = V_4 - 4V_3 V_1 + 6V_2 V_1^2 - 3 V_1^4$

The central moments are useful in measuring 'skewness' and 'kurtosis' of curves.

6.2   SKEWNESS

Sometimes, even if the two measures like 'Mean' and 'standard deviation' are same for the distributions still the shape of the two curves may differ. For example, one curve may be symmetric and the other may be asymmetric. We shall define here 'symmetric' and 'asymmetric' curves for the Uni-modal frequency distribution.

6.2.1 Symmetric Curve

A symmetric curve is one where the shape of the curve on either side of the mean is identical. That is, in a frequency distribution, the frequencies on either side of a mean should be equal. In this curve, the mean, median and mode coincide at one point and the ordinate drawn from peak of the curve to mean on the X-axis would, bifurcate the area under the curve into two equal halves. The skewness (or bending) in this case is zero. The symmetric curve is depicted in Fig. 6.1.

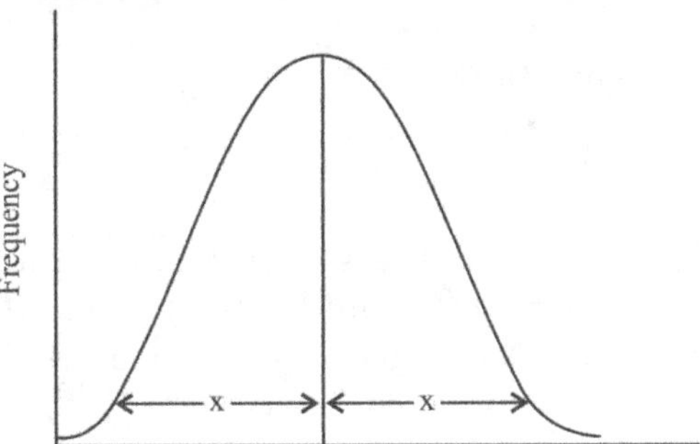

Fig. 6.1 Symmetric curve.

6.2.2 Asymmetric Curve

A curve which is not symmetric is known as 'Asymmetric' or 'skewed' curve. In this case, the peak of the curve may bend towards right or towards left with respect to mean.

The curve bending towards right from the mean and having a long tail on left is said to be negatively skewed. This curve is shown in Fig. 6.2.

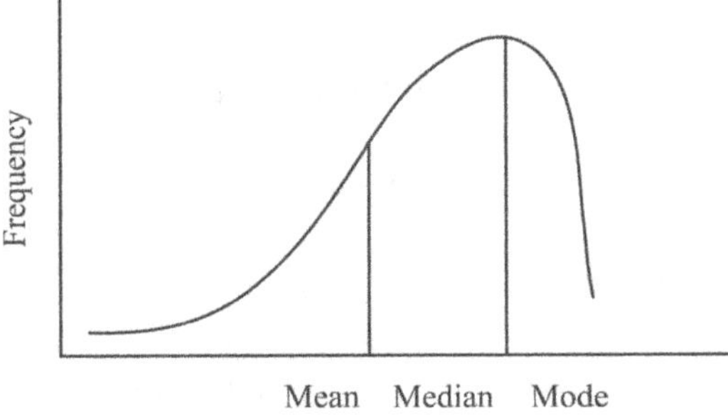

Fig. 6.2 Negative skewness.

The curve bending towards left from the mean and having long tail towards right is said to be positively skewed. The curve is shown in Fig. 6.3.

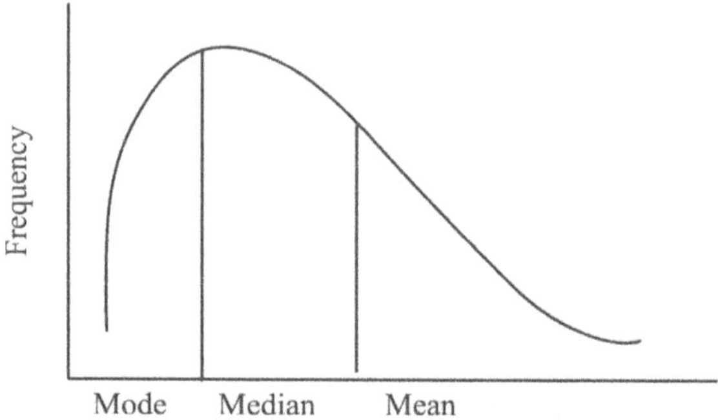

Fig. 6.3 Positive skewness.

The different measures of coefficient of skewness are given by

1. Pearson's Coefficient of skewness

$$= \frac{(\text{Mean–mode})}{\text{S.D}} \qquad \ldots\ldots(6.3)$$

This measure is due to K. Pearson. In Fig.6.2. Mode will be greater than Mean and hence (Mean-Mode) is negative. Therefore, the coefficient of skewness is negative. In Fig. 6.3, Mode will be less than Mean and hence (Mean–Mode) is positive. Hence, the coefficient of skewness is positive since the denominator in the formula is always positive. In Fig. 6.1, Mean is equal to Mode and hence (Mean–Mode) is zero. In the formula of coefficient of skewness, S.D. is used in order to make the coefficient independent of units so as to facilitate the comparison of two or more distributions. Median always lies between Mean and Mode in Figs. 6.2 and 6.3.

2. Quartile coefficient of skewness $= \dfrac{(Q_3 - Q_2) - (Q_2 - Q_1)}{(Q_3 - Q_1)}$ $\quad\ldots\ldots(6.4)$

where Q_1, Q_2 and Q_3 are the 1st, 2nd and 3rd quartiles respectively. This coefficient always lies between -1 and $+1$. In this case also the denominator is taken to make the coefficient independent of units.

3. Moment coefficient of skewness, $\sqrt{\beta_1} = \dfrac{\mu_3}{\mu_2^{3/2}}$ $\quad\ldots\ldots(6.5)$

where μ_2 and μ_3 are 2nd and 3rd central moments respectively. In this formula μ_3 measures the excess of negative deviations over positive deviations and excess of positive deviations over negative deviations in Fig. 6.2 and Fig. 6.3, respectively. Here also the denominator is used to make the coefficient independent of units.

6.3 KURTOSIS

The shape of the Vertex of the curve is known as Kurtosis. The measure of Kurtosis is known as coefficient of Kurtosis and is denoted β_2.

6.3.1 Platykurtic

The peak or Vertex of the curve is more flat and the tails on both sides are long compared to normal curve (chapter 9). Here $\beta_2 < 3$.

6.3.2 Mesokurtic

The peak of the curve is normal and the tails on both sides are also normal. Here $\beta_2 = 3$.

6.3.3 Leptokurtic

The peak of the curve is narrow and sharp and the tails on both sides are short compared to normal curve. Here $\beta_2 > 3$.

The above three curves are depicted in Fig. 6.4. $(\beta_2 - 3)$ is taken as the departure from normality. This quantity would be negative for Platykurtic, zero for Mesokurtic and positive for Leptokurtic. The measure of Kurtosis is denoted by coefficient of Kurtosis and is given by the formula,

$$\beta_2 = \frac{\mu_4}{\mu_2^{\,2}} \qquad\qquad \ldots\ldots(6.6)$$

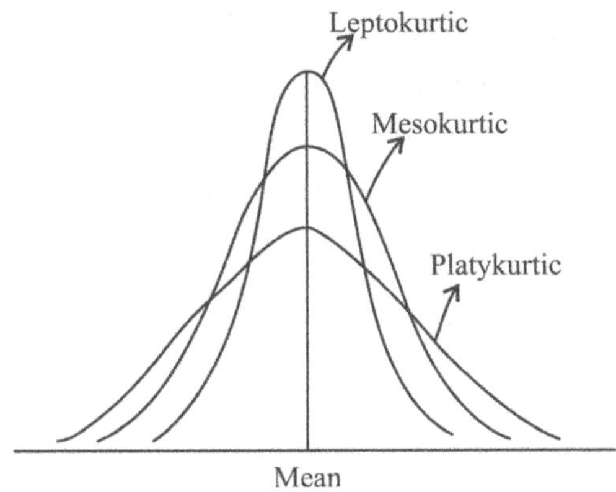

Fig. 6.4 Kurtosis.

Sometimes it is convenient to express the coefficients of skewness and kurtosis in terms of γ_1 and γ_2 respectively, where

$$\gamma_1 = \beta_1, \ \gamma_2 = (\beta_2 - 3).$$

EXAMPLE: Compute the different coefficients of skewness and kurtosis for the following data on milk yield.

TABLE 6.1

Milk yield (kgs)	No. of cows (f_1)	X_i	d_i	$f_i d_i$	$f_i d_i^2$	$f_i d_i^3$	$f_i d_i^4$
				Mid-value			
4-6	8	5	−3	−24	72	−216	648
6-8	10	7	−2	−20	40	−80	160
8-10	27	9	−1	−27	27	−27	27
10-12	38	11	0	0	0	0	0
12-14	25	13	1	25	25	25	25
14-16	20	15	2	40	80	160	320
16-18	7	17	3	21	63	189	567
	135			0	15	307	51	1747

$$d_1 = \frac{X - A}{C} \text{ where } A = 11, C = 2$$

$$V_1 = C/N \ \Sigma f_i \ d_1 = 2 \times 15/135 = 0.222$$

$$V_2 = C^2/N \ \Sigma f_i \ d_1^2 = 4 \times 307/135 = 9.096$$

$$V_3 = C^3/N \ \Sigma f_i \ d_1^3 = 8 \times 51/135 = 3.022$$

$$V_4 = C^4/N \ \Sigma f_i \ d_1^4 = 16 \times 1747/135 = 207.052$$

$$\mu_2 = V_2 - V_1^2 = 9.096 - (0.222)^2 = 9.047$$

$$\mu_3 = V_3 - 3V_2 V_1 + 2V_1^3$$

$$= 3.022 - 3(9.096)(0.222) + 2(0.222)3 = -3.014$$

$$\mu_4 = V_4 - 3V_3 V_1 + 6V_2 V_1^4$$

$$= 207.052 - 4(3.022)(0.222) + 6(9.096)(0.222)^2 - 3 \times (0.222)^4$$

$$= 207.051$$

Moment coefficient of skewness, $\sqrt{\beta_1} = \dfrac{\mu_3}{\mu_2^{3/2}}$

$$= -3.014/(9.047)^{3/2} = -0.1108$$

Mean = 11.222,   Mode = 10.917,   S.D. = 3.008

Pearson's coefficient of skewness $= \dfrac{11.222 - 10.917}{3.008} = 0.1014$

$Q_1 = 9.185$, $Q_2 = 11.211$, $Q_3 = 13.480$

Quartile coefficient of skewness

$$= \frac{(13.480 - 11.211) - (11.211 - 9.185)}{(13.480 - 9.185)} = 0.0566$$

Coefficient of Kurtosis, $\beta_2 = \dfrac{\mu_4}{\mu_2^2}$

$$= 207.051/(9.047)^2 = 2.530$$

The coefficients of skewness obtained by Pearson's coefficient of skewness and quartile coefficient of skewness are positive whereas the Moment coefficient of skewness gave negative value. The coefficient of Kurtosis indicates that the curve is platy kurtic since $(\beta_2 - 3)$ i.e., $(2.529-3)$ is negative.

EXERCISES

1. In a frequency distribution, size 701, range 30-150 divided into 8 class intervals of equal width, the first three moments measured in terms of the scale units u from the midpoint of the fourth class interval from the top are $\Sigma fu = -150, \Sigma fu^2 = 1532, \Sigma fu^3 = -750$. Determine A.M. and the values of the first three moments from the mean in terms of the original unit. Calculate the standard deviation. *(B.Sc Madras, 1944)*

2. If the first three moments about an arbitrary mean 4 are 2, 15, and 84. Calculate the mean, variance and third moment about mean.

3. Define skewness and arrange median, mode and mean in ascending order of their magnitude for positively skewed curve.

4. Define r-th moment about mean and give the formula for 4-th moment about mean in terms of the moments about arbitrary mean.

5. For any two groups of data A and B the statistical constants are

	A	B
Median	19.64	24.46
Lower quartile	13.46	15.64
Upper quartile	25.94	37.76

Comment on the dispersion and skewness of A and B.

6. Define the various measures of dispersion and discuss their relative advantages.

Find the standard deviation and the Pearson coefficient of skewness for the following distribution.

Percentage ash Content	3.0-3.9	4.0-4.9	5.0-5.9	6.0-6.9	7.0-7.9	8.0-8.9	9.0-9.9	10.0-10.9
Frequency	3	7	28	78	84	45	28	7

7. Find the 3rd and 4th central moments given the following raw moments

$$V_4 = 40,\ V_3 = 10,\ V_2 = 5 \text{ and } V_1 = 2.$$

8. Obtain the 'quartile coefficient of skewness' given the following data on minimum temperatures [celcius] on 11 days in December month in a city.

$$4,\ 5,\ 6,\ 7,\ 10,\ 3,\ 1,\ 5,\ 12,\ 18,\ 16.$$

9. Find the coefficient of Kurtosis given the following raw moments on rain fall data and also specify the type of kurtosis.

$$V_1 = 2,\ V_2 = 6,\ V_3 = 8 \text{ and } V_4 = 20.$$

10. Find the 'coefficient of skewness' given the following on yields of wheat in a region

$$\text{Mean} = 40,\ \text{Mode} = 50\ \text{Variance} = 6$$

11. Obtain 3rd and 4th central moments for the following data on disease effected birds in 10 poultry farms

$$6, 13, 10, 2, 21, 6, 150, 96, 74, 65$$

12. Compute coefficients of 'skewness' and 'kurtosis' for the following data on rainfall (cm) in the month of August at a Regional Agricultural Research Station in Coastal Andhra Pradesh.

$$2, 0, 6, 8, 13, 7, 2, 2, 4, 10, 11, 2, 0, 5, 11, 0, 0, 8, 7, 9, 6,$$
$$5, 12, 3, 4, 5, 0, 6, 1, 2, 4$$

13. Find the coefficients of 'skewness' and Kurtosis for the following data on iron content (percentage) in samples of leafy vegetable sold in markets and also draw diagrams.

Iron content (%)	0-2	2-4	4-6	6-8	8-10
Samples	6	12	18	9	5

CHAPTER 7

ELEMENTARY PROBABILITY

We use the word 'probability' several times in our daily life such as 'It may probably rain today', 'India may probably win the Davis Cup tie against Australia', 'Probably this year we may have a good harvest'. We frame these statements according to chances of happening of a particular event. The statements mentioned above· are no longer subjective since the development of probability theory.

7.1. DEFINITION OF PROBABILITY

$$\text{Probability of an event} = \frac{\text{No of favourable cases to that event}}{\text{Total No. of equally likely cases}} \quad\quad .....(7.1)$$

This definition is also called *a priori probability* since the probability is obtained prior to happening of an event. It is difficult to enumerate or imagine all the equally likely cases in some situations for evaluating probability by this method.

For example, we take the statement 'It may probably rain today'. Following the above definition, the total number of equally likely cases can, at the most, be two and the number of favourable cases to the event is one. Therefore, the probability that it may rain today is $\frac{1}{2}$. But this may not be correct since the probability of raining is more in monsoon season compared to other seasons. In certain cases, we can safely say that there will be no rain today. Therefore, the probability is $\frac{1}{2}$ no longer true. Similarly, the statement like 'a man will die tomorrow is $\frac{1}{2}$', is no longer holds though it is obtained strictly based on mathematical definition of probability. Therefore, one should know the conditions affecting a particular event before finding the probability of happening a particular event.

If an event can happen in 'a' ways and fail to happen in 'b' ways and all these are equally likely then the probability of happening an event is $\frac{a}{a+b}$

and probability of not happening an event is $\dfrac{b}{a+b}$.

$$\text{Let } p = \dfrac{a}{a+b} \text{ and } q = \dfrac{b}{a+b}, \text{ then } p + q = 1 \qquad \qquad .....(7.2)$$

P always lies between 0 and 1. If p = 1 and q = 0, the event will certainly happen. If p = 0 and q = 1 the event will certainly not happen. p/q refers to the odds in favour of the event and q/p as the odds against the event.

EXAMPLE: (1) Find the probability of drawing an ace from a pack of cards.

Total No. of favourable cases for drawing an ace = 4

Total No. of equally likely cases=52.

$$\text{Probability} = \dfrac{4}{52} = \dfrac{1}{13}$$

EXAMPLE: (2) An urn contains 10 red, 15 black and 20 white balls. A ball is drawn at random from the urn. What is the probability that being a black ball ?

Total No. of favourable cases for drawing a black ball = 15

Total No. of equally likely cases = 45

Probability = 15/45 = 1/3.

7.2   MUTUALLY EXCLUSIVE EVENTS

Events are said to be mutually exclusive and exhaustive if the occurrence of any one of the events excludes the occurrence of any other event at a particular occasion.

Suppose an urn contains 10 white, 5 black and 15 red balls. A ball is drawn at random and it is required that the ball should be either white or black. Here there are two events since either white or black ball can come in a single draw. If white ball comes first then black ball cannot come and *vice versa*. Hence these two events are mutually exclusive. Similarly in a lottery a bowl contains numbers from 1 to 100. A number is drawn at random from the bowl. Supposing that number 5 and a number ending with 5 comes in this draw, the owner of that number will be awarded a prize. Here any one of the numbers 5, 15, 25,..., 95 can come in the draw. If the number 5 comes first it excludes the occurrence of other numbers. Therefore, these are the mutually exclusive events.

7.2.1 Addition Rule

Let $A_1, A_2, \dots , A_N$ be N mutually exclusive events then the probability of occurrence of any one of the events is the sum of the probabilities of the separate events.

$$p\,(A_1 \text{ or } A_2 \text{ or,}\dots A_N) = p\,(A_1) + p\,(A_2) + \dots + p\,(A_N) \qquad .....(7.3)$$

EXAMPLE: What is the probability of drawing a card either of a spade or a heart from a pack of cards?

Pack of cards contains 4 suits each consists of 13 cards. Probability of drawing a card of spade = 13/52, probability of drawing a card of heart = 13/52.

$\therefore$ Probability of drawing either spade or heart card

$$= \frac{13}{52} + \frac{13}{52} = \frac{1}{2}$$

7.3   MUTUALLY INDEPENDENT EVENTS

Events are said to be mutually independent, if the occurrence of any one of the events does not effect the occurrence of any other event.

Drawing two cards of heart simultaneously from two packs of cards, drawing one black and one white ball simultaneously from two urns containing black, white, red balls are some of the examples of mutually independent events.

7.3.1 Multiplication Rule

Let A_1, A_2... , A_N be N mutually independent events then the probability of the simultaneous occurrence of all the events is the product of all the probabilities of the individual events.

$$p(A_1, A_2, ..., A_N) = p(A_1)\, p(A_2) ... p(A_N) \qquad .....(7.4)$$

EXAMPLE: What is the probability of obtaining two heads successively in tossing a coin two times?

Probability of obtaining 'Head' from the first trial is $\frac{1}{2}$ and the probability of obtaining 'Head' in second trial also is $\frac{1}{2}$, since the outcome of the second trial is independent of the first and two events are mutually independent. Therefore, the probability of getting two 'Heads' successively in two trials is

$$\frac{1}{2}.\frac{1}{2} = \frac{1}{4}$$

EXAMPLE: One card is drawn from each of the two packs of cards. What is the probability that both of them are Aces?

Probability of drawing an 'Ace' from the first pack = 4/52, Probability of drawing an 'Ace' from the second pack = 4/52. Since the above two events are mutually independent, probability of drawing an 'Ace' from each of the two packs=1/13×1/13 = 1/169.

7.3.2 Not mutually exclusive events

Suppose that from a pack of cards two cards are to be drawn which contain one ace. Here the first card is replaced before drawing the second. The first event is that ace comes first or and the second event is that ace comes in the second but it also happens that the events may happen together. That is, the two Aces may come from both the draws. In that case the two events are not mutually exclusive.

Rule: Let A_1, A_2 are two events which are not mutually exclusive, then:

$$p(A_1 \text{ or } A_2 \text{ or both}) = p(A_1) + p(A_2) - p(A_1A_2) \text{ or } p(A_1UA_2) = p(A_1) + p(A_2) - p(A_1A_2) \quad \ldots\ldots(7.5)$$

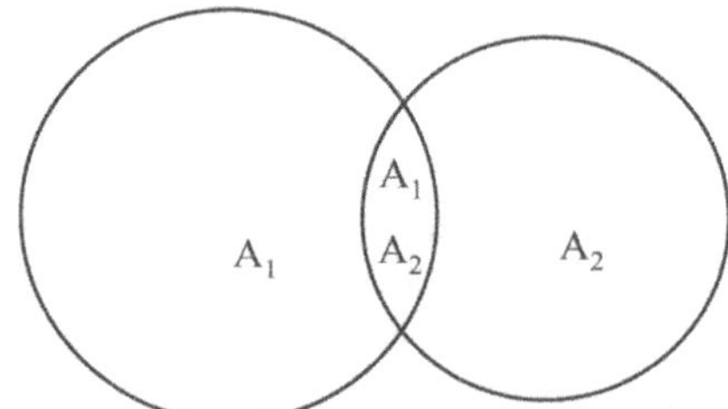

Fig. 7.1 Not mutually exclusive events.

In the case of three events, we have

$$p(A_1UA_2UA_3) = p(A_1) + p(A_2) + p(A_3) - p(A_1A_2) - p(A_2A_3) - p(A_3A_1) + p(A_1A_2A_3) \quad \ldots\ldots(7.6)$$

7.4   DEPENDENT EVENTS

The events are said to be dependent if the occurrence of any one of them depends on the occurrence of any other event.

Let A_1, A_2 are two dependent events then the probability of the simultaneous occurrence of the two events is given by the relation,

$$p(A_1A_2) = p(A_1) P(A_2/A_1) \text{ or } p(A_2/A_1) = p(A_1A_2) / p(A_1) \quad \ldots\ldots(7.7)$$

Also $p(A_1A_2) = P(A_2) p(A_1/A_2)$ or $p(A_1/A_2) = p(A_1A_2)/p(A_2)$

It may be noted that the multiplication rule holds good when the events are dependent or independent.

EXAMPLE: An urn contains 5 white, 10 black and 6 red balls. Two balls are drawn in succession that being 1 white and 1 black: Find the probability that *(i)* If the first ball is not replaced, *(ii)* if the first ball is replaced.

Case *(i)* *(a)* If the first ball is not replaced and it is white. Probability of drawing white ball =5/21.

Now, one white ball is taken out, and the remaining balls in the urn are 20.

Probability of drawing black ball if the first ball is white = 10/20.

Probability of drawing 1 white and I black, ball = 5/21 × 10/20 = 5/42.

***(b)* If the first ball drawn is black, the probability of drawing a black bail = 10/21.**

Now one black ball is taken out and the remaining balls in the urn are 20.

Probability of drawing a white ball if the first ball is black = 5/20.

Probability of drawing 1 white and 1 black ball= 10/21 × 51/20 =5/42.

Since *(a)* and *(b)* are two mutually exclusive events, the probability of drawing 1 white and 1 black ball is 5/42 + 5/42 = 5/21.

Case *(ii) (a)* If the first ball is replaced and it is white.

Probability of drawing a white ball = 5/21. As the first ball is replaced, the total number of balls in the urn remains same.

Therefore, the probability of drawing black ball 10/21.

Since these two events are independent, the probability of drawing 1 white and 1 black ball is 5/21 ×10/21= 50/441.

***(b)* If the first ball is replaced and it is black**

Probability of drawing black ball= 10/21

Probability of drawing,white ball when and black ball is $= \dfrac{5}{12}$

Pr obability of drawing, 1 white and 1 black balls is $\dfrac{10}{21}, \times \dfrac{5}{21}, = \dfrac{50}{441}$

Since *(a)* and *(b)* events are mutually exclusive, the probability of drawing 1 white and 1 black ball $= \dfrac{50}{441} + \dfrac{50}{441} = \dfrac{100}{441}$

7.5   SUB-POPULATIONS

A population of n units or elements consists of $\binom{n}{r}$ different sub-populations of size $r \le n$.

Let A, B, C be three letters. These three letters can be arranged in 3! ways, as ABC, ACB, BCA, BAC, CAB and CBA. These are also called as permutations. Now out of these three letters, two letters are to be selected at random irrespective of the order of the two letters. This can be done in 3 ways as AB, AC and BC. These are called as combinations. These can be obtained by the formula $\binom{3}{2} = \dfrac{3!}{2!(3-2)} = 3$. It may be noted that 0! = 1.

Similarly r units can be arranged in rl ways and the population of n units can be sub-divided into $\binom{n}{r}$ sub-populations each of size r where $r \leq n$ and the order of the units in the sub-populations is not taken into account

EXAMPLE: In how many ways the pack of cards can be divided into groups of 13 in each hand in a game of bridge?

Since the order of the cards in each hand is immaterial, the total number of ways 52 cards can be divided into groups of 13 each is $\binom{5\ 2}{1\ 3}$.

7.6   PROBABILITY BASED ON BINOMIAL DISTRIBUTION

Let P_r be the probability of exactly r successes out of n trials with p, the probability of success and q = (1–p), probability of failure, then

$$p_r = \binom{n}{4} p^r q^{n-r}.$$

EXAMPLE: (1) If a coin is tossed 4 times and the turning up of head is taken as success, the probability of exactly 2 successes out of '4' trials is obtained as follows.

The two heads may turn up in any of the four trials as H H T T, H T H T, T H H T, H T T H, T H T H, T T H H *i.e.,* in 6 ways

The probability of getting Head is $\dfrac{1}{2}$ and the probability of getting two heads is $\left(\dfrac{1}{2}\right)^2$. Probability of getting tail is $\dfrac{1}{2}$ and the probability of getting two tails is $\left(\dfrac{1}{2}\right)^{4-2}$. As these are dependent events, the probability of getting two heads and two tails is $\left(\dfrac{1}{2}\right)^2 \left(\dfrac{1}{2}\right)^{4-2}$.

The probability for each of the 6 ways is $\left(\dfrac{1}{2}\right)^2 \left(\dfrac{1}{2}\right)^{4-2}$.

Since all these six events are mutually exclusive, we have

$$\left(\dfrac{1}{2}\right)^2 \left(\dfrac{1}{2}\right)^{4-2} + \ldots + \left(\dfrac{1}{2}\right)^2 \left(\dfrac{1}{2}\right)^{4-2} = \binom{4}{2}\left(\dfrac{1}{2}\right)^2 \left(\dfrac{1}{2}\right)^{4-2}.$$

EXAMPLE: (2) Find the probability of getting at least 3 Heads if a coin is tossed 7 times.

The probability of getting 3 heads out of 7 trials is $\binom{7}{3}\left(\dfrac{1}{2}\right)^3\left(\dfrac{1}{2}\right)^4$;

Similarly, the probability of getting 4 heads, 5 heads, 6 heads and 7 heads are $\binom{7}{4}\left(\dfrac{1}{2}\right)^4\left(\dfrac{1}{2}\right)^3$, $\binom{7}{5}\left(\dfrac{1}{2}\right)^5\left(\dfrac{1}{2}\right)^2$ $\binom{7}{6}\left(\dfrac{1}{2}\right)^6\left(\dfrac{1}{2}\right)$ and $\left(\dfrac{1}{2}\right)^7$ respectively.

Hence the probability of obtaining at least 3 heads out of 7 trails is

$$\binom{7}{3}\left(\dfrac{1}{2}\right)^3\left(\dfrac{1}{2}\right)^4+\binom{7}{4}\left(\dfrac{1}{2}\right)^4\left(\dfrac{1}{2}\right)^3+\binom{7}{5}\left(\dfrac{1}{2}\right)^5\left(\dfrac{1}{2}\right)^2+\binom{7}{6}\left(\dfrac{1}{2}\right)^6\left(\dfrac{1}{2}\right)+\left(\dfrac{1}{2}\right)^7 \text{ or}$$

$$1-\text{prob of obtaining at most 2 heads} = 1-\sum_{i=0}^{2}\binom{7}{i}\left(\dfrac{1}{2}\right)^i\left(\dfrac{1}{2}\right)^{7-i}.$$

EXAMPLE: (3) Two throws are made, the first with three dice and the second with two. The faces of each die are numbered from 1 to 6. What is the probability that the total in the first throw is not less than 16 and at the same time the total in the second throw is not less than 10.

First throw is done with three dice. The following are the number of ways in which the total in the throw with three dice can be from 16 to 18. Each of the combinations 466, 565, 566 will have three permutations each and 666 will have one and hence the total number of favourable cases for obtaining total 16 to 18 is 10. The total number of equally likely cases is $6 \times 6 \times 6 = 6^3$ Therefore, the probability $= 10/6^3$.

Similarly with two dice the combinations 46 and 56 will have two permutations each and the combinations 55 and 66 will have one permutation each and the total number of favourable ways for obtaining total 10 to 12 is 6. The total number of equally likely cases is $6 \times 6 = 6^2$. Therefore the probability is $6/6^2 = 1/6$.

Since the two events are independent the probability of happening the two events simultaneously is $10/6^3 \times 1/6 = 10/6^4$.

EXAMPLE: (4) There are two urns. One of them contains 6 red, 8 white and 10 blue balls and the other contains 10 red, 6 white and 12 blue balls, One-ball is transferred from the first urn to the second urn and a ball is drawn from the second urn. What is the probability that it is a blue ball ?

Case (i) If a red ball is transferred from the first urn to the second, then there will be 11 red, 6 white and 12 blue balls in the second urn. The probability of drawing a red ball from 1st urn is 6/24. The probability of drawing a blue ball from 2nd urn is 12/29. Hence the probability of drawing a ball from the 2nd urn when a red ball is placed from the 1st urn to 2nd urn is 6/24 × 12/29.

Case (ii) If a white ball is transferred from the first urn to the second urn, then there will be 10 red 7 white and 12 blue balls in the second urn. The probability of drawing a white ball from 1st urn is 8/24. The probability of drawing a blue ball from 2nd urn is 12/29. The probability of drawing a blue ball from 2nd urn when white ball is transferred from 1st urn to 2nd urn is 8/24 × 12/29.

Case (iii) If a blue ball is transferred from the first to the second urn, then there will be 10 red, 6 white and 13 blue balls In the second urn. The probability of drawing a blue ball from the 1st urn is 10/24. The probability of drawing a blue ball from 2nd urn is 13/29. The probability of drawing a ball from 2nd urn when a blue ball is transferred from 1st urn to 2nd urn is 10/24×13/29. Since the above three cases are mutually exclusive and exhaustive the required probability = 6/24 × 12/29 + 8/24 × 12/29 + 10/24 × 13/29 = 298/696.

7. 7   BAYES THEOREM

An event A can be described by a set of n exhaustive and mutually exclusive hypotheses $B_1, B_2,..., B_n$. Given

(i) a priori probabilities $P(B_1), P(B_2)...P(B_n)$; $P(B_i) \neq 0$ for $i = 1, 2,...,n$ corresponding to a total absence of a knowledge regarding the occurrence of A and

(ii) conditional probabilities $P(A/B_1), P(A/B_2)...P(A/B_n)$

The a posteriori probabilities $P(B_i/A)$ are obtained as

$$P(B_i/A) = \frac{P(B_i)\,P(A/B_i)}{\sum_{i=1}^{n} P(B_i)\,P(A/B_i)}$$

EXAMPLE: A factory has 3 machines A, B and C. A produces 4000 injection needles, B produces 5000 and C produces 3000 per day. Machine A produces 2%, B produces 1% and C produces 3% defective needles. A needle is checked at random at the end of a day and is found to be defective. What is the probability that it is produced by machine B.

Let P(A), P(B) and P(C) be the probabilities of production of injection needles by machines A, B and C, respectively.

$$P(A) = \frac{4000}{4000 + 5000 + 3000} = \frac{1}{3}$$

$$P(B) = \frac{5000}{4000 + 5000 + 3000} = \frac{5}{12}$$

$$P(C) = \frac{3000}{4000 + 5000 + 3000} = \frac{3}{12} = \frac{1}{4}$$

Let P(D/A) P(DIB) and P(D/C) be the probabilities of producing defective needles from machines A, B and C, respectively.

$$P(D/A) = \frac{2}{100} = 0.02, \ P(D/B) = \frac{1}{100} = 0.01, \ P(D/C) = \frac{3}{100} = 0.03$$

$$P(B/D) = \frac{P(B).P(D/B)}{P(A).P(D/A) + P(B).P(D/B) + P(C).P(D/C)}$$

$$= \frac{\dfrac{5}{12} \times 0.01}{\dfrac{1}{3} \times 0.02 + \dfrac{5}{12} \times 0.01 + \dfrac{3}{12} \times 0.03} = \frac{0.05}{0.22} = 0.23$$

EXERCISES

1. Four balls are drawn from a bag containing 5 red 6 black and 10 white balls. Find the probability that they are 2 black and 2 red balls.

2. There are two bags one of which contains 5 red and 8 black balls and the other 7 red and 10 black balls and a ball is to be drawn from one or other of two bags. Find the chance of drawing a red ball.

   (M.U.,1946)

3. A coin is tossed 10 times. What is the probability of getting heads exactly as many times in the first seven throws as in the last three?

   (Tr.Uni.,1946)

4. A and B throw two dice in turn. 1 hose who throw 8 will be declared as winner. What is the probability of A winning if he starts first?

5. A bridge player and his partner have 8 diamonds between them. What is the probability that the other pair have 5 diamonds in the ratio 4: I?

6. A seed production firm produces seeds and sells in packets each containing 500 seeds. Packets are inspected by taking 20 seeds from each packet. If no seed is found defective in each packet then it is accepted otherwise it is rejected. What is the probability that a packet is accepted with 5 defective seeds?

7. A box contains 3 red and 7 white balls. One ball is drawn at random, and in its place a ball of the other colour is put in the box. Now one ball is drawn at random from the box. Find the probability that it is red.

8. Assuming that the ratio of male children to be $\frac{1}{2}$, find the probability that in a family with 6 children (i) all children will be of the same sex (ii) the 4 oldest children will be girls.

9. Four balls are drawn from a bag containing 5 black, 6 white 2 red and 7 blue balls. Find the probability that the balls drawn are all different colour.

10. A bag contains 6 white and 8 black balls. If 4 balls are drawn at random, find the probability that (i) 2 are black (ii) not more than 2 are black.

11. Find the probability of obtaining at least 4 successes out of 6 trials by tossing a coin 6 tines,

12. What is the probability of selecting either medium or short duration variety out of 10 short, 15 medium and 6 long duration varieties available in paddy from 'seed' shop.

13. There are two baskets A, B. A contains 20 Apples, 30 Mangoes and 40 Oranges. B contains 30 Apples, 20 Mangoes and 50 Oranges. A fruit is transferred from A to B and a fruit is drawn at random from B. What is the probability that it is Apple?

14. What is the probability of not obtaining 'HEAD' by tossing a coin 4 times?

15. Three seed farms A, B and C belong to same corporation produce seed sample packets each containing 200 seeds in the ratio 2: 3: 5. A seed packet is found defective if it contains 10 or less defective seeds. A, B and C farms produce defective sample packets with 4, 5 and 1 percent respectively. A seed sample packet is drawn from the mix at random and it is found defective. What is the probability it is from farm B?

16. Three dairy farms A, B and C produce milk with 30, 25 and 45 percent respectively. A milk packet is found defective if it contains fat percentage 3 or less. Three farms A, B and C produce defective milk packets in the ratio 2: 3: 5. A milk packet is drawn at random and it is found defective and what is the probability that it is from farm C ?

CHAPTER 8

BINOMIAL AND POISSON DISTRIBUTIONS

8.1 BINOMIAL DISTRIBUTION

Bernoulli trials: Repeated independent trials with two possible outcomes at each trial are known as Bernoulli trials. The probability of success (failure) at each of the outcomes is assumed to be same throughout the experiment.

For example, if a coin is tossed ten times, at each trial the outcome may be either Head or Tail. If Head turns up it may be called a success and Tall for failure. Therefore, in ten trials, there will be ten outcomes which consist of either success or failure or both. The outcome of each trial is independent of the other. The probability of each success is equal and the probability of each failure is equal throughout the experiment. Similarly, there are 20 seed samples each sample consists of say 100 seeds. A sample may be accepted if 10 or less seeds do not germinate otherwise rejected. There are 20 outcomes in which each sample may be either rejected or accepted. These 20 outcomes may be considered as 20 Bernoulli trials.

The probability of r successes out of n trials or at least or utmost r successes out of n trials is obtained as follows.

If the probability of success is denoted by p and the failure by q then

$p + q = 1$. There will be $\binom{n}{r}$ ways in which r successes can occur in n trials. The probability of each combination of r successes and (n–r) failures is $p^r q^{n-r}$. Since each trial is independent from the other, the $\binom{n}{r}$ combinations are mutually exclusive and therefore their probabilities are added. Let b_r be the probability of r successes out of n trials, where

$$b_r = \binom{n}{r} p^r q^{n-r}$$

The probabilities in Table 8.1 are the (n+1) terms of a well known expansion called Binomial expansion for $(q + p)^n$, where

TABLE 8.1

Success	Probability	Frequency
0	q^n	$N.\,q^n$
1	$\binom{n}{1} pq^{n-1}$	$N.\,\binom{n}{1} pq^{n-1}$
2	$\binom{n}{2} p^2 q^{n-2}$	$N.\,\binom{n}{2} p^2 q^{n-2}$
.	.	.
.	.	.
r	$\binom{n}{r} p^r q^{n-r}$	$N.\,\binom{n}{r} p^r q^{n-r}$
.	.	.
.	.	.
N	p^n	$N.\,p^n$
Total	1	N

$$(q + p)^n = q^n + \binom{n}{1} p\, q^{n-1} + \ldots + \binom{n}{r} p^r p^{n-r} + \ldots + p^n$$

Therefore, the distribution of successes given in Table 8.1 is known as Binomial distribution. Since the numbers of successes are the values of a discrete variable, Binomial distribution is a discrete distribution. The different frequencies of successes can be obtained by multiplying the respective probabilities with the total frequency and are presented in 3rd column of Table 8.1 where N is the total number of trials or total frequency. This frequency distribution can be represented by diagram in Fig. 8.1

8.2 PROPERTIES OF BINOMIAL DISTRIBUTION

Mean $= np$; $\mu_2 = npq$; $\sigma = \sqrt{npq}$; $\mu_3 = npq\,(q-p)$

$$\mu_4 = 3p^2 q^2 n^2 + pqn\,(1 - 6\,pq)$$

Coefficient of skewness $= \sqrt{\beta_1} = (q - p) / \sqrt{npq}$

Coefficient of Kurtosis $= \beta_2 = 3 + (1 - 6\,pq)/npq$

The central term is obtained in order to obtain successive probabilities after knowing the probability of preceding number of success as follows.

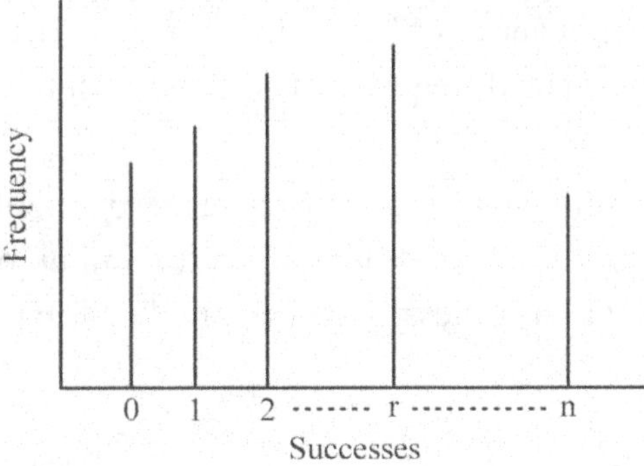

Fig. 8.1 Discrete distribution (Binomial).

The central term is given as

$$b_r / b_{r-1} = \binom{n}{r} p^r q^{n-r} / \left(n_{r-1}\right) p^{r-1} q^{n-r+1}$$

$$= \left\{ \frac{n-r+1}{r} \right\} p/q$$

$$b_r = \left\{ \frac{n-r+1}{r} \right\} p/q, \; b_{r-1} = \left\{ 1 + \frac{(n+1)p - r}{rq} \right\} b_{r-1}$$

If $r < (n + 1) \, p$; $b_r > b_{r-1}$

If $r > (n + 1) \, p$; $b_r < b_{r-1}$

If $r = (n + 1) \, p$; $b_r = b_{r-1}$

Therefore there exists exactly one integer 's' such that $(n + 1) \, p - 1 < s$, $< (n + 1)p$.

As r takes values from 0 to n, the value of b_r increases monotonically first and reaches maximum when r = s except when $b_r = b_{r-1}$ for $s = (n + 1)p$ and decreases monotonically later on. As n increases, p remains fixed, the binomial distribution tends to the normal distribution which we discuss in chapter 9.

In this distribution 'n' and 'p' are the parameters to be estimated.

8.3   FITTING OF THE BINOMIAL DISTRIBUTION

If 'p' is estimated from the sample data then the expected frequencies can be obtained.

Further the significance of the difference between the observed frequencies and the expected frequencies is tested with the help of chi-square test.

EXAMPLE: The following is the frequency distribution of successes obtained by throwing 10 dice 200 times. The turning up of an even number on a die is considered as a success. Fit a Binomial distribution.

TABLE 8.2

Success (r)	Frequency	$(f_i X_r)$	b	Expected frequency
0	3	0	.000798	0.1596
1	7	7	.008304	1.6608
2	13	26	.038893	7.7786
3	18	54	.106688	21.3376
4	30	120	.194324	38.8648
5	45	225	.242707	48.5414
6	38	228	.210511	42.1022
7	20	140	.125202	25.0404
8	16	128	.048867	9.7734
9	9	81	.011303	2.2606
10	1	10	.001190	0.2380
	200	1019		

$$np = \frac{\Sigma f_{ir}}{\Sigma f_i} = 1019/200 = 5.095$$

$p = np/n = 5.095/10 = 0.5095 = 0.51, q = 0.49$

$b_0 = q^n = (0.49)^{10} = 0.0007980, Nb_0 = 200 \times 0.0007980 = 0.1596$

Using central term.

$$b_1 = \frac{n-1+1}{1} \cdot \frac{p}{q} \, b_0 = 0.008304$$

Similarly all the expected probabilities $b_2, \ldots, b_{10}$ were calculated and furnished in column (4) of Table 8.2. The expected frequencies are obtained by multiplying each value in column (4) by the total frequency and furnished in column (5) of Table 8.2.

8.4   POISSON DISTRIBUTION

In the Binomial distribution as n increases and the probability of success decreases and np remains fixed, the Binomial distribution tends to Poisson distribution.

Using central term in binomial distribution

$$b_r = (n - r + 1) / r \times p/q \times b_{r-1}$$
$$= [np - (r - 1)p] / rq \times b_{r-1}$$

as $n \to \infty$ and $np = m$ (fixed), we have

$$b_r \approx m/r \, b_{r-1}$$
$$b_0 = q^n = (1-p)^n = (1-m/n)^n \text{ since } p = m/n$$

$$\text{Log } b_0 = n \text{ Log } (1-m/n) = -m - m^2/2n - \ldots\ldots$$

$$b_0 \approx e^{-m}$$

For $r = 1$, $b_1 = me^{-m}$; $r = 2$, $b_2 = m^2/2! \, e^{-m}$; .....

$$r = k, \; bk = \frac{m^k}{k!} e^{-m}$$

By induction process the probability of r successes in poisson distribution is given by $b_r = (m^r / r!) \, e^{-m}$. This distribution was first developed by S.D. Poisson in 1837. Since the numbers of successes are the values of a discrete variate the poisson distribution is a discrete distribution.

8.5 PROPERTIES OF POISSON DISTRIBUTION

Mean $= m$; $\mu_2 = m$; $\sigma = \sqrt{m}$; $\mu_3 = m$; $\mu_4 = 3m^2 + m$.

It may be noted that the first three moments of a poisson distribution are same. Coefficient of skewness $\sqrt{\beta_1} = m/m^{3/2} = 1 / \sqrt{m}$; coefficient of Kurtosis $= 3 + 1/m$, m is the parameter of this distribution.

This distribution is found to be much useful in the cases where the probability of success is small. The number of deaths due to car accidents in a day, number of printing mistakes in a page of a book, the number of plants infested with a particular disease in a plot of a field, number of defective screws manufactured in a day by a particular factory, number of deaths of centenarians in a year, bacteria counts on a Petrie plate where the plate is divided into small squares, number of weed plants of a particular species in different plots of a field, are some of the common examples which follow poisson distribution.

8.6 FITTING OF A POISSON DISTRIBUTION

We would like to know sometimes whether the sample data follows the poisson distribution or not. For that the value of the parameter, m has to be estimated. Further, e^{-m} has to be obtained. For different values of r, the different expected (theoretical) frequencies are obtained using the formula N. $e^{-m} \, m^r / rl$ where $N = \Sigma f_1$. The difference between the observed and theoretical frequencies are tested with the help of the chi-square test for its significance.

EXAMPLE: The following is the frequency distribution of weed plants of a particular species in 100 plots of a field. Fit a poisson distribution to the sample data.

TABLE 8.3

No. of weed plants of a particular (r)	No.pf plots (f₁)	$f_1 \times r$	Expected frequencies $N.e^{-m} \dfrac{m^r}{r!}$
0	4	0	2.73
1	9	9	9.84
2	12	24	17.70
3	20	60	21.24
4	25	100	19.12
5	18	90	13.77
6	8	48	8.26
7	3	21	4.21
8	1	8	1.90
	100	360	98.77

$$m = \Sigma f_1 \times r / \Sigma f_1 = 360/100 = 3.6$$

$$\text{Log } e^{-m} = \text{Log } e^{-3.6} = -3.6 \times 0.4343 = -1.5635$$

$$e^{-m} = 0.02732$$

The central term is $p_r = m/_r \times p_{r-1}$

$$P_o = e^{-m} m^o/0! = e^{-m} = 0.02732$$

Expected frequency = $N p_o = 100 \times 0.02732 = 2.73$

$r = 1$ $p_1 = 3.6/1 \times p_o = 3.6 \times 0.02732 = .09835$

$Np_1 = 100 \times 3.6 \times 0.0732 = 9.84$

$r = 2$ $Np_2 = 100 \times [3.6/2] \times .09835 = 17.70$

$$\vdots \qquad\qquad \vdots \qquad\qquad \vdots$$

$r = 8$ $Np_8 = 100 \times [3.6/8 \times .0421 = 1.90$

These expected frequencies are furnished in column (4) of Table 8.3.

The total of the theoretical frequency should be equal to the total of observed frequency. For testing the significance of the differences between observed and theoretical frequencies the reader is advised to refer the chapter on chi-square distribution.

EXERCISES

1. What is the probability of getting two Aces in a hand of bridge?

2. In a cytrus orchard, it is found that 10% of them are affected by a particular disease. Obtain the probability that exactly 15 trees were affected out of 100 trees?

3. 10 dice are thrown 300 times. If 1 and 3 appear on a dice it was considered a success. The following is the frequency distribution of successes.

Success	0	1	2	3	4	5	6	7	8	9	10
Frequency	3	10	22	40	60	72	54	30	6	2	1

   Obtain the theoretical frequencies assuming that it follows a binomial distribution.

4. From a pack of cards, 4 cards are drawn at random for100 times. The number of red cards are recorded from each of the 4 cards. Obtain the theoretical frequencies for 0, 1, 2, 3 and 4 red cards and also the expected number of red cards in all the drawings.

5. 100 seed samples of 10 each were tested for germination. The following are the number of seeds germinated with different frequencies.

No. of seeds germinated	0	1	2	3	4	5	6	7	8
No. of samples ...		2	5	8	20	30	24	6	4	1

   Fit a Binomial distribution to the above data.

6. The number of noxious and weed seeds were recorded in160 samples.

No. of noxious and weed seeds per sample	0	1	2	3	4	5
No. of samples	41	35	34	17	23	10

   Fit a Poisson distribution to the above data.

7. The following were the results obtained by conducting an experiment of chromosome interchanges induced by X-ray Irradiation of first experiment (Data are due to D.G. Catche side, D.E. Lea and J.N. Thoday).

Cells with K interchanges	0	1	2	3
Observed No. of cells	753	266	49	5

   Fit a Poisson distribution to the above data.

8. The following are the observed number of squares on a Petri plate with K dark spots (counts of Bacteria) in first experiment (Due to J. Neyman).

K counts of bacteria	0	1	2	3	4	5	6
Observed no. of squares	5	19	26	26	21	13	8

Fit a Poisson distribution to the above data.

9. A coin is tossed 6 times in succession and a person will get 1, 2, 3, 4, 5 and 6 rupees respectively, if head turns on the 1st, 2nd, 3ed, 4th, 5th and 6th occasion.

If tail appears he will have to give the same amounts. Find his expected gain and the variance.

(B.Sc. Madras 1968 Sept.)

10. The number of accidents in a year to taxi-drivers in a city follows a poisson distribution with mean equal to 3. Out of 1000 Taxi-drivers, find approximately the number of drivers with

    (i)   no accident in a year

    (ii) more than 3 accidents in a year.

11. The following is the frequency distribution of 'printing errors' per page

No. of printing errors	0	1	2	3	4	5	6
No. of pages	60	50	16	10	9	2	1

Fit a poisson distribution to the data.

12. In certain district the incidence of rinderpest disease in cattle was found to be 8 percent in a dairy farm consisting of 210 animals. Find the average number of animals effected with the disease, standard deviation, coefficient of skewness and coefficient of Kurtosis assuming that the incidence of disease follows Binomial distribution.

13. The incidence of nutritional disorder in children of tribal population was found to be 3 percent. Assuming that the incidence follows poisson distribution, find the expected average number of children effected with nutritional disorder in a population of 12,000, the standard deviation, coefficient of skewness and coefficient of kurtosis.

14. Find the 'coefficient of skewness' and coefficient of Kurtosis in Binomial distribution gives the following

N = 200, p = 0.8

State also the type of 'skewness' and 'kurtosis'

15. The incidence of 'white fly' attack in cotton farms was found to be 15 percent in Prakasam district of Andhra Pradesh. In a Mandal consisting of 1000 farms in Prakasam district find the number of farms effected with 'white fly' incidence, standard deviation, all central moments, coefficient of skewness, coefficient of Kurtosis assuming that white fly attack on cotton crop follows Binomial distribution.

16. The following is the distribution of male births in different hospitals in a city on a particular day.

Male birth	0	1	2	3	4	5
Hospitals	6	8	12	9	7	8

Fit a binomial distribution to the above distribution.

CHAPTER 9

NORMAL DISTRIBUTION

In the case of Binomial distribution, the probability of r successes in n trials is given by

$$b_r = \frac{n!}{r!(n-r)!}p^r q^{n-r} \qquad \dots\dots(9.1)$$

Using Sterling's approximation as $n! \sim \sqrt{2\pi}\, n^{\frac{n+1}{2}} e^{-n}$ for the factorials in the above expression as $n \to \infty, r \to \infty$ and p is fixed, we have

$$b_r \sim \frac{1}{\sqrt{2\pi npq}} e^{\frac{-(r-np)^2}{2npq}} \qquad \dots\dots(9.2)$$

where np, and npq are the mean and variance respectively in the Binomial distribution and ' $\sim$ ' denotes approximation. The proof is not dealt here as it is beyond the scope of this book. The expression can be rewritten as

$$f(X) = \frac{1}{\sigma\sqrt{2\pi}} e^{\frac{-(X-\mu)^2}{2\sigma^2}} \qquad \dots\dots(9.3)$$

where f(X) is the density function of a continuous distribution called Normal distribution. This is also known as Gaussian distribution. The mean 'μ' and standard deviation 'σ' are the parameters of a normal distribution. The curve for f(X) is also called as normal probability curve and normal curve of error. This was first developed by A. Demoivre in 1718, later independently worked out by Laplace in 1812 and Gauss (1777–1855).

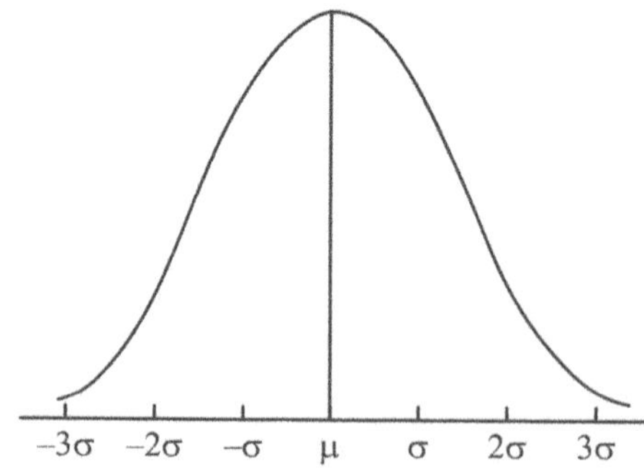

Fig. 9.1 Normal curve.

The area under the normal curve for the expression (9.3) between the ordinates at -3σ and 3σ in Fig. 9.1 is 99.7% of the total area.

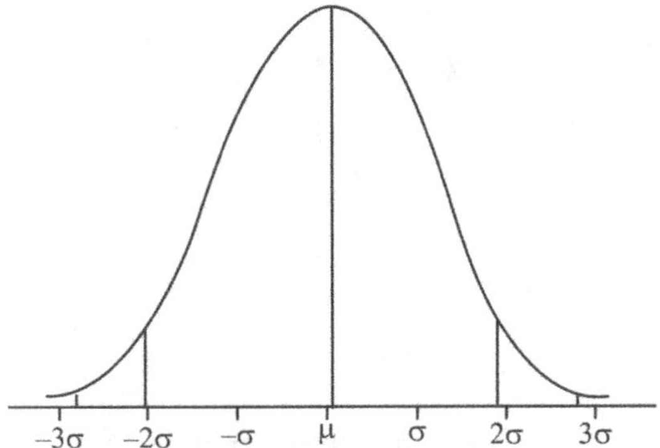

Fig. 9.2 Normal curve.

The area under the normal curve between the ordinates at -2σ and 2σ in Fig. 9.2 is 95.5% of the total area. The area under the normal curve between the ordinates at $-\sigma$ and σ is 68.3% of the total area as shown in Fig. 9.3.

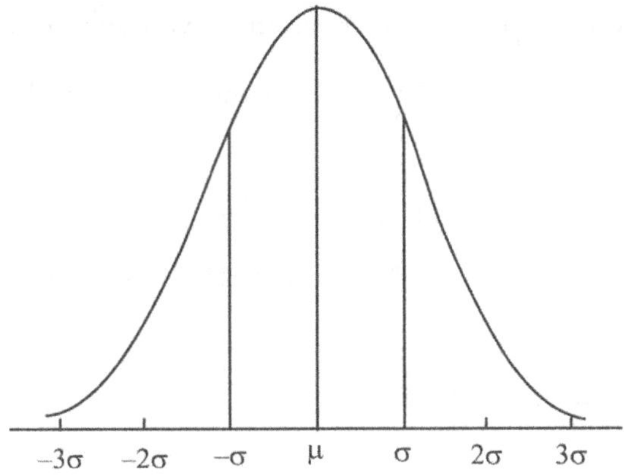

Fig. 9.3 Normal curve.

9.1   STANDARD NORMAL DISTRIBUTION

If X is the normal variate $Z=(X-\mu)/\sigma$ is called the standard normal variate or standard normal deviate which is distributed normally with mean zero and S.D. Unity.

$$f(Z) = \frac{1}{\sqrt{2\pi}} e -\frac{1}{2}Z^2 \qquad \qquad .....(9.4)$$

and
$$\int_{-\infty}^{\infty} f(Z) = 1$$

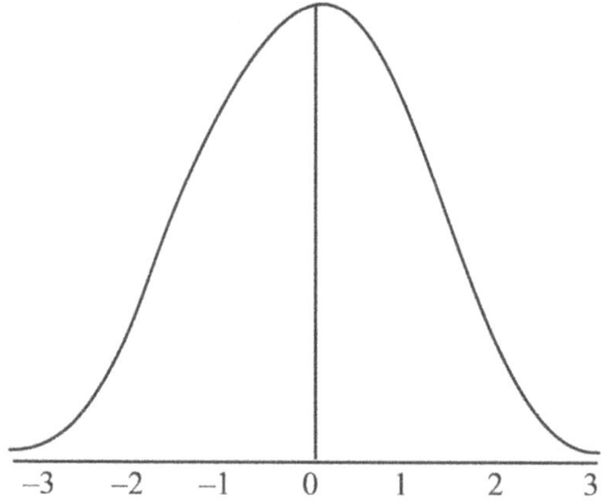

Fig. 9.4 Standard normal curve.

Here f(Z) is the density function of a standard normal variate. The curve of f(Z) which is shown in Fig. 9.4 is bell shaped and symmetric on either side of the mean. The total area under the curve is 1 sq unit. The maximum ordinate is at mean which is equal to $1/\sqrt{2\pi}$. Therefore, mode and mean coincide. This ordinate divides the curve into two equal halves each has area $\dfrac{1}{2}$ sq unit. Therefore, median also coincides with mean and mode.

9.2   FREQUENCY FUNCTION

If N is the total frequency of a Normal distribution, then the frequency function of X is given b

$$f(X) = \frac{N}{\sqrt{2\pi}\sigma} e^{\frac{-(X-\mu)^2}{2\sigma^2}} \qquad\qquad .....(9.5)$$

9.3   PROPERTIES OF NORMAL DISTRIBUTION

(1) The curve is bell shaped and tails off symmetrically on both sides of the mean, (2) The curve extends $-\infty$ to ∞, from (3) The second moment about mean (or variance) $\mu_2 = \sigma^2$, (4) The only maximum ordinate is at $\mu = N\Big/\sqrt{2\pi}\sigma$ (5) The odd moments about mean are zero, i.e., $\mu_1 = \mu_3 = ... = 0$.

(6) The fourth moment $\mu_4 = 3\sigma^4$, (7) Coefficient of skewness, $\sqrt{\beta_1} = \mu3/\mu_2^{3/2} = 0$ and (8) Coefficient of Kurtosis, $\beta_2 = \mu_4/\mu_2^2 = 3$.

9.4   DISTRIBUTION FUNCTION

$$F(X) = \int_{-\infty}^{X} \frac{1}{\sqrt{2\pi}\sigma} e^{\frac{-(X-\mu)^2}{2\sigma^2}} \qquad\qquad .....(9.6)$$

Is called the distribution function of X. In other words F(X) represents the area from $-\infty$ to X in the normal curve.

If $\qquad Z = (X-\mu)/\sigma$, we have

$$F(Z) = \int_{-\infty}^{Z} \frac{1}{\sqrt{2\pi}} e^{-\frac{1}{2}z^2} \qquad\qquad .....(9.7)$$

For different values of Z, the values of F(Z) are provided in the table called the Normal probability integral table. The values of F(Z) are given in proportions and the corresponding frequencies are obtained by multiplying F(Z) with N.

The area to the right of Z can be obtained by subtracting F(Z) from 1 as shown in Fig. 9.5.

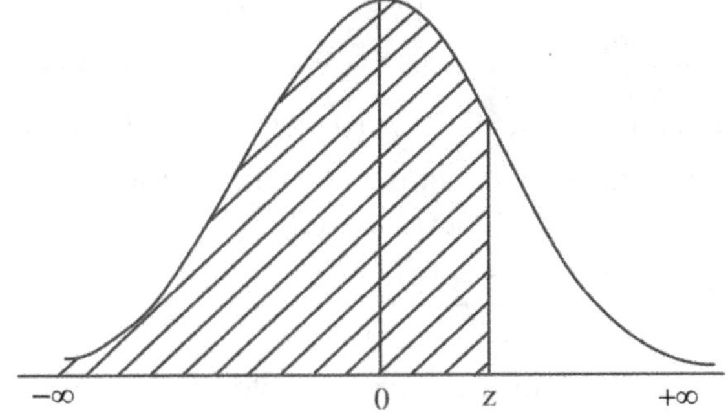

Fig. 9.5 Normal curve

EXAMPLE: If the marks of 1000 students in B.Sc. (Exam.) follow normal distribution with mean mark 40 and standard deviation 8, find the number of students *(i)* between 50 and 59 marks *(ii)* below 35 marks and *(iii)* 60 and above.

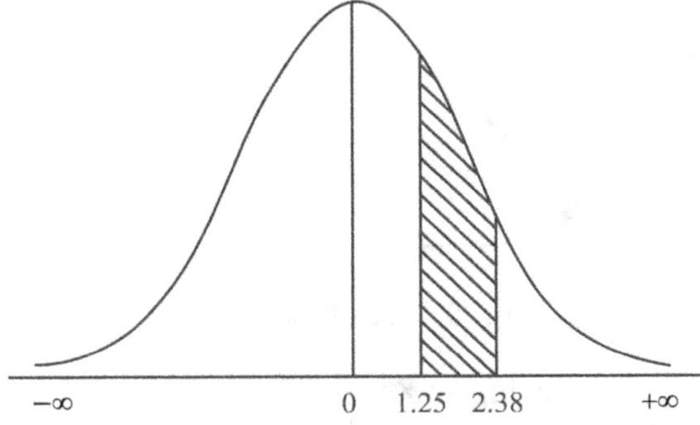

Fig. 9.6 Normal curve.

(i)   Here $\mu = 40$, $\sigma = 8$.

Let $X = 50$, then $Z = \dfrac{X_1 - \mu}{\sigma} = \dfrac{50 - 40}{8} = 1.25$

Let $X2 = 59$ then $Z_2 = \dfrac{59 - 40}{8} = 2.38$

The areas to the left of 1.25 and 2.38 can be obtained by interpolation from normal probability integral table as

		Z value	**Area to the left of Z**
		1.2	0.88493
		1.3	0.90320
By interpolation	...	1.25	0.89406
		2.3	0.98928
		2.4	0.99180
By interpolation	...	2.38	0.99130

Area between 2.38 and 1.25 is (0.99130–0.89406) =0.09724.

Hence the number of students scored in the range from 50 to 59 is 1000 X 0.09724 = 97 (approximately).

(ii)   Let $X = 35$, $Z = \dfrac{35 - 40}{8} = -\,0.63$

In the normal probability integral table, the areas are given only for positive values of Z. The areas to the left of negative value of Z are obtained as shown in Fig. 9.7.

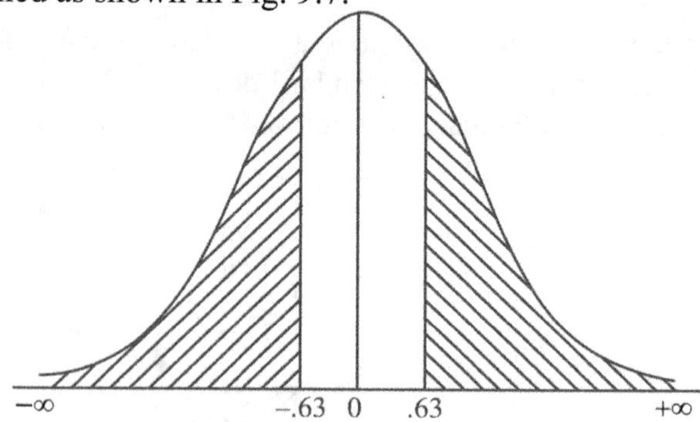

Fig. 9.7 Normal curve.

Hence the area to the left of Z= –0.63 is equal to the area to the right of Z = + 0.63 where 1 is the total area under the normal curve.

From table we have

	Z Values	Areas to the left of Z
	0.6	0.72575
	0.7	0.75804
By interpolation	0.63	0.73544

The area to the right of $0.63 = 1 - 0.73544 = 0.26456$ which is the area to the left of -0.63

Therefore, the number of students below 35 marks $= 0.26426 \times 1000 = 265$ (approximately)

(iii) Let $X = 60$, $Z = (60 - 40)/8 = 2.5$

Since the area to the right of 2.5 is equal to the I – area to the left of 2.5, we have from the table $1 - 0.99379 = .00621$.

Therefore, number of students who scored 60 and above $= 1000 \times 0.00621 = 6$ (approximately).

9.5   FITTING OF THE NORMAL DISTRIBUTION

The frequency distribution based on sample data can be fitted to the normal distribution by computing the estimates of two parameters μ and σ from the sample. The method of fitting is illustrated here with an example.

EXAMPLE: The following is the frequency distribution of marks obtained in an examination out of 100 by 900 students. Fit a normal distribution to the data.

Marks	No. of students	Limits (X)	$Z = \dfrac{X - \mu}{\sigma}$	Area to the left of Z	Area between the limits	Theoretical frequencies
1	2	3	4	5	6	7
0-10	13	$-\infty$	$-\infty$	0.000000	0.016586	14.9
10-20	42	10	−2.13	0.016586	0.016422	41.8
20-30	93	20	−1.53	0.063008	0.110602	99.5
30-40	175	30	−0.95	0.173610	0.189560	170.6
40-50	220	40	−0.35	0.363170	0.231660	208.5
50-60	196	50	0.24	0.594830	0.201900	181.5
60-70	88	60	0.83	0.796730	0.125466	112.9
70-80	45	70	1.42	0.922196	0.056112	50.5
80-90	23	80	2.02	0.978308	0.017165	15.5
90-100	5	90	2.61	0.995473	0.004527	4.1
	900	∞	∞	1.000000		900.0

$$\text{A.M.} \quad (\mu) = 45.94$$
$$\text{S.D.,} \quad \sigma = 16.9$$

The mean and S.D. were calculated in the usual way. Column (3) was obtained by identifying the data with normal distribution with lowest value equal to $-\infty$ instead of 0 and the highest values as $+\infty$ in place of 100. The standard normal deviate values (Z) were obtained in Col. (4). The area to the left of Z values were posted in Col. (5) by entering into the normal probability integral table. These values can be directly obtained from the Fisher & Yates table without resorting to interpolation for in between 'Z' values. Column (6) gives the areas lying in between two limits and were obtained by subtracting the values from the succeeding values of Col. (5). In other words these areas are the proportions of candidates who scored in between the corresponding two limits. Column (7) was obtained by multiplying each value in Col. (6) by the total frequency which are the theoretical frequencies in each group.

The differences between the observed and theoretical frequencies can be tested with the help of chi-square test for goodness of fit. If the differences are not significant then the fit is said to be good otherwise not.

EXERCISES

1. The height of barley plants in a field is assumed to: follow normal distribution with mean height 35" and standard deviation 4.0". A sample of 150 plants was selected from a plot of the field. Find the number of plants *(i)* having height more than 40" *(ii)* between the heights 32" and 38" *(iii)* below the heights 30".

2. The daily milk yield of 800 goats assumes to follow normal distribution with standard deviation 0.85 kg. There are100 goats which give daily milk below 3 kg. Find the mean yield of a goat and also the probability that the yield per day exceeds 5.5 kgs.

3. The following is the distribution of daily milk yield for a herd of cows

Milk yield (kgs.)	1-3	3-5	5-7	7-9	9-11	11-13	13-15	15-17
No. of cows	10	95	150	375	260	110	85	15

Fit a normal distribution to the above data.

4. The following is the data based on the bead weights of Drosophila Melanogoster.

Head weight(mgs)	No. of insects	Head weight(mgs)	No. of insects
0.01-0.02	20	0.05-0.06	134
0.02-0.03	75	0.06-0.07	96
		0.07-0.08	38
0.03-0.04	120	0.08-0.09	17
0.04-0.05	185	0.09-0.10	15

Test the assumption of normality of the above data by finding the proportion of insects whose head weights lie between $\mu \pm \sigma, \mu \pm 2\sigma, \mu \pm 3\sigma$.

5. The life of a calgas cylinder is 45 days with a standard deviation of 6 days. If 15000 cylinders are issued, find *(i)* how many need replacement after 40 days *(ii)* How many need replacement within 20 days and *(iii)* If it is to have a probability of not less than 0.9 of having a life between 42 and 50 days, what is the highest allowable standard deviation ?

   (Assume that the life of a calgas cylinder follows a Normal Law).

6. The breaking strength X (in pounds) of a type of rope has a normal distribution with mean 98.5 and standard deviation 4.5. Each 100 feet coil of rope fetches a profit of Rs. 3.75 if X>90. If X<90, the coil fetches a profit of Rs. 1.80.

   Determine the expected profit realised per coil.

7. Fit a normal distribution for the following distribution of yields of wheat in a locality.

Yield[quintals/hectare]	3-8	8-13	13-18	18-23	23-28	28-33	33-38
Farms	10	32	64	75	46	40	33

8. If the yields of 500 maize farms follow normal distribution with mean yield 8 quintels/hectare and standard deviation as 3.4, find the number of farms (i) between 5 and 9. (ii) below 4 quintals and (iii) 10 quintals and above.

9. The monthly income of a head of a family in a locality of 1000 families follow normal distribution with mean income as Rs 2,000 with a standard deviation as Rs 350. Find the number of families (i) between Rs 2000 and 3,500 per month (ii) below Rs1000/- and (iii) Rs 4,000 and above.

10. If the egg production of 200 poultry farms follow normal distribution with annual egg production per bird was found to be 224 with a

standard deviation of 14 eggs, find the number of farms (i) between 200 and 210 (ii) below 190 and (iii) 240 and above.

11. Fit a normal distribution for the following distribution of body weights of goats in a farm.

Body weight (kg)	8-10	10-12	12-14	14-16	16-18	18 and above
No. of animals	5	18	25	35	12	4

12. The following is the data on rain fall (mm) in two rainy months July and August in a year at an Agricultural Research Station,

10,	15,	9,	8,	13,	0,	7,	12,	22,	15,	1,
35,	40,	38,	26,	20,	0,	5,	19,	14,	13,	6,
13,	18,	17,	24,	21,	29,	32,	3,	10,	13,	6,
5,	8,	9,	1,	3,	4,	13,	11,	5,	24,	15,
19,	20,	0,	13,	6,	2,	3,	8,	1,	4,	42,
6,	5,	16,	22,	9,	3,	5				

Verify the above data on rainfall whether it follows normal distribution pattern using all the properties of normal distribution.

13. The mean and standard deviation of marks obtained by 800 students in an examination are respectively 50 and 10. Assuming the normality of distribution, find the number of students scored marks between 55 and 75.

14. The following is the distribution of weights of students admitted in college of agriculture in a particular year.

Weight (Kgs)	40-45	45-50	50-55	55-65	65-75
Students	9	20	35	28	8

Fit a normal distribution to the above distribution.

CHAPTER 10

TESTS OF HYPOTHESES

10.1 INTRODUCTION

The estimate based on sample values do not equal to the true value in the population due to inherent variation in the population. The samples drawn will have different estimates compared to the true value. It has to be verified that whether the difference between the sample estimate and the population value is due to sampling fluctuation or real difference. If the difference is due to sampling fluctuation only it can be safely said that the sample belongs to the population under question and if the difference is real we have every reason to believe that sample may not belong to the population under question.

Type I error: Rejecting the hypothesis when it ought to be accepted.

Type II error: Accepting the hypothesis when it ought to be rejected.

10.1.1 Statistical Significance

The probability of the difference between sample estimate and the true value taken with standard error compared with observed difference is very small then we say that there is significant difference between the sample estimate and the population value. That is the probability is less for the difference between sample estimate and the true value (or population value) not due to sampling fluctuation but due to real difference.

If the above probability is very large then we say that there is no significant difference between the sample estimate and the true value. The difference so obtained is due to sampling fluctuation only.

10.1.2 Levels of Significance

The maximum probability at which we would be willing to risk a type I error is known as the level of significance. In general 5 per cent and 1 per cent are taken as 'levels of significance' thereby indicating that on an average we may go wrong 5 out of 100 cases and 1 out of 100 cases respectively. To say that 5 per cent level of significance there is 95 per cent confidence in the result with a margin of error 5 per cent. The 5 per cent level of significance is shown in Fig. 10.1.

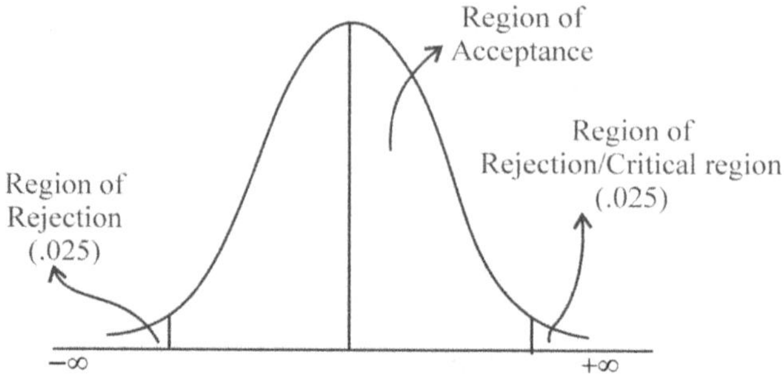

Fig. 10.1 Normal curve.

10.1.3 Degrees of Freedom

It is defined as the difference between the total number of items and the total number of constraints.

If n is the total number of items and k, the total number of constraints then the degrees of freedom (d.f.) is given by d.f. $= n - k$

Suppose we want to select 10 values with a restriction that the total of the ten values should be equal to 100. Thus, we can select only 9 items at our choice but the tenth item must be chosen in such a way that the total of them should be equal to 100. Therefore, degrees of freedom of selecting 10 items is only 9 with one constraint.

10.1.4 Null Hypothesis

Null hypothesis is the statement about parameters which is likely to be rejected after testing. We start with the hypothesis that the two items are equal.

10.1.5 Standard Normal Deviate Tests

If X follows normal distribution with mean μ and S.D., σ then $\overline{X}$ also follows normal distribution with mean μ and S.D. $\dfrac{\sigma}{\sqrt{n}}$. This can be denoted by

$$\overline{X} \to N\left(\mu, \dfrac{\sigma}{\sqrt{n}}\right) \text{ i.e., } \dfrac{\overline{X} - \mu}{\dfrac{\sigma}{\sqrt{n}}} \to N(0,1)$$

the expression on the left hand side is called as standard normal variate (or standard normal deviate) which follows normal distribution with mean zero and standard deviation unity. The test of hypothesis based on this deviate is called standard normal deviate test. The confidence limits for the population mean can be obtained as

$$\overline{X} \pm 1.96 \frac{\sigma}{\sqrt{n}}$$

i.e., the population mean, lies between $\overline{X} - 1.96 \frac{\sigma}{\sqrt{n}}$ and $\overline{X} + 1.96 \frac{\sigma}{\sqrt{n}}$ where

1.96 is the tabulated value of normal distribution at 5 percent level of significance. These are also called as lower and upper limits respectively.

10.2   ONE SAMPLE TEST: CASE (I)

Assumptions: 1. Population is normal.

2. The sample is drawn at random.

Conditions: 1. Population S.D., σ is known.

2. Size of the sample may be small or large.

Null hypothesis: H_o: $\mu = \mu_0$, H_1: $\mu \neq \mu_0$, where H_1 is alternate hypothesis.

$$Z = \frac{\left|\overline{X} - \mu_0\right|}{\sigma / \sqrt{n}}$$

we know that 'Z' values follow normal distribution with zero mean and unit S. D. The values of Z on either side of normal curve corresponding to the areas 0.025 and 0.025 are

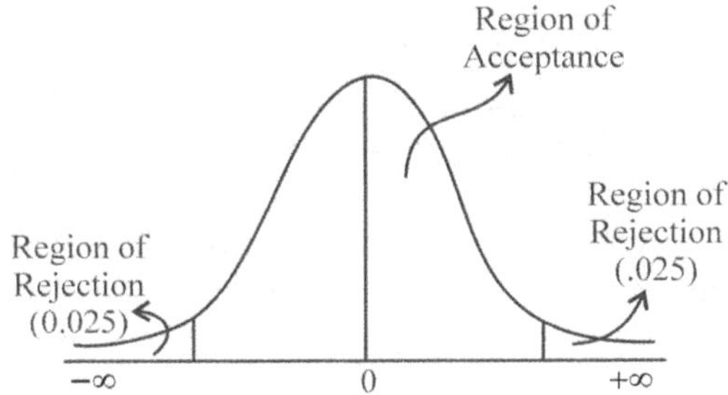

Fig. 10.2 Standard normal curve.

−1.96 and 1.96 respectively. That is, the total area of the region of rejection is 0.05 out of the total area I sq units which is shown in Fig. 10.2. For 1 per cent level of significance, the regions of rejection on either side of the standard normal curve comprises area of 0.005 and the corresponding Z values are −2.576, and 2.576 which is shown in Fig. 10.3.

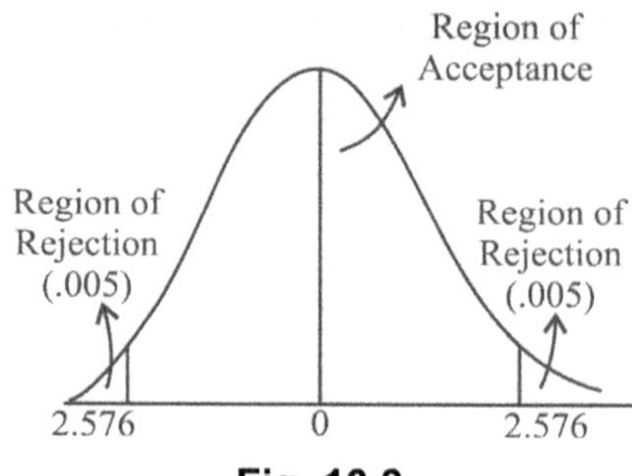

Fig. 10.3

Conclusion: If Z (calculated) $\leq$ Z (tabulated), at 5 per cent (or 1 per cent) level of significance, the null hypothesis is accepted i.e., there is no significant difference between sample mean and population mean.

Otherwise, the null hypothesis is rejected. That is, there is significant difference between sample and population means. In other words, the sample may not belong to the population having given mean with respect to character under consideration.

Here, we are comparing with positive values of tabulated Z since we are taking modulus in S.N.D. test as the Z values on either side of mean are identical.

EXAMPLE: The average number of mango fruits per tree in a particular region was known from a considerable experience as 520 with a standard deviation 4.0. A sample of 20 trees gives an average number of fruits 450 per tree. Test whether the average number of fruits per tree selected in the sample is in agreement with the average production in that region?

Null hypothesis: H$_o$: $\mu = \mu_0$, H$_1$: $\mu \neq \mu_o$

$$Z = \frac{|4500 - 520|}{4 / \sqrt{20}} = 78.26$$

Conclusion: Z (calculated) > Z (tabulated), 1.96 at 5 per cent level significance. Therefore it can be concluded that there is significant difference between sample mean and population mean with respect to average performance.

10.2

Case (ii): If the S.D. in the population is not known still we can use the standard normal deviated test.

Assumptions: 1.   Population is normal

              2.   Sample is drawn at random

Conditions: 1. σ is not known

              2.   Size of the sample is large (say > 30)

Null hypothesis: $H_o: \mu = \mu_0$, $H_1: \mu \neq \mu_0$

$$Z = \frac{\left|\overline{X} - \mu_0\right|}{S/\sqrt{n}}$$

where $$S = \sqrt{\frac{1}{n}\sum_{i=1}^{n}(X_1 - \overline{X})^2}$$

where $\overline{X}$ = mean of sample

n = size of the sample

Conclusion: Same as in the case (i).

EXAMPLE: The average daily milk production of a particular variety of buffalo was given as 12 ltrs. The distribution of daily milk yield in a dairy farm was given as follows:

TABLE 10.1

Daily milk yield (ltrs)	6-8	8-10	10-12	12-14	14-16	16-18
No. of buffaloes	9	20	35	42	17	7

Test whether the performance of dairy farm was in agreement with the record.

Null hypothesis: $H_o: \mu = \mu_0$, $H_1: \mu \neq \mu_0$,

Using the formula given in Section 4.1.3 (b), we have

$$\overline{X} = 11.91$$

Using the formula given in Section 5.4.2; we have

$$S = 2.49$$

Therefore, $$Z = \frac{|11.91 - 12|}{\dfrac{2.49}{\sqrt{130}}} = 0.41$$

Conclusion: The Z (calculated) < Z (tabulated), 1.96 at 5 per cent level of significance, Therefore, the null hypothesis is accepted. That is there is no significant difference between the average daily milk yield of dairy farm and the previous record.

10.3 TWO SAMPLE TEST: CASE (I)

Assumptions: 1. Populations are normal

2. Samples are drawn independently and at random.

Conditions: 1. σ is known.

2. Sizes of samples may be small or large.

Null hypothesis: H_o: $\mu_1 = \mu_2$ H_1: $\mu_1 \neq \mu_2$ where μ_1, μ_2 are the population means for the 1st and 2nd populations respectively.

$$Z = \frac{\left|\overline{X}_1 - \overline{X}_2\right|}{\sqrt{\sigma^2\left(\dfrac{1}{n_1} + \dfrac{1}{n_2}\right)}}$$

where $\overline{X}_1$, $\overline{X}_2$ are the means of 1st and 2nd samples with sizes n_1, n_2 respectively.

Conclusion: If Z (calculated) > Z (tabulated), the null hypothesis is rejected. There is significant difference between two sample means; In other words, the two samples have come from two different populations having two different means. Otherwise, the null hypothesis is accepted.

10.3

Case (ii): In this case the common population S.D. is not known.

Assumptions: 1. Populations are normal.

2. Samples are drawn independently and at random.

Conditions: 1. σ is not known.

2. Sizes of samples are large.

Null hypothesis: H_o: $\mu_1 = \mu_2$, H_1: $\mu_1 \neq \mu_2$

$$Z = \frac{\left|\overline{X}_1 - \overline{X}_2\right|}{\sqrt{\dfrac{S_1^2}{n_1} + \dfrac{S_2^2}{n_2}}}$$

where $\quad S_1^2 = \dfrac{1}{n_1}\Sigma\left(X_1 - \overline{X}_1\right)^2, S_2^2 = \dfrac{1}{n_2}\Sigma\left(X_2 - \overline{X}_2\right)^2$

and $\overline{X}_1$, $\overline{X}_2$ are the means of 1st and 2nd samples with sizes n_1, n_2 respectively.

Conclusion: If Z (calculated) > Z (tabulated) at chosen level of significance, the null hypothesis is rejected. Otherwise, it is accepted.

EXAMPLE: A random sample of 90 poultry farms of one variety gave an average production of 240 eggs per bird/year with a S.D. of 18 eggs. Another random sample of 60 poultry farms of another variety gave an average production of 195 eggs per bird/year" with a S,D. of 15 eggs. Distinguish between two varieties of birds with respect to their egg production.

Null hypothesis: H_0: $\mu_1 = \mu_2$, H_1: $\mu_1 \neq \mu_2$

$$Z = \frac{|240 - 195|}{\sqrt{\dfrac{(18)^2}{90} + \dfrac{(15)^2}{60}}} = 16.61$$

Conclusion: Z (calculated) > Z (tabulated), 1.96 at 5 per cent level of significance. Here there is significant difference between two varieties of birds with respect to egg production.

10.4 STUDENT'S t-DISTRIBUTION

In small samples drawn from a normal population, the ratio of the difference between the sample and population means to its estimated standard error follows a distribution known as t-distribution, where

$$t = \frac{|\overline{X} - \mu|}{\dfrac{s}{\sqrt{n}}} \quad \text{where} \quad s^2 = \frac{1}{n-1} \Sigma \left(X_i - \overline{X} \right)^2$$

10.4.1 Properties of t-distribution

This distribution is symmetrical about the origin and unimodel and extends from $-\infty$ to $+\infty$ in both directions. It is known as student's t-distribution, the name 'Student' being the Pen name of W.S. Gosset who has developed this distribution.

For large n, this distribution tends to standard normal distribution having zero mean and unit variance.

The moments μ_r^1 of the distribution exist only for r < (n–I). All odd order moments are zero by symmetry. The even moments are given by

$$\mu_{2r} = \frac{(n-1)^r \sqrt{r + 1/2} \sqrt{\dfrac{n-1}{2} - r}}{\sqrt{\dfrac{1}{2}} \sqrt{\dfrac{n-1}{2}}} \; ; 2r < (n-1)$$

The Skewness and Kurtosis coefficients are

$$\gamma_1 = 0, \; \gamma_2 = 6/(n-5), \; (n-1) > 4,$$

The values of 't' were given at different levels of significance and presented in the t-table. The tabulated values of 't' would differ for different degrees of freedom unlike in the case of Normal distribution. For normal distribution, the tabulated value would be entered into t-table at ∞ degrees of freedom.

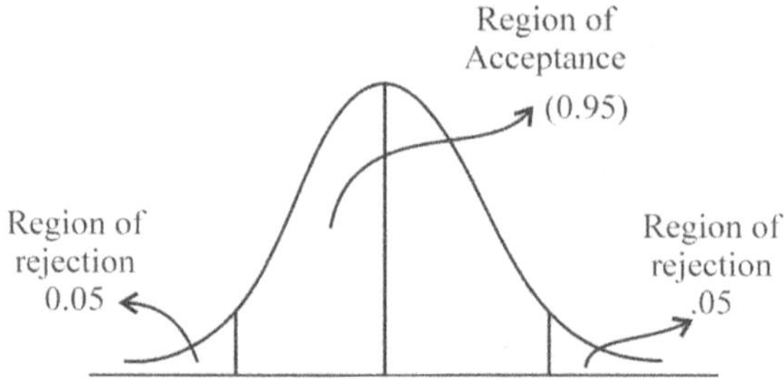

Fig. 10.4 Student's t-distribution.

Just as in the case of standard normal deviate test, student's t-test also plays an important role in tests of hypothesis in the case of small samples when the S.D. in the population is not known.

$$\text{Student's t} = \frac{\text{A standard normal deviate}}{\sqrt{\dfrac{\text{A chi-square variate}}{\text{d.f.}}}}$$

$$t = \frac{\dfrac{\left|\overline{X}-\mu\right|}{\dfrac{\sigma}{\sqrt{n}}}}{\sqrt{\dfrac{(n-1)s^2}{\sigma^2}}\Big/(n-1)}$$

The numerator follows normal distribution with zero mean and unit S.D. and the denominator' follows chi-square distribution with (n–l) d.f. where n is the size of the sample.

Simplifying, we have

$$t = \frac{\left|\overline{X}-\mu\right|}{\dfrac{s}{\sqrt{n}}} \quad \text{where} \quad s = \sqrt{\frac{1}{n-1}\Sigma(X_1 - \overline{X})^2}$$

The confidence limits for the population mean, μ based on students t-test is given as

$$\overline{X} \pm t_{tab}\,(n-1)\,d.f. \times \frac{s}{\sqrt{n}}$$

These limits can easily be derived from the expression of 't' by solving for 'μ'

10.5 ONE SAMPLE t-TEST

Assumptions: 1. Population is normal,

                 2. Sample is drawn at random,

Conditions: 1. σ is not known.

2. Size of the sample is small.

Null hypothesis: H_o: $\mu = \mu_0$, H_1: $\mu \neq \mu_o$

$$t = \frac{\left|\overline{X} - \mu_0\right|}{s / \sqrt{n}} \quad \text{where} \quad s^2 = \frac{1}{n-1}\Sigma(X_i - \overline{X})^2$$

and is the unbiased estimate of σ^2

n = size of the sample.

Conclusion: 1f t (calculated) < t (tabulated) with (n–l) d.f. at chosen. level of significance, the null hypothesis is accepted. That is, there is no significant difference between sample mean and population mean. Otherwise, the null hypothesis is rejected.

EXAMPLE: The heights of plants in a particular field were assumed to follow normal distribution. A random sample of 10 plants were selected and whose heights (in cms) were recorded as 96, 100, 102, 99, 104, 105, 99, 98, 100 and 101. Discuss in the light of the above data the mean height of plants in the population is 100.

Null hypothesis: H_o: $\mu = \mu_o$, H_1: $\mu \neq \mu_o = 100$

TABLE 10.2

X	d_i	d_i^2
96	–4	16
100	0	0
102	2	4
99	–1	1
104	4	16
105	5	25
99	–1	1
98	–2	4
100	0	0
101	1	1
	4	68

$d_i = (X_i-A)$ where $A = 100$

$$\overline{X} = A + \Sigma\frac{d_1}{n} = 100 + \frac{4}{10} = 100.4$$

$$s = \sqrt{\frac{1}{n-1}\left[\Sigma d_1^2 - \frac{(\Sigma d_1)^2}{n}\right]}$$

$$= \sqrt{\frac{1}{9}(68 - 16/10)} = 2.72$$

$$t = \frac{\left|100.4 - 100\right|}{2.72 / \sqrt{10}} = 0.46$$

Conclusion: t (calculated) (0.46) < t (tabulated), (2.262) with 9 d.f. at 5 per cent level of significance. Therefore, the null hypothesis is accepted. In other words, the sample may belong to the population whose mean height is 100 cm.

SPSS DATA ANALYSIS PROCEDURE

1. ONE-SAMPLE t- TEST;

    Data entry

    X

    13

    10

    1 7

    X(wt./kg.)variable   will   be   entered   with   data   in   column   as
    shown above. Click ANALYZE from the TOOL BAR and take option

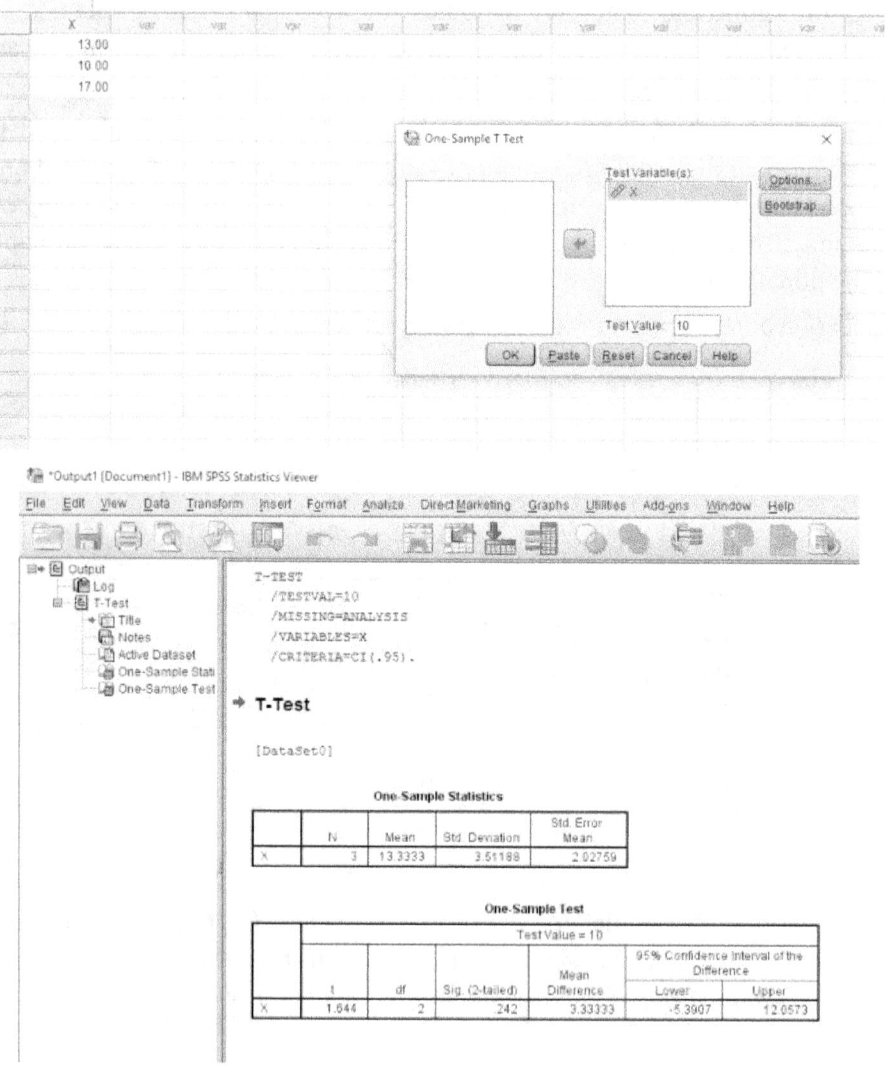

'COMPARISON OF MEANS' AND OPT FOR 'ONE-SAMPLE t-test'. In the DISPLAY table take X variable to the 'TEST VARIABLE' AND CLICK 'TEST VALUE' BAR and then enter population mean value with which we want to test say $\mu = 0$ or 10 or any value. Then click 'O.K'. option. We get the results   display table which sample mean, S.D, S.E. t(calculated) value and probability value under sign. Head. The decision or conclusion will be drawn based on the value of significance.

10.6 TWO SAMPLE t-TEST

Assumptions: 1. Populations are normal,

2. Samples are drawn independently and at random.

Conditions: 1. S.D.'s in the populations are same and not known.

2. Sizes of the samples are small.

Null hypothesis: H_o: $\mu_1 = \mu_2$, H_1: $\mu_1 \neq \mu_2$

where μ_1, μ_2 are the means of 1st and 2nd populations respectively.

$$t = \frac{\left| \overline{X}_1 - \overline{X}_2 \right|}{\sqrt{s_c^2 \left(\dfrac{1}{n_1} + \dfrac{1}{n_2} \right)}}$$

where $$s_c^2 = \frac{(n_1 - 1)s_1^2 + (n_2 - 1)s_2^2}{n_1 + n_2 - 2} = \frac{\Sigma(X_{1i} - \overline{X}_1)^2 + \Sigma(X_{2i} - \overline{X}_2)^2}{n_1 + n_2 - 2}$$

and $$s_1^2 = \frac{1}{n_1 - 1}\Sigma(X_{1i} - \overline{X}_1)^2 \text{ and } s_2^2 = \frac{1}{n_2 - 1}\Sigma(X_{2i} - \overline{X}_2)^2$$

Conclusion: 1f t (calculated) <t (tabulated) with ($n_1 + n_2 - 2$) d.f. at chosen level of significance, the null hypothesis is accepted. That is there is no significant difference between the two samples, means. Otherwise, the null hypothesis is rejected.

EXAMPLE: Two types of diets were administered to two groups of pre-school going children for increase in weight and the following increases in weight (100 gm) were recorded after a month.

Increases in weight

Diet A	4	3	2	2	1	0	5	6	3
Diet B	5	4	4	2	3	2	6	1	

Test whether there is any significant difference between the two diets with respect to increase in weight.

Null hypothesis: H_0: $\mu_1 = \mu_2$, H_1: $\mu_1 \neq \mu_2$

TABLE 10.3

X_1	X_2	X_1^2	X_2^2
4	5	16	25
3	4	9	16
2	4	4	16
2	2	4	4
1	3	1	9
0	2	0	4
5	6	25	36
6	1	36	1
3	1	9	
26	27	104	111

$$\overline{X}_1 = 2.89, \overline{X}_2 = 3.38$$

$$s_c^2 = \frac{\left[104 - \frac{(26)^2}{9}\right] + \left[111 - \frac{(27)^2}{8}\right]}{9 + 8 - 2} = 3.25$$

$$t = \frac{|2.89 - 3.38|}{\sqrt{3.25(1/9 + 1/8)}} = 0.56$$

Conclusion: Here t (calculated) value i.e., 0.56 < t (tabulated), value (2.131) with 15 d.f, at 5 per cent level of significance. Therefore, the null hypothesis is accepted. Hence, there is no significant difference between the two diets with respect to increases in weight.

SPSS DATA ANALYSIS PROCEDURE

1. Two-SAMPLE t- TEST

   DATA ENTRY

X	Y
1	12
1	9
1	12
1	15
2	21
2	27
2	19
2	24

Here 'Gouping variable' is taken as X and 'TEST variable' is taken as Y(gain in wt/kg.). Click on 'GROUPING VARIABLE' then it will ask 'DEFINE VARIABLE'. For VARIABLE (1)/GROUP 1 give '1' and VARIABLE(2)/GROUP 2 give '2' .Then click 'CONTINUE ' OPTION . Then on display table. Click 'O.K.' OPTION. Then we get result displaying table with two means, S,D.s , S.E.s and t(calculated) and t(tabulated) values   for equal and unequal variances

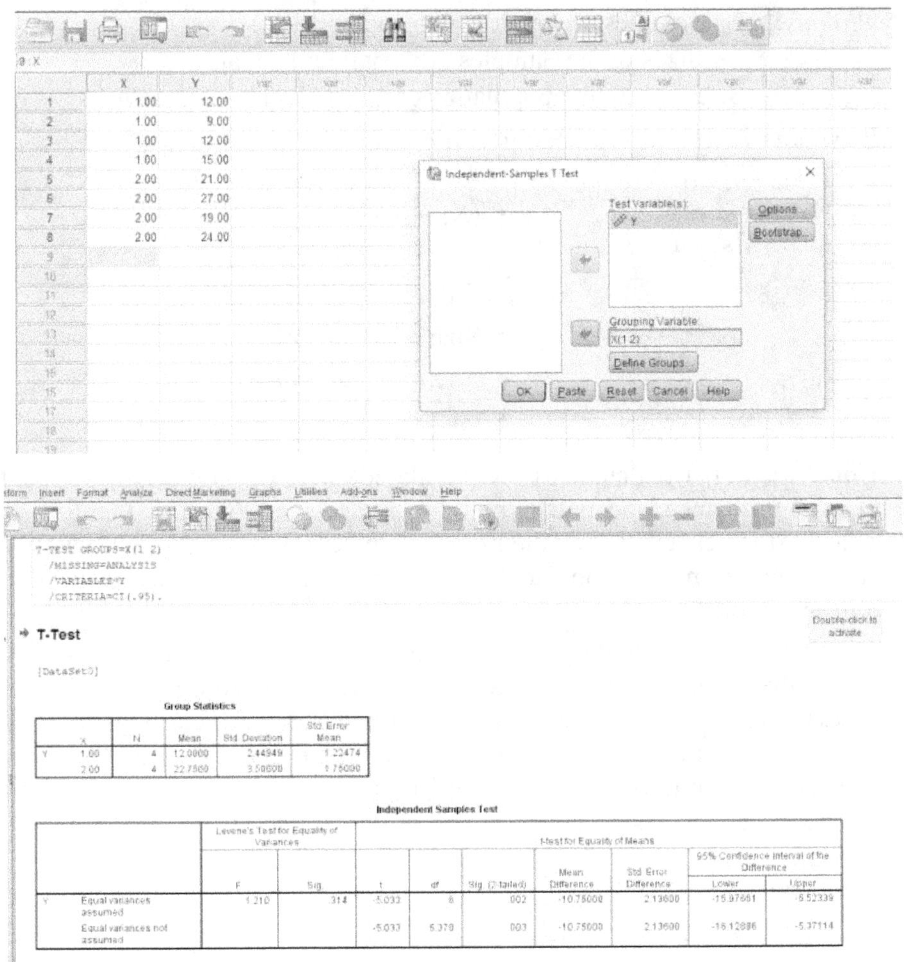

10.7 PAIRED t-TEST

When the two small samples of equal size are drawn from two populations and the samples are dependent on each other then the paired t-test is used in preference to independent t-test. The same patients for the comparison of two drugs with some time interval; the neighbouring plots of a field for comparison of two fertilizers with respect to yield assuming that the neighbouring plots will have the same soil composition; rats from the same litter for comparison of two diets; branches of same plant for comparison of the nitrogen uptake, etc., are some of the situations where paired t-test can be used.

In the paired t-test the testing of the difference between two treatments means was made more efficient by keeping all the other experimental conditions same.

Assumptions: 1. Populations are normal.
                   2. Samples are drawn independently and at random.

Conditions: 1. Samples are related with each other.
                   2. Sizes of the Samples are small and equal.
                   3. S.D.'s in the populations are equal and not known.

Null hypothesis: H_o: $\mu_1 = \mu_2$, H_1: $\mu_1 \neq \mu_2$

$$t = \frac{|\bar{d} - 0|}{\sqrt{s_d^2 / n}} \qquad d_i = (X_{1i} - X_{2i})$$

$$\bar{d} = \sum d_1 / n$$

$$n = \text{Size of the sample.}$$

$$s_d^2 = \frac{1}{n-1}\left[\sum d_{1^2} - \frac{(\sum d_1)^2}{n}\right]$$

Conclusion: If t (calculated) < t (tabulated) with (n–1) d.f. at 5 per cent level of significance, the null hypothesis is accepted. That is, there is no significant difference between the means of the two samples, In other words, the two samples may belong to the same population. Otherwise, the null hypothesis is rejected.

EXAMPLE: The following is the experiment conducted on Agronomy farm in the year 1969-70 at S.K.N. College of Agriculture, Jobner (Rajasthan) for comparing two types of grasses on neighbouring. plots of size 5×2 meters in each replication. The weights of' grasses per plot (in kgs) at the harvesting time were recorded on 7 replicates:

	1	2	3	4	5	6	7
Cenchrus cilioris (Grass I)	1.96	2.10	1.64	1.78	1.95	1.70	2.00
Losirus Sindicus Grass (II)	2.13	2.10	2.14	2.08	2.20	2.12	2.05

Test the Significant difference between the two grasses with respect to their yield.

Null-hypothesis: H_o: $\mu_1 = \mu_2$,

                H_1: $\mu_1 \neq \mu_2$

TABLE 10.4

$X_{1\iota}$	$X_{2\iota}$	d_ι	d_ι^2
1.96	2.13	−0.17	0.0289
2.10	2.10	0	0
1.64	2.14	−0.50	0.2500
1.78	2.08	−0.30	0.0900
1.95	2.20	−0.25	0.0625
1.70	2.12	−0.42	0.1764
2.00	2.05	−0.05	0.0025
		−1.69	0.6103

$$\bar{d} = -0.24$$

$$s_d^2 = \frac{1}{(7-1)}\left[0.6103 - \frac{(-1.69)^2}{7}\right] = 0.0337$$

$$t = \frac{|0.24|}{\sqrt{0.0337/7}} = 3.46$$

Conclusion: Here t(calculated) value i.e., 3.46 > t (tabulated), value (2.447) with 6 d.f. at 5 per cent level of significance. The null hypothesis is rejected. There is significant difference between the two grasses with respect to yield.

SPSS DATA ANALYSIS PROCEDURE

    1. PAIRED t – TEST

       Data entry

X	Y
6	12
8	15
9	20

X (Loss of vitamin-B) and Y(Loss of vitamin-B) variables be entered with data as columns underneath in two columns and click 'ANALYZE' ON THE TOOL BAR AND OPT FOR 'COMPARISON OF MEANS' and OPT for 'PAIRED t-test'. In the display table enter 'X 'for variable 1 and 'Y' for variable 2 as per indicator marker. Then click 'O.K.' We will get the display table of analysis wherein the means of two variables X and y and also in the last table the calculated value of t and probability of significance. Based on the value of probability the decision will be drawn about the significance level.

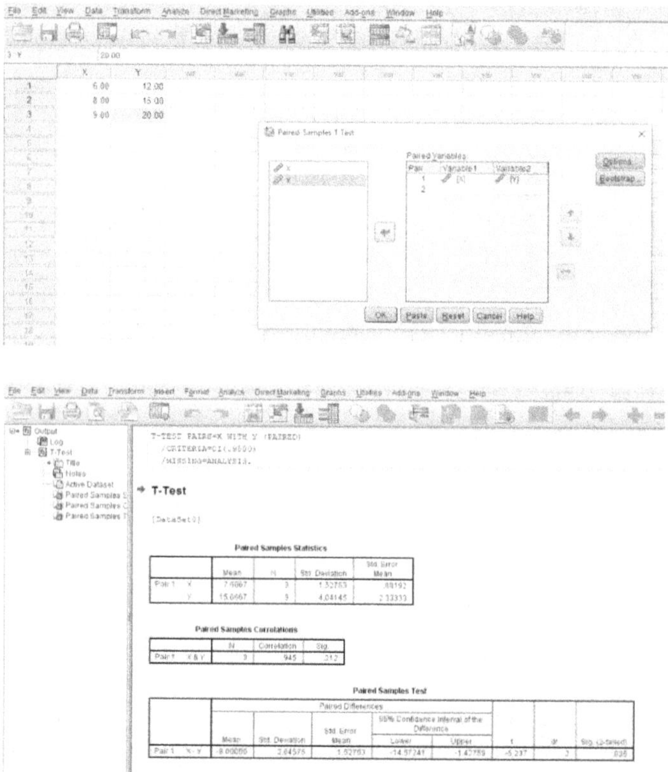

10.8 S.N.D. TEST FOR PROPORTIONS

Sometimes there is need to have the tests of hypothesis for proportion of individuals (or objects) having a particular attribute. For example, to know whether the proportion of disease infected plants in the sample is in confirmity with the proportion in the whole field (or population).

Here the number of plants in the sample is identically equal to the n independent trials with constant probability of success, p. The probabilities of 0, 1, 2,... successes are the successive terms of the binomial expansion $(q+p)^n$ where $q=(1-p)$. For the Binomial distribution the first and second moment of the number of successes are 'np' and 'npq' respectively.

Mean of proportion of successes=P

S.E of the proportion of successes $= \sqrt{PQ / n}$

10.8.1   One Sample Test

Assumptions: 1. Population-is normal,

2. Sample is drawn at random without replacement if it is a finite population.

Conditions: 1. P is known in the population.

2. Size of the sample is large.

Null hypothesis: H_o: $P = P_o$, H_1: $P \neq P_o$

$$Z = \left| \frac{X}{n} - P_0 \right| \sqrt{\frac{P_0 \, Q_0}{n}}$$

where P_0 = Given proportion in the population

$\qquad Q_0 = 1 - P_0$

Conclusion: If Z (calculated) < Z (tabulated) at chosen level of significance, the null hypothesis is accepted. That is, there is no significant difference between the proportions in the sample and population. In other words, the sample may belong to the given population. Otherwise, the null hypothesis is rejected.

EXAMPLE: For a particular variety of wheat crop it was estimated that 5 per cent of the plants attacked with a disease. A sample of 600 plants of the same variety of wheat crop was observed and found that 50 plants were infected with a disease. Test whether the sample results were in confirmity with the population.

Null hypothesis: H_o: $P = P_o$, H_1: $P \neq P_o$

$$Z = \frac{\left| 50 / 600 - 0.05 \right|}{\sqrt{\dfrac{0.05 \times 0.95}{600}}} = 3.74$$

Conclusion: Here Z (calculated), value i.e. 3.74 > Z (tabulated), 1.96 at 5 per cent level of significance, the null hypothesis is rejected. Therefore, there is significant difference between the proportion of diseased plants in the sample and the population.

10.8.2 Two Sample Test: Case (i) P is known

There are two population of individuals (or objects) haying the same proportion, P of possessing a particular character. Let $P_1 = X_1/n_1$ arid $P_2 = X_2/n_2$ are the two proportions of individuals possessing the same attribute in the samples of sizes n_1 and n_2 respectively. It is to test whether the proportions in the samples are significantly different from each other or not.

Assumptions: 1. Populations are normal.

$\qquad\qquad\quad$ 2. Samples are drawn independently and at random.

Conditions: 1. P is known.

$\qquad\qquad\quad$ 2. Sizes of the samples are large.

Null hypothesis: H_o: $P = P_o$, H_1: $P \neq P_o$

$$Z = \frac{\left| p_1 - p_2 \right|}{\sqrt{p_0 Q_0 \left(\dfrac{1}{n_1} + \dfrac{1}{n_2} \right)}}$$

Conclusion: If Z (calculated) < Z (tabulated) at chosen level of significance, the null hypothesis is' accepted, i.e., there is no significant difference between the two proportions in the samples. In other words, their populations are having the same proportion, P_o.

EXAMPLE: In an investigation, it was found that 4 per cent of the farmers accepted the improved seeds for a barley crop in a particular state. On conducting a survey in two Panchayat Samithi, 340 farmers accepted out of 1,500 in the first samithi and 200 out of 1,000 in the second samithi. Test whether the difference between the two samithi is significant.

Null hypothesis: H_o: $P = P_o$, H_1: $P \neq P_o$

$$P_0 = 4/100 = 0.40 \qquad Q_0 = 1 - 0.04 = 0.96$$
$$P_1 = 340/1500 = 0.23 \qquad p_2 = 200/1000 = 0.20$$

$$Z = \frac{|0.23 - 0.20|}{\sqrt{0.4 \times 0.96(1/1500 + 1/1000)}} = 1.19$$

Conclusion: Here Z (calculated) value i.e., 1.19 < Z (tabulated), 1.96 at-5 per cent level of significance. Therefore, the null, hypothesis is accepted. That is, there is no significant difference between the proportions of the two Samithis with regard to acceptability of the improved seeds.

10.8.2 Two Sample Test: Case (ii) P is not known

When the proportion in the populations is same and is not known then it is estimated from the proportions of the two samples.

If p_1 and p_2 are the proportions having an attribute in the two samples of sizes n_1 and n_2 respectively then p, its proportion having the same attribute in the populations is estimated by taking the weighted average of p_1 and p_2

$$p = \frac{n_1p_1 + n_2p_2}{n_1 + n_2} \text{ and } q = 1 - p$$

$$Z = \frac{|p_1 - p_2|}{\sqrt{pq(1/n_1 + 1/n_2)}}$$

Conclusion: If Z (calculated) $\leq$ Z (tabulated) at chosen level of significance, then the null hypothesis is accepted. Otherwise, the null hypothesis is rejected.

EXAMPLE: In the example of 10.8.2 case (i) if P is not known, test the significance of the difference between the proportions of the two samples.

Null Hypothesis: H_o: $P_1 = P_2$, H_1: $P_1 \neq P_2$

where P_1, P_2 are the proportions in the 1st and 2nd populations respectively.

$$p = \frac{1500 \times 0.23 + 1000 \times 0.20}{1500 + 1000} = 0.22, q = 1 - 0.22 = 0.78$$

$$Z = \frac{|0.23 - 0.20|}{\sqrt{0.22 \times 0.8(1/1500 + 1/1000)}} = 1.75$$

Conclusion: Here Z (calculated) value i.e., 1.75 < Z (tabulated), 1.96 at 5 per cent level of significance. There is no significant difference between the two smithies with regard to proportions of farmers accepting the improved seeds.

10.9 ONE-TAILED TESTS

Sometimes we confront with one-sided hypothesis of the type, say, brothers are taller than sisters, or the yield of rice crop for nitrogen at 20 kg/acre is less than the yield for nitrogen at 40 kg/acre, etc.

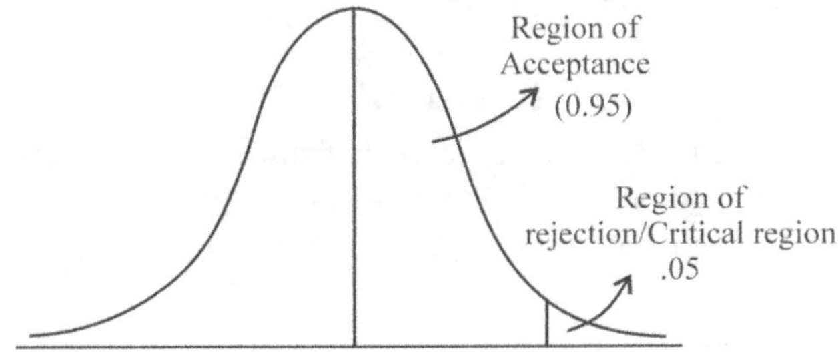

Fig.10.5 Normal curve.

When the hypothesis to be tested is one-sided as $\mu_1 > \mu_2$ or $\mu_1 < \mu_2$ instead of $\mu_1 = \mu_2$, the test of hypothesis will slightly change since we have to consider only one of the sides of standard normal curve or t-distribution because the differences based on the samples may be either +ve or − ve at a time. The one-sided hypothesis is shown in Fig. 10.5.

Here the calculated value of Z (or t) is compared with tabulated value of Z (or t) respectively at 2.5 per cent level of significance which corresponds to 5 per cent level of significance in the two-sided case. Similarly 1 per cent level of significance corresponding to 2 per cent level of significance in the two-sided case.

EXAMPLE: A group of 200 boys in the first year course of B.Sc. (Ag.) have the mean height 65" with a S.D. 3.2" and a group of 150 boys in the final year course have the mean height 67" with a S.D. 2.5". Test whether the final year boys are taller than the first year boys. Obtain also the fiducial limits for the difference in the two population means.

Null hypothesis: $H_0 : \mu_2 > \mu_1;$

$$H_1 : \mu_2 \ngtr \mu_1$$

$$Z = \frac{|65 - 67|}{\sqrt{\dfrac{(3.2)^2}{200} + \dfrac{(2.5)^2}{150}}} = 6.56$$

Conclusion: The calculated value of Z i.e. 6.56 is greater than the tabulated value of Z (1.645) at 10 per cent level of significance. The null hypothesis is rejected. The final year boys may not be taller than the first year boys.

10.9.1

Fiducial limits for the difference between the two population means are given as:

$$\left(\overline{X}_1 - \overline{X}_2\right) \pm 1.96 \sqrt{\frac{S_1^2}{n_1} + \frac{S_2^2}{n_2}}$$

$$2 \pm 1.96 \times .3047, \text{ i.e., } (1.4, 2.6)$$

EXERCISES

1. The average life of an electric bulb from a manufacturing company was estimated' to be of 3,600 hrs: A sample of 100 bulbs was tested in order to confirm the above hypothesis and the results were given as follows:

Average life (hours)	3200-3300	3300-3400	3400-3500	3500-3600	3600-3700
No. of bulbs	13	18	22	32	15

   Test whether the results obtained from a sample are in agreement with the hypothesis.

2. A random sample of 500 pine apples was taken from a large consignment and 65 were found to be bad. Show that S.E. of the proportion of bad ones in a sample of this size is 0.015 and deduce that the percentage of bad pine apples in the consignment almost certainly lies between 8.5 and 17.5

3. A pot experiment was conducted to test the quality of water on the yield of the crop. Saline water and distilled water were used for comparison on ten pots each having 3 soybean plants and the weight of seeds per plant were recorded in gms at the time of harvesting.

	1	2	3	4	5	6	7	8	9	10
Saline water	19	18	20	23	20	25	25	19	17	16
Distilled water	25	20	21	24	26	23	27	25	24	25

   Test whether there is any significant difference between saline and distilled waters.

4. An experiment was conducted in Agronomy Farm, College of Agriculture, Jobner (Rajasthan) to test the difference between two fertilizers, Suphala (N: P.=20 : 20) and urea with superphosphate (U_1S_1, N=20/S=20) on the straw yield.

   The fertilizers are applied independently on 8 plots each of size $4 \times 3 = 12$ sq metres. The results were recorded in the following table.

Suphala 20 : 20	4.15	3.80	5.00	2.15	3.40	3.65	2.50	4.00
Urea with super-phoshate $U_1 S_1$ 20 : 20	3.50	2.80	3.40	2.15	2.75	3.50	2.70	2.10

   Test whether the two fertilizers give the same straw yield?

5. The average number of seeds set per pod in lucerne were determined for top flowers and bottom flowers in 10 plants,

   The values observed were as follows:

Top flowers	4.2	5.0	5.4	4.3	4.8	3.9	4.2	3.1	4.4	5.8
Bottom flower	4.6	3.5	4.8	3.0	4.1	4.4	3.6	3.8	3.2	2.2

   Test whether there is any significant difference between the top and bottom flowers with respect to average numbers of seeds set per pod.

6. In a random sample of 400 adults and 600 teenagers who watched a certain television programme 100 adults and 300 teenagers indicated that they liked it. Construct *(a)* 95 per cent, and *(b)* 99 per cent confidence limits for the difference in proportion of all adults and all teenagers who watched the programme and liked it.

   (B.Sc. Madras, 1967 Sept.)

7. Observations on 320 families having exactly five children gave the following data, X representing the number of boys in the family and f the corresponding frequency.

X	0	1	2	3	4	5
f	42	126	105	29	15	3

   Test the hypothesis that boy and girl births are equally likely in families having 5 children

   (B.Sc. Madras, 1967 April)

8. The average thermal efficiency for a particular brand of stove was quoted by a firm as 48%. The different thermal efficiencies of stoves of the same brand in a survey of 25 households are given as 44, 42,

46, 52, 50, 41, 39, 49, 41, 39, 38, 37, 45, 48, 42, 40, 35, 32, 30, 45, 46, 35, 31, 30 and 36.

Test whether there is any significant difference between the quoted percentage and the percentage obtained from sample data.

9. The thermal efficiencies of wick and non-wick stoves are given as follows:

Samples

Wick stove	42	46	49	35	45	48	37	39
Non-wick stove	53	50	51	48	46	57	45	

Test whether there is any significant difference between wick and Non-wick stoves with respect to thermal efficiency.

10. Two methods of preparation of same green leafy vegetable are to be compared for percentage calcium contents gm). 10 pairs of samples were studied.

	1	2	3	4	5	6	7	8	9	10
Method I	0.25	0.30	0.27	0.24	0.21	0.26	.024	0.27	0.29	0.30
Method II	0.27	0.32	0.23	0.27	0.28	0.29	0.25	0.21	0.23	0.26

Test whether there is any significant difference between two methods of preparation with respect to calcium content.

11. Fruit dropping for a particular variety of mango was estimated as 4 percent per tree. In a garden of 240 trees the fruit dropping was found to be 6 percent/tree. In another garden of 420 trees the fruit dropping was found to be 4.5 percent/tree. Test whether there is any significant difference between the two gardens with respect to fruit dropping at 5 per cent level of significance.

12. The milk consumption per family in a month in a particular town was estimated as 34 litres. A sample survey conducted in a locality of that town obtained the following distribution of consumption of milk. Test whether there is any significant difference between the averages of town and locality with respect to milk consumption.

Milk Consumption	15-20	20-25	25-30	30-35	35-40	40-45
Families	42	35	30	27	17	19

13. In a college of Home science, the average O.G.P.A scored by 120 final B.SC (Hom.Sc.) students in the year 2004 is 8.20 out of 10.0 and the average O.G.P.A scored by 132 final B.Sc (Hom. Sc.)

students of 2005 batch in the same college is 8.48 out of 10.00. The standard deviation of OGPAs' in the said college was obtained as 2.16 from records. Test whether there is any significant difference between 2004 and 2005 batches at 1 per cent level of significance.

14. An average expenditure for food per family in a city was obtained from records as 32 percent of the income with a standard deviation of 9 per cent. From the survey conducted in a locality of 240 families, the average expenditure on food per family was found as 38 percent. Test whether there is any significant difference between city and locality with respect to average expenditure on food at 5 per cent level of significance.

15. The following are the yields of 1st and 3rd tillers of 10 paddy hills.

1st (10 gm)	10	12	9	8	7	13	18	20	26	19
3rd (10 gm)	7	8	6	6	5	8	9	11	13	10

Test whether there is any significant difference between the yields of 1st and 3rd tillers at 1 per cent level of significance.

16. The following are the daily yields (kg) of 1st and 2nd lactation of 10 Murrah buffaloes in a dairy farm.

Lactation

1 st	10.2	9.8	6.9	14.6	11.9	17.9	11.4
2nd	12.4	10.6	7.0	16.5	10.8	20.4	13.8
1 st	7.5	10.6	15.7				
2nd	8.2	13.0	18.2				

Test whether 2nd lactation yields are significantly superior to 1st lactation yields at 1 percent level of significane.

17. The following are the yields (tons)/hectare of sugarcane for the first and ratoon crops in farms at 10 sugarcane farmers fields.

Crop	1	2	3	4	5	6	7	8	9	10
1st	35	28	40	18	54	50	38	48	38	27
Ratoon	46	32	63	25	68	74	46	64	60	40

Test whether ratoon crop yield is significantly superior to first crop yields with probability at 0.05.

18. The yields (bales) of cotton in 1st and 3rd pickings in 12 cotton farms in cotton growing area are given as follows

Picking	1	2	3	4	5	6	7	8	9	10	11	12
First	40	32	37	56	41	50	46	52	39	60	51	59
Third	28	50	44	50	39	57	50	40	30	52	49	60

Test whether first picking yields are significantly superior to third picking yields at 5 percent level of significance.

19. In a fertilizer factory the average daily production of Ammonia was recorded as 26 tons. On random inspection of 10 days the daily production were found to be 24, 32, 28, 20, 34, 32, 18, 25, 40 and 27. Test whether sample production is in conformity with the record at probability level 0.0 I.

20. A corporate company has two fertilizer factories A and B. The standard deviation of daily production of urea of these factories was computed as 2.8 tons over a long period. On inspection of 12 days the average daily production of factory A was observed as 30 tons and the average production of factory B was 38 tons based on 15 days. Test whether there is any significant difference between the output of two factories at 5 percent level.

21. The daily production (tons) of two pesticide factories are recorded as follows

Factory	1	2	3	4	5	6	7	8
A	12	16	9	13	15	17	20	18
B	8	13	21	16	20	12		

Test whether there is any significant difference between the two factories at probability level 0.05.

22. The following are the data on 'total iron' and 'total available iron' in different food products of 'Ragi' millet.

Product	1	2	3	4	5	6
Total Iron	16	25	18	20	16	13
Total available iron	9	19	11	14	7	5

Test the significant difference between 'Total' and 'Available' Iron at 5% level.

23. The following are the I.Q. scores of 'Abnormal' children 'before' and 'after' psychological treatment.

	1	2	3	4	5	6	7	8
Before	14	9	20	30	21	16	11	12
After	22	11	29	34	28	22	19	22

Test the significant difference between 'Before' and 'After' scores after psychological treatment at 5% level.

CHAPTER 11

CHI-SQUARE DISTRIBUTION

11.1 CHI-SQUARE DISTRIBUTION

Let X_1, X_2,..., X_n be n independent normal variates each is distributed normally with mean zero and S.D. unity then $X^2_1 + X^2_2 + ... + X^2_n = \Sigma X^2_1$ is distributed as Chi-square (χ^2) with n degrees of freedom (d.f.) where n is large. The probability that χ^2 lies in the interval $d\chi_2$ is given by

$$f(\chi^2) = \frac{1}{2^{n/2}\sqrt{n/2}} e^{-\chi^{2/2}} \left(\chi^2\right)^{n/2-1} d\left(X^2\right), 0 < \chi^2 < \infty \qquad .....(11.1)$$

The expression (11.1) is called the chi-square distribution with n.d.f. This was first given by Helmert in 1875, later it was independently given by Karl Pearson in 1900 along with chi-square test of goodness of fit. The chi-square probability curves for d.f., n. = 1,2, ... , 6 are given in Fig. 11.1

The areas under the chi-square curve for different fixed value X^2 on the X-axis are tabulated.

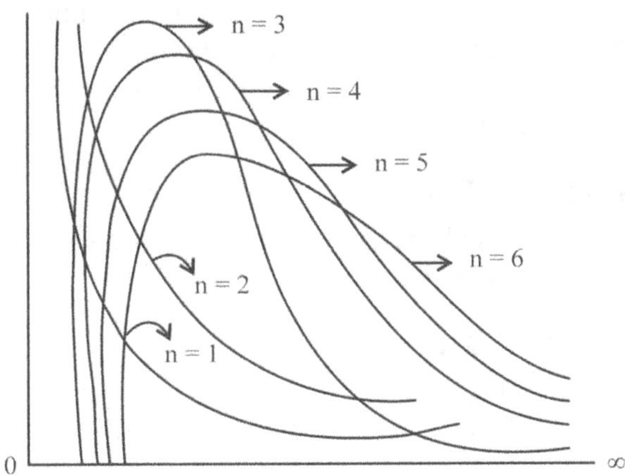

Fig. 11.1 Chi-square distribution.

11.2 PROPERTIES

1. For n > 2, the χ^2–distribution has mode at (n – 2). For n = 2, the distribution is L-shaped with maximum ordinate at zero, while for 0 <

$n < 2$ the distribution is L-shaped and has infinite ordinate at the origin.

2. Chi-square distribution is having additive property.

   If $\chi^2_1, \chi^2_2 ..., \chi^2_\kappa$ are k independent chi-square variates with $n_1, n_2, ...,$ n_k d.f. respectively, then $\chi^2_1 + \chi^2_2 + ... + \chi^2_\kappa$ is a χ^2 – variate with $n_1 + n_2 + ... + n_k$ d.f.

3. The different central moments are $\mu_2 + 2n$, $\mu_3 = 8n$, $\mu_4 = 48n+12n^2$

4. The coefficients of skewness and kurtosis are given as $\gamma_1 = \left(\dfrac{8}{n}\right)^{\frac{1}{2}}$,

   $\gamma_2 = \beta_2 - 3 - 12/n$

5. As n tends to infinity, the chi-square distribution tends to normality. The distribution function is an incomplete Gamma-function. Fisher showed that when n is large $\sqrt{2\chi^2}$ is approximately normally distributed with mean $\sqrt{2n-1}$ and variance unity. Wilson and Hilferty showed that $(\chi^2/n)^{1/3}$ is approximately normally distributed with mean, $1-2/9n$ and variance, $2/9n$ The latter one is 'more accurate approximation.

6. The ratio of two chi-square variates χ^2_1 and χ^2_2 with n_1 and n_2 d.f. respectively is a β-variate, i.e., $\beta(n_1/2, n_2/2)$

11.3 CHI-SQUARE TEST OF GOODNESS OF FIT

Chi-square test is used to know whether the given sampling distribution is in agreement with the known theoretical distribution or whether the given objects are segregating in a theoretical ratio or whether the two attributes are independent in a contingency table, etc.

11.3.1   Measurement Data

The data obtained by actual measurement is called measurement data. For example height, weight, age, income, area etc.

11.3.2   Enumeration Data

The data obtained by enumeration or counting is called enumeration data. For example number of blue flowers, number of intelligent boys, number of curled leaves etc.

11.3.3

χ^2-test is used for enumeration data which generally relate to discrete variable whereas t-test and standard normal deviate tests are used for measurement data which generally relate to continuous variable.

The expression for χ^2-test for goodness of fit is

$$\chi^2 = \sum_{i=1}^{k} \frac{(O_i - E_i)^2}{E_1} = \sum \frac{O_i^2}{\Sigma_i} - N \qquad \text{.....(11.2)}$$

where O_i = observed frequency of i-th cell, E_i = expected frequency of i-th cell, and k, the number of cells.

Here the term 'cell' is used for class interval in the case of frequency districution and compartment (or category) in the case of enumeration data.

Conclusion: If χ^2 (calculated) $\leq \chi^2$ (tabulated) then the hypothesis that the observed frequencies are in agreement with the expected frequencies is accepted. Otherwise, the hypothesis is rejected.

The expression (11.2) has been obtained from the expression $\sum_{ij} \alpha^{ij} \phi_i \phi_j$

where f_i is the likelihood function of the multinominal probability function, α^{ij} is the inverse of the covariance matrix,

11.4 2 × 2 CONTINGENCY TABLE

When the individuals (objects) are classified into two categories with respect to each of the two attributes then the table showing frequencies distributed over 2×2 classes is called 2×2 contingency table,

EXAMPLE: Suppose the individuals are classified according to two attributes colour (B) and intelligence (A). The distribution of frequencies over cells is shown in Table 11.1

TABLE 11.1

A B	A_1	A_2	Total
B_1	a	b	R_1
B_2	c	d	R_2
Total	C_1	C_2	N

where a, b, c and d are frequencies of the different cells. R_1, R_2, C_1 and C_2 are the respective marginal totals, and N is the grand total.

Null Hypothesis: H_o: The two attributes are independent. H_1: The two attributes are not independent.

$$\chi^2 = \sum_i \sum_j \frac{(O_{ij} - E_{ij})^2}{E_{ij}} \qquad \text{.....(11.3)}$$

where O_{ij} = Observed frequency of $(i, j)^{th}$ cell

E_{ij} = Expected frequency of $(i, j)^{th}$ cell

If the colour is not dependent on intelligence, then

$$\frac{a}{C_1} = \frac{b}{C_2} = \frac{R_1}{N} \quad \text{or}$$

$$\frac{c}{C_1} = \frac{d}{C_2} = \frac{R_2}{N}$$

Similarly if the intelligence is nothing to do with colour,

then $\quad a/R_1 = c/R_2 = C_1/N$

and $\quad b/R_1 = d/R_2 = C_2/N$

The expected frequencies are obtained as follows

$$E(a) = \frac{C_1 R_1}{N}$$

$$E(b) = \frac{C_2 R_1}{N} \quad \text{or} \quad E(b) = R_1 - C_1 R_1/N$$

$$E(c) = \frac{C_1 R_2}{N} \quad \text{or} \quad E(c) = C_1 - C_1 R_1/N$$

$$E(d) = \frac{C_2 R_2}{N} \quad \text{or} \quad E(d) = C_2 - E(b)$$

where, $E(a)$, $E(b)$, etc., are the expected frequencies of the cells containing observed frequencies 'a', 'b', etc. Since the marginal totals are fixed we need to compute only one expected frequency in 2 x 2 contingency table and the rest are obtained by subtraction as shown in R.H.S. above.

The degrees of freedom for testing the X^2 value is derived as total number of items–No. of restrictions from rows–No. of restrictions from the grand total, i.e., $4 - (2 - 1) - (2 - 1) - 1 = 1$.

Substituting the expected frequencies in (11.3), we have

$$\chi^2 = \frac{\left(a - \dfrac{C_1 R_1}{N}\right)^2}{\dfrac{C_1 R_1}{N}} + \frac{\left(b - \dfrac{C_2 R_1}{N}\right)^2}{\dfrac{C_2 R_1}{N}} + \frac{\left(c - \dfrac{C_1 R_2}{N}\right)^2}{\dfrac{C_1 R_2}{N}} + \frac{\left(d - \dfrac{C_2 R_2}{N}\right)^2}{\dfrac{C_2 R_2}{N}} \quad \dots(11.4)$$

$$\chi^2 = \frac{(ab - bc)^2 . N}{R_1 R_2 C_1 C_2} \quad \dots(11.5)$$

Conclusion: If χ^2 (calculated) $\leq \chi^2$ (tabulated) with $(2 - 1) \times (2 - 1)$ d.f. at chosen level of significance, the null hypothesis is accepted, i.e., the two attributes are independent. Otherwise, the null hypothesis is rejected.

EXAMPLE: 100 individuals of a particular race were tested with an intelligence test and classified into two classes. Another group of 120 individuals belong to another race were administered the same intelligence test and classified into the same two classes. The following are the observed frequencies of the two races:

Race	Intelligence Intelligent	Non-intelligent	
Race I	42	58	100
Race II	55	65	120
	97	123	220

Test whether the intelligence is anything to do with the race.

Null Hypothesis: Intelligence and Race are two independent attributes. Using (11.3), we have

$$\chi^2 = \frac{(42 \times 65 - 58 \times 55)^2 \times 220}{100 \times 220 \times 97 \times 123} = 0.33$$

Conclusion: χ^2 (calculated) $< \chi^2$ (tabulated), (3.481) with $(2-1)(2-1)$ d.f. at 5 per cent level of significance. Therefore, there is evidence to conclude that race and intelligence may be independent.

CHI-SQUARE TEST (2 × 2 CONTINGENCY TABLE)

SPSS DATA ANALYSIS PROCEDURE

Vaccination	Recovered	Not-recovered	Total
Vaccinated	8	2	10
Not-vaccinated	4	8	12
Total	12	10	22

Enter the data of above 2 × 2 contingency table in the SPSS Excel sheet column wise in the following manner.

M   V
1
1   1   }
1   1   } Repeat   8 times equivalent to the number in n_{11}
                  (1st row and 1st column)

2 }
1   2   } Repeat   2 times equivalent to the number in n_{12}
                  (1st row and 2nd column)

2   1   } Repeat   4 times equivalent to the number in n_{21} (2nd
                  row and 1st column)

2   2   } Repeat 8 times equivalent to the number in n_{22} (2nd row
                  and 2nd column0

After entering the data in the above manner then CLICK 'ANALYZE' FROM THE TOOL BAR. Then OPT FOR 'DISCRIPIVE STATISTICS' AND CHOOSE ' CROSS TABS'. Enter 'M' in columns and 'V' in rows. After obtaining the displayed table 'CHOOSE ' STATISTICS' AND OPT FOR ' CHI-SQUARE' & "CONTINUE'. THEN CLICK 'O.K.

We get the Table containing results of chi-square test , Yates-correction , Fisher's exact test along with probabilities of significance.

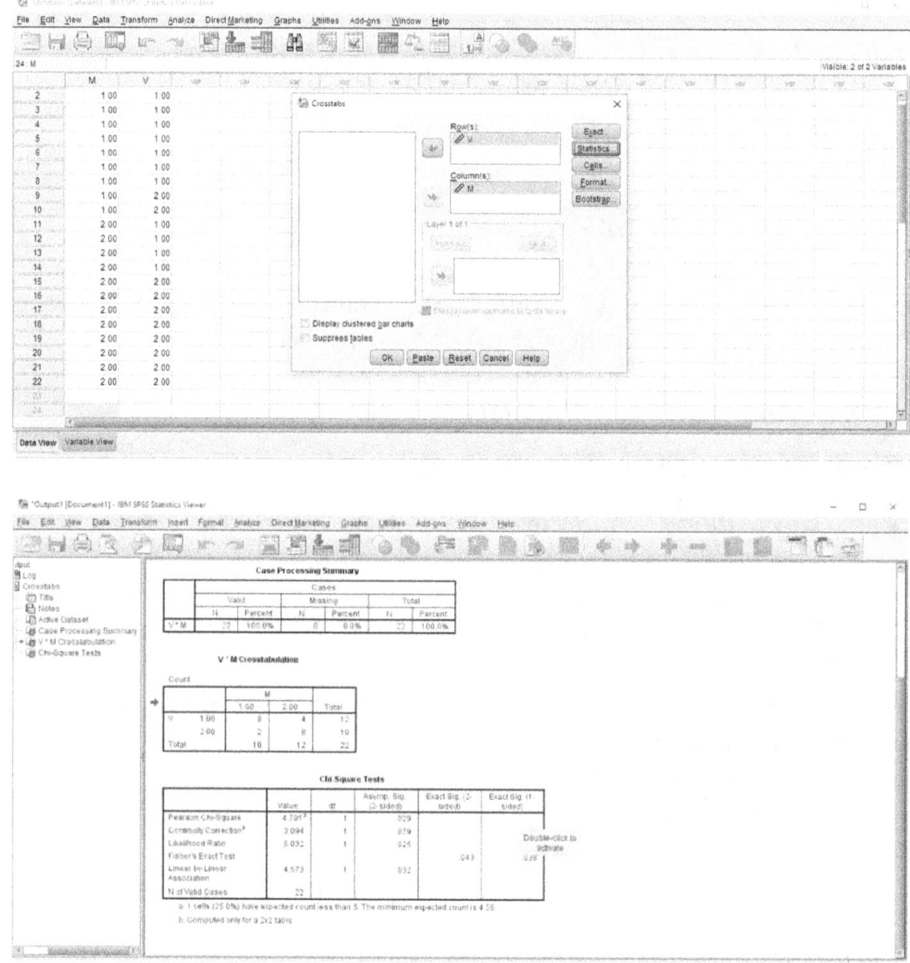

11.4.1   Yates Correction for Continuity

In the 2×2 contingency table, if the expected cell frequencies are large, the discrete distribution of the probabilities of the cell frequencies approximate to normal distribution and hence χ^2-statistics is distributed as χ^2 distribution with 1 d.f. On the other hand, if the expected frequencies are small (less than 5) the distribution of χ^2 cannot be used.

The corrected value of χ^2 can also be obtained directly using the expression

$$\chi^2 = \frac{\left(|ad - bc| - N/2\right)^2}{R_1 R_2 C_1 C_2} \times N \qquad \ldots\ldots(11.6)$$

11.4.2   V.M. Dandekar's Method

If χ_0^2 is the χ^2 value obtained from original observed frequencies, χ_{-1}^2 is the one obtained by increasing the smallest frequency by unity keeping the marginal totals fixed and χ_{+1}^2 is the value of χ^2 obtained by decreasing the smallest frequency by one keeping the marginal totals constant, then,

$$\chi^2 = \chi_0^2 - \frac{(\chi_0^2 - \chi_{-1}^2)}{\chi_{+1}^2 - \chi_{-1}^2}\left(\chi_{+1}^2 - \chi_0^2\right) \qquad \ldots\ldots(11.7)$$

The value of χ^2 is to be obtained from χ^2 and is to be compared with tabulated value at 10 per cent level of significance of normal distribution.

EXAMPLE: The following were the data of 40 individuals classified according to smoking and residential background. Test whether the smoking habit is inherent with the residential background.

	Smokers	Non-smokers	Total
Rural	15	3	18
Urban	15	7	22
Total	30	10	40

Yates method: Here the expected frequency of cell containing observed frequency 3 is less than 5. Hence Yates correction for continuity is applied since 15 x 3 < 15 × 7, subtracting $\frac{1}{2}$ from 15 and 7 and adding $\frac{1}{2}$ to 15 and 3, we have

	Smokers	Non-smokers	Total
Rural	14.5	3.5	18
Urban	15.5	6.5	22
Total	30.0	10.0	40

Null Hypothesis: H_o: Smoking and residential background are independent.

$$\chi^2 = \frac{(14.5 \times 6.5 - 15.5 \times 3.5)^2}{30 \times 10 \times 18 \times 22} \times 40 = 0.5387$$

$$\chi = 0.7340$$

CONCLUSION: The calculated value of χ is less than the tabulated value 1.645 at 10 per cent level of significance of normal table. Therefore, there is evidence to say that the smoking habit is independent of the residential background.

V.M. Dandekar's method:

$$\chi_0^2 = \frac{(105 - 45)^2 \times 40}{30 \times 10 \times 18 \times 22} = 1.2121$$

$$\chi_{-1}^2 = \frac{(14 \times 6 - 16 \times 4)^2 \times 40}{30 \times 10 \times 18 \times 22} = 0.1347$$

$$\chi_{+1}^2 = \frac{(16 \times 8 - 14 \times 2)^2 \times 40}{30 \times 10 \times 18 \times 22} = 3.3670$$

$$\chi^2 = \chi_0^2 - \frac{\chi_0^2 - \chi_{-1}^2}{\chi_{+1}^2 - \chi_{-1}^2}(\chi_{+1}^2 - \chi_0^2)$$

$$\chi^2 = 1.2121 - \frac{(1.2121 - 0.1347)}{3.3670 - 0.1347}(3.3670 - 1.2121)$$

$$= 0.4938$$

$$\chi = 0.7027$$

Conclusion: The X (calculated) i.e., $0.7027 < X$ (tabulated), (1.645) at 10 per cent level of significance of normal table. Therefore, null hypothesis is accepted. The two attributes smoking and residential background may be considered independent.

11.5 r × s CONTINGENCY TABLE

When the number of individuals (or objects) are classified according to 'r' categories A_1, A_2, ..., A_r with respect to attribute A and 's' categories B_1, B_2 ... B_s with respect to attribute B, then the arrangement of the frequencies in r × s cells is called as r × s contingency table.

TABLE 11.2

A \ B	B_1	B_2	..	B_1	..	B_2	Total
A_1	O_{11}	O_{12}	..	O_{1j}	..	O_{1s}	R_1
A_2	O_{21}	O_{22}	..	O_{2j}	..	O_{2s}	R_2
$\overline{A}_1$	$\ddot{O}_{i1}$	$\ddot{O}_{i2}$	..	$\ddot{O}_{ij}$	..	$\ddot{O}_{is}$	R_i
$\ddot{A}_r$	$\ddot{O}_{r1}$	$\ddot{O}_{r2}$	..	$\ddot{O}_{rj}$	..	$\ddot{O}_{rs}$	R_r
Total	C_1	C_2	, ,	C_j	, ,	C_s	N

Null Hypothesis: H_o: The two attributes A and B are independent

$$\chi^2 = \sum_{i=1}^{r}\sum_{j=1}^{s}(O_{ij} - E_{ij})^2 / E_{ij} \qquad\qquad .....(11.8)$$

The expected frequency of each cell can be obtained in the same way as in 2×2 contingency table, For example.

$$E_{ij} = E(O_{ij}) = \frac{R_i \times C_j}{N}$$

where E (O_{ij}) is the expected frequency of cell containing observed frequency O_{1j}. The expected frequencies and observed frequencies will be substituted in the expression (11.8) for arriving at the calculated value of χ^2.

Conclusion: If χ^2 (calculated) $> \chi^2$ (tabulated) with (r – 1) (s – 1) d.f. at chosen level of significance, the null hypothesis is rejected.

CHI-SQUARE TEST FOR r × s CONTINGENCY TABLE

SPSS DATA ANALYSIS PROCEDURE

Literacy	More progressive	Progressive	Less pregressive	Total
Post-graduate	8	2	4	14
College	4	8	6	18
High school	6	10	5	21
Total	18	20	15	53

Here the data will be entered in the SPSS Excel sheet column wise as follows;

P (Row)   L Column)

1	1	Repeat 8 times as in n_{11} (1st row and 1st column)
1	2	Repeat 2 times as in n_{12} (1st row and 2nd column)
1	3	Repeat 4 times as in n_{13} (1st row and 3rd column)
2	1	Repeat 4 times as in n_{21} (2nd row and 1st column)
2	2	Repeat 8 times as in n_{22} (2nd row and 2nd column)
2	3	Repeat 6 times as in n_{23} (2nd row and 3rd column)
3	1	Repeat 6 times as in n_{31} (3rd row and 1st column)
3	2	Repeat 10 times as in n_{32} (3rd row and 2nd column)
3	3	Repeat 5 times as in n_{33} (3rd row and 3rd column)(

After entering the data in the above manner then CLICK 'ANALYZE' FROM THE TOOL BAR. Then OPT FOR 'DISCRIPIVE STATISTICS' AND CHOOSE ' CROSS TABS'. Enter 'P' in rows and 'L' in columns. After obtaining the displayed table 'CHOOSE ' STATISTICS' AND THEN   OPT FOR ' CHI-SQUARE' and then' CONTINUE' and then CLICK 'O.K. 'We get the Table containing results of chi-square test , Yates-correction etc.,

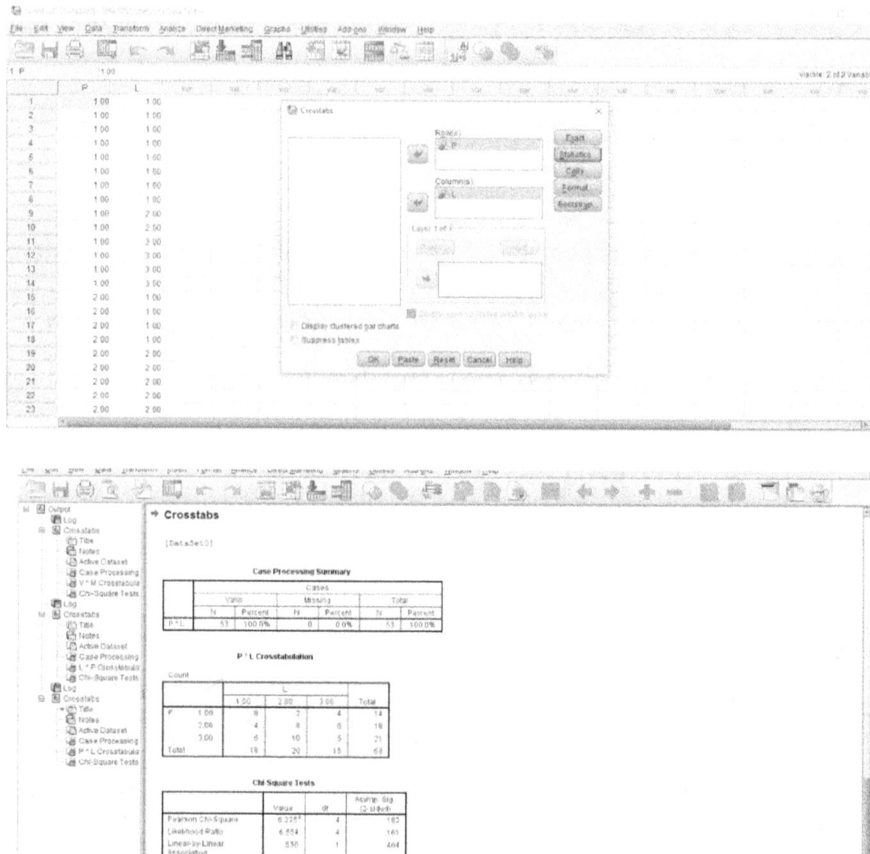

11.6 χ^2-TEST FOR GENETIC PROBLEMS

Genetic theory states that different genes segregate independently and in a particular phenotypic ratios depending upon the cross. A dominant gene segregates from its recessive allele in the ratio 3:1 in F_2 generation and in the ratio 1:1 in a back cross. If there are two genes responsible for a particular character then there are four type of phenotype combinations in F_2 generation and back cross. They are $(3:1)\times(3:1) = 9:3:3:1$ and $1:1\times 1:1 = 1:1: 1.1.$ for F_2 generations and back cross respectively.

EXAMPLE: A cross between the two varieties of Sorghum one giving high yield and the other for high amount of fodder was made. The number of plants in F_2 generation were observed as 79, 160, and 85. Test whether this sample data is in agreement with the Mendelian ratio 1: 2:1 or not.

Null Hypothesis: H_o: The sample ratio is in agreement with 1: 2:1 ratio. H_1: The sample ratio is not in agreement with 1:2:1 ratio.

Observed Freq. (O_i)	Expected Freq. (E_i)	(O_i-E_i)	$\dfrac{(O_i-E_i)^2}{(E_i)}$
79	$324 \times 1/4 = 81$	-2	0.0494
160	$324 \times 2/4 = 162$	-2	0.0247
85	$324 \times 1/4 = 81$	4	0.1975
324	324		0.2716

Conclusion: χ^2 (calculated). $<\chi^2$ (tabulated), (5.991) with (3–1) d.f. at *5* per cent level of significance. Therefore, the null hypothesis is accepted, i.e., the plants are segregating according to Mendelian ratio, 1: 2:1 in F_2 generation.

11.6.1

If there are two or more families, first we compute the χ^2 individually with 1 d.f. for testing the deviation from the theoretical ratio and pool the different χ^2 values for obtaining the total heterogeneity. This χ^2 will be partitioned into two Chi-squares, one due to deviation from theoretical ratio for the totals of the observed frequencies and the other χ^2 due to heterogeneity among families. The different observed frequencies are shown in Table 11.3.

If the families segregate into e_1: e_2 theoretical' ratio, then

$$\chi_1^2 = \frac{\left[(O_{11})e_2 - (O_{12})e_1\right]^2}{R_1 e_1 e_2}$$

$$\chi_2^2 = \frac{\left[(O_{21})e_2 - (O_{22})e_1\right]^2}{R_2 e_1 e_2}$$

$$\vdots \qquad\qquad \vdots$$

$$\chi_k^2 = \frac{\left[(O_{k1})e_2 - (O_{k2})e_1\right]^2}{R_k e_1 e_2}$$

$$\chi_t^2 = \chi_1^2 + \chi_2^2 + \ldots + \chi_k^2$$

TABLE 11.3

Family	Observed Frequencies		
1	O_{11}	O_{12}	R_1
2	O_{21}	O_{22}	R_2
$\vdots$	$\vdots$	$\vdots$	$\vdots$
k	O_{k1}	O_{k2}	R_k
	C_1	C_2	N

The χ_t^2 would be partitioned into two components, one due to deviation from $e_1 : e_2$ ratio and another due to heterogeneity between the families *as* $\chi_t^2 = \chi_d^2 + \chi_h^2$, where

$$\chi_d^2 = \frac{(C_1e_2 - C_2e_1)^2}{Ne_1e_2} \quad \text{and}$$

$$\chi_h^2 = \chi_t^2 - \chi_d^2$$

The calculated values of χ_d^2 and χ_h^2 can be compared' with tabulated values with 1 and $(k-1)$ d.f respectively.

EXAMPLE: In the rabi crop the segregation of the plants (phenotype) in F_2 generation for different parents with regard to one pair of genes (colour) is given as follows:

Parents	Purple	Green	Total
I	320	100	420
II	280	85	365
III	315	110	425
IV	270	84	354
V	295	105	400
Total	1480	484	1964

Test whether the F_2 generation is segregating in the 3:1 ratio as a whole and test whether there exists homogeneity among different pairs of parents.

TABLE 11.4

Parent	χ^2 (calculated)	d.f.	χ^2 (tabulated)	Remarks
I	0.3175	1	3.841	Not significant
II	0.5708	1	,,	-do-
III	0.1765	1	,,	-do-
IV	0.3051	1	,,	-do-
V	0.3333	1		-do-
Total	1.7032	5	11.070	

TABLE 11.5

	χ^2 (calculated)	d.f.	χ^2 (tabulated)	Remarks
Total Deviation From 3 : 1 ratio	1.7032	5	11.070	Not significant
	0.1331	1	3.841	-do-
Hetero-Geneity	1.5701	4	9.488	-do-

Conclusion: It is observed from the Tables 11.4 and 11.5 that χ^2 values are not significant within each of the pairs of parents indicating thereby that phenotype frequencies are segregating in the ratio 3:1 and it is also seen that the F_2 generation for all the parents is segregating in the 3:1 ratio. The calculated value of χ^2 for the heterogeneity between the families is obtained by subtracting the χ^2 value of deviation from 3:1 ratio from the total value of χ^2. The calculated value of χ^2 for heterogeneity is also found not significant and hence it can be inferred that the heterogeneity between the families is not significant.

11.7 X^2-TEST FOR LINKAGE PROBLEMS

Linkage is defined as the tendency of genes belong to the same chromosome or linkage group to enter the gametes in the parental combinations.

The alternative to linkage is cross over. In this case the genes would tend to enter the gametes other than the parental combinations. These are called recombination.

11.7.1

When the two pairs of genes (or characters) are under consideration whether the F_2 data (phenotype frequency) is segregating according to expected theoretical ratio or not can be tested with the help of chi-square test. If there is a significant departure from the theoretical ratio, the reason for this departure may be either due to the individual characters (pair of genes) may not be segregating into theoretical ratio or it may be due to significant linkage.

If χ^2_t is the departure of F_2 data from theoretical ratio (say) 9:3:3:1 for the two characters A and B with 3 d.f. and χ^2_A is the departure from the theoretical ratio 3:1 for the character A with 1 d.f. and χ^2_B is the departure from the theoretical ratio 3:1 for the character B with 1 d.f., we have

$$\chi^2_1 = \chi^2_t - (\chi^2_A + \chi^2_B) \text{ where } \chi^2_1 \text{ is the value of } \chi^2 \text{ for linkage}$$

TABLE 11.6

	B	b	
A	AB	Ab	R_1
a	aB	ab	R_2
	C_1	C_2	N

The different chi-square values are presented in Table 11.7.

TABLE 11.7

Source	χ^2 Value	d.f.	χ^3 (tabulated)
Character			
A	χ^2_a	1	3.841
B	χ^2_b	1	-do-
Linkage	χ^2_1	1	-do-
Total	χ^2_1	3	

EXAMPLE: In an experiment on sweet peas purple and long pollen grains were crossed with red and round pollen grains and progeny in F_2 generation were observed as follows. Purple and long pollen grains were 220, purple and round pollen 105, red and long pollen 90 and red and round pollen 17. Test departure of F_2 generation from the theoretical ratio 9:3:3:1 and also the linkage and the individual characters in the ratio 3:1.

Null Hypothesis: H_o: Observed frequencies are in the theoretical ratios. H_1: Observed frequencies are not in the theoretical ratios.

TABLE 11.8

	Long	Round	
Purple	220	105	325
Red	90	17	107
	310	122	432

$$\chi^2_c \text{ (Colour)} = \frac{(325 - 3 \times 107)^2}{3 \times 432} = 0.012$$

$$\chi^2_s \text{ (Shape)} = \frac{(310 - 3 \times 122)^2}{3 \times 432} = 2.420$$

$$\chi^2_l \text{ (Linkage)} = \frac{(220 - 3 \times 105 - 3 \times 90 + 9 \times 17)^2}{9 \times 432} = 11.56$$

TABLE 11.9

Observed (O_i)	Expected (E_i)	$\dfrac{(O_i - E_i)^2}{(E_i)}$
220	243	2.177
105	81	7.111
90	81	1.000
17	27	3.704
432	432	13.992

$$\chi^2 \text{ (Total)} = 13.992$$

TABLE 11.10

Source	χ^2 (calculated)	d.f.	χ^2 (tabulated)	Remarks
Colour (Purple vs Red)	0.012	1	3.841	Not significant
Shape (long vs Round)	2.420	1	3.841	-do-
Linkage	11.560	1	3.841	Significant
Total	13.992	3	7.815	Significant

Conclusion: Here the values of χ^2 for colour and shape are not significant, thereby indicating that the individual characters are segregating in the theoretical ratio 3:1. But the $\chi^2 t$ value is significant indicating that both the pairs of genes are not assorting independently. Therefore, the discrepancy may be due to linkage. The χ^2_1 is found significant which confirms the earlier statement.

11.7.2

If there are three pairs of genes A, a, B, b, C, c segregating in a back cross (Aa Bb Cc × aa bb cc) and each gene is expected to show a 1: 1 ratio. If there is no linkage the three pairs of genes segregate independently. There are in all eight classes ABC, ABc, AbC, aBC, Abc, aBc, abC, and abc. The expected frequencies will be same in all the classes i.e., 1:1:1:1:1:1:1:1 in a back cross, The total χ^2 will have 7 d.f. for the eight classes as a whole. The coefficients for the different pairs of comparisons can be obtained with the help of Fisher's evens Vs Odds rule.

TABLE 11.11

		ABC	ABc	AbC	aBC	Abc	aBc	abC	Abc
Main effects	A	+	+	+	−	+	−	−	−
	B	+	+	−	+	−	+	−	−
	C	+	−	+	+	−	−	+	−
Interaction	AB	+	+	−	−	−	−	+	+
	BC	+	−	−	+	+	−	−	+
	AC	+	−	+	−	−	+	−	+
Second order Interaction	ABC	+	−	−	−	+	+	+	−

Since in the back cross, the ratio of segregation for the pairs of genes in F_2 generation is 1: 1: 1: 1: 1: 1: 1: 1, the divisor for each comparison is equal to N where N is the total frequency.

$$\chi^2_A = [(ABC) + (ABc) + (AbC) + Abc - (aBC) - (aBc) - (abC) - (abc)]^2 / N$$

The total χ^2 with 7 d.f. will be partitioned into different $\chi^2 s$ for each of the 7 comparisons with 1d.f. each and these will be tested as follows,

TABLE 11.12

Item (Source)	χ^2 (calculated)	d.f.	χ^2 (tabulated)
A	χ^2_A	1	3.841
B	χ^2_B	1	,,
C	χ^2_C	1	,,
Interaction			
Linkage			
AB	χ^2_{AB}	1	,,
BC	χ^2_{BC}	1	,,
AC	χ^2_{AC}	1	,,
ABC	χ^2_{ABC}	1	,,
Total	χ^2_t	7	14.067

EXERCISES

1. A group of school children were classified according to intelligence level (I) and economic level (E) and the results were as under

2.

	E_1	E_2	E_3	E_4
I_1	85	216	165	206
I_2	144	305	320	152
I_3	120	185	160	45

Test for the independence of the two-factors at .one percent level.

3. The following table gives the number of literates and criminals in the three cities A, B and C. Compare the degree of association between criminality and illiteracy in each of the three cities.

		A	B	C
Total number (in ten thousands)	. .	246	185	228
Literates	(,,) . .	42	47	32
Literate criminals	(,,) . .	4	2	3
Illiterate criminals	(,,) . .	41	22	24

3. The probability for an animal to catch a particular infection is 0.20. In an experiment 60 animals were treated with a new vaccine, and 5 of

them caught the infection. Applying the χ^2-test, arrive at a decision regarding the efficiency of the new vaccine at the 5 per cent level of significance.

4. In an experiment on chillies, the following results were obtained

TABLE

Shape	Pungent	Not-pungent
Long	48	27
Short	32	73

Test whether there is any association between taste and shape of chillies at 5 per cent level of significance.

5. To prevent eye disease in children, an experimental diet was recommended. In an investigation the following results were obtained.

	Prevented the disease	Not-Prevented the disease
Experimental diet	10	6
Control diet	4	14

Test whether experimental diet has any effect in preventing eye disease in children.

6. Experimental plots were classified according to duration of the variety and infestation of particular pest and the following results were obtained in an investigation.

Duration	High infested	Medium infested	Low infested
Long	10	18	27
Medium	14	22	34
Short	38	20	12

Test whether duration of a variety has any effect in controlling the pest at 5 per cent level.

7. Tobacco leaves were classified according to shape and quality in a survey and the results are presented as follows.

	Shape		
Quality	Large	Medium	Narrow
Good	34	28	9
Satisfactory	30	16	11
Not-good	14	20	38

Test whether there is any association between "quality" and 'shape' of Tobacco leaves.

8. The following table gives the number of children vaccinated against polio disease. Test whether vaccination has any effect in controlling the disease at 1 per cent level of significance.

	Effected	Not-effected
Vaccinated	2	22
Not- vaccinated	18	10

9. In an experiment conducted on Tomatoes Red and round were crossed with Pink and elongated and the progeny in F_2 generation were observed as follows. Red and round tomatoes were 40, Red and elongated 84, Pink and round 69 and Pink and elongated 46. Test departure of F_2 generation from the theoretical ratio 9:3:3:1 and also the linkage and individual characters in the ratio 3: 1.

10. The following are the number of seeds germinated in each pot when 10 seeds are sown in glass house experiment on sunflower crop.

Pot

Pot	1	2	3	4	5	6	7	8	9	10
Seeds germinated	8	6	3	8	6	4	5	7	9	4

Test whether seeds germinated uniformly in all the pots at 5 per cent level of significance.

11. The following are the observed and expected frequencies obtained in fitting Binomial distribution.

Observed	6	10	19	28	13	9	5
Expected	4	12	20	27	14	7	6

Test whether expected frequencies are in close agreement with observed frequencies at 1 per cent level of significance.

12. The following is the classification of 'Ergonomic' score according to age of women working in package section of Pharmaceutical Competency.

	0-20	20-30	30-40	40 and above
Age < 30	15	8	6	2
30-45	10	18	22	18
45-55	9	24	30	32
55 and above	8	25	32	40

Test the association between 'Age' and Ergonomic score' at 5% level.

CHAPTER 12

CORRELATION AND REGRESSION

12.1 INTRODUCTION

Here we consider the behaviour of two related variables. If one variate changes in sympathy with another variate then it can be said that there exists some association between these two variates. The degree of association ship (or the extent of relationship) is known as "coefficient of correlation". It is known that height and weight of different individuals are related with each other. In other words, if height changes weight also changes. This type of relationship may be denoted as 'correlation'. The amount of this correlation is known as 'Correlation coefficient'. This relationship can be observed from a graph shown in Fig. 12.1 corresponding to heights and weights of different individuals. The diagram shown in Fig. 12.1 is called Scatter diagram. If the points cluster together and indicate a straight line, then the relationship is termed as linear relationship, otherwise it is known as 'Non-linear relationship'.

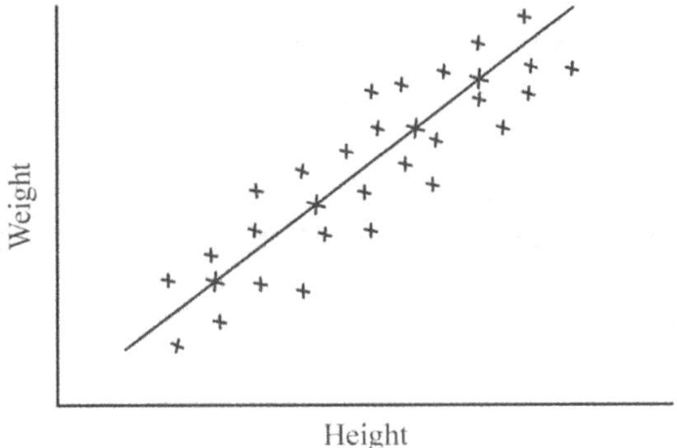

Fig. 12.1 Scatter diagram.

The linear correlation coefficient between X and Y variables based on sample data is denoted by 'r', where

$$r = \frac{\text{Covariance (XY)}}{\sqrt{\text{Variance (X). Variance (Y)}}}$$

$$r = \frac{\Sigma(X_i - \overline{X})(Y_i - \overline{Y})}{\sqrt{\Sigma(X_i - \overline{X})^2 . \Sigma(Y_i - \overline{Y})^2}}$$

Simplifying

$$r = \frac{\Sigma X_i Y_i - \dfrac{(\Sigma X_i)(\Sigma Y_i)}{n}}{\sqrt{\left[\Sigma X_i^2 - \dfrac{(\Sigma X_i)^2}{n}\right]\left[\Sigma Y_i^2 - \dfrac{(\Sigma Y_i)^2}{n}\right]}} \qquad \dots(12.1)$$

where n is the number of observations for each variate. 'r' is also called as 'simple correlation coefficient'. r lies between −1 and +1 and it is independent of units. If r = 0, it indicates no correlation between two variates. If r = − 1, there exists perfect negative correlation between the two variates. If r = + 1, there exists perfect positive correlation between the two variates. These cases are shown in Fig. 12.2 (a), (b) and (c) respectively.

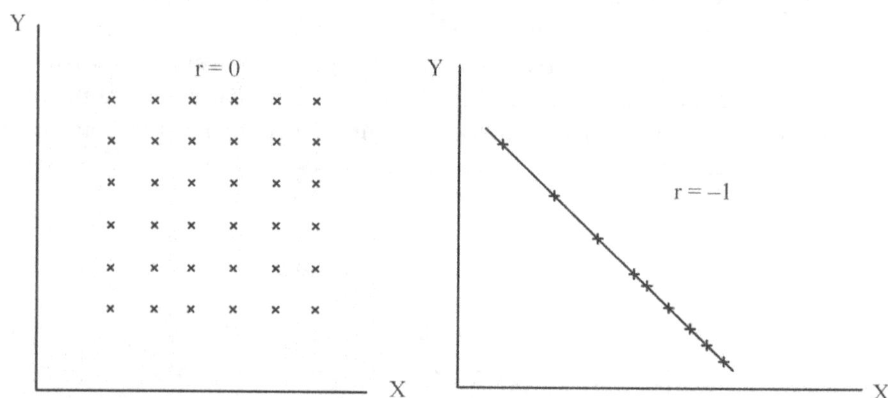

Fig. 12.2(a) Zero correlation.   **Fig. 12.2(b)** Perfect negative correlation.

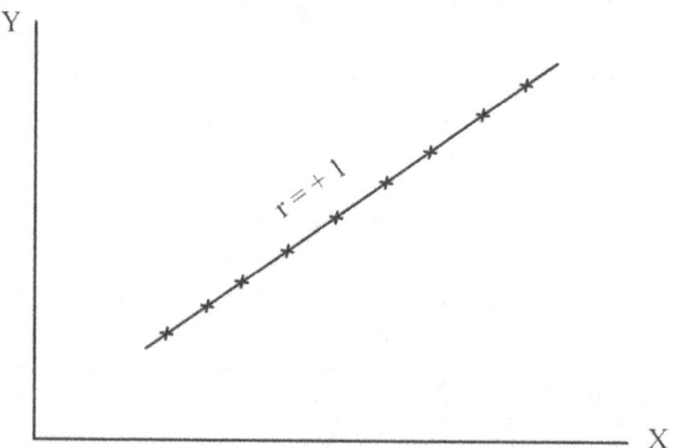

Fig. 12.2(c) Perfect positive correlation.

Let $d_1 = \dfrac{X-A}{C_1}$ and $d_2 = \dfrac{Y-B}{C_2}$ where A and B, Are arbitrary means and

C_1 C_2 are the common divisors respectively for X and varieties, then

$$r = \frac{\Sigma d_1 d_2 - \dfrac{(\Sigma d_1)(\Sigma d_2)}{n}}{\sqrt{\left[\Sigma d_1^2 - \dfrac{(\Sigma d_1)^2}{n}\right]\left[\Sigma d_2^2 - \dfrac{(\Sigma d_2)^2}{n}\right]}} \qquad .....(12.2)$$

12.2 TEST OF SIGNIFICANCE OF SIMPLE CORRELATION COEFFICIENT

Null Hypothesis: H_o: $\rho = 0$, H_1: $\rho \neq 0$

where 'ρ' is called the population correlation coefficient.

$$t = \frac{|r - 0|}{\sqrt{1 - r^2}} \sqrt{n - 2}$$

Conclusion: 1f t (calculated) > t (tabulated) with (n–2) d.f. at chosen level of significance, the null hypothesis is rejected. That is, there may be significant correlation between the two variates. Otherwise, the null hypothesis is accepted.

EXAMPLE: The following is the data of head and body weights of 10 insects (drosophila melanaogoster).

| Head weight (mg) | 20 | 22 | 25 | 27 | 31 | 32 | 35 | 38 | 39 | 40 |
| Body weight (mg) | 60 | 64 | 72 | 80 | 84 | 86 | 92 | 96 | 97 | 102 |

Compute the correlation coefficient between head and body weights and test its significance.

TABLE 12.1

X_i	Y_i	$d_{1i}=(X_i-31)$	$d_{2i}=(Y_i-86)$	$d_{1i}d_{2i}$	d^2_{1i}	d^2_{2i}
20	60	−11	−26	286	121	676
22	64	−9	−22	198	81	484
25	72	−6	−14	84	36	196
27	80	−4	−6	24	16	36
31	84	0	−2	0	0	4
32	86	1	0	0	1	0
35	92	4	6	24	16	36
38	96	7	10	70	49	100
39	97	8	11	88	64	121
40	102	9	16	144	81	256
		−1	−27	918	465	1909

Using formula (12.2), we have

$$r = \frac{918 - \dfrac{(-1)(-27)}{10}}{\sqrt{\left[465 - \dfrac{(-1)^2}{10}\right]\left[1909 - \dfrac{(-27)^2}{10}\right]}} = 0.99$$

Test of Significance

Null Hypothesis: H_o: $\rho = 0$, H_1: $\rho \neq 0$

$$t = \frac{r}{\sqrt{1-r^2}}\sqrt{n-2} = \frac{0.99}{\sqrt{1-(.99)^2}}\sqrt{(10-2)} = 19.84$$

CONCLUSION: Here t (calculated) > t (tabulated) i.e, 2.306 with 8 d.f. at 5 per cent level of significance. Hence the null hypothesis is rejected. That is, there may be significant correlation between the two variates.

CORRELATION COEFFICIENT

SPSS DATA ANALYSIS PROCEDURE

X	Y
7	12
11	19
14	23
17	26
19	32

Take X(ht.cm) and Y(wt.kg) as related variables and we have to compute correlation coefficient. Enter the data column wise underneath X and Y. Click 'ANALYZE' from TOOL BAR and opt for 'CORRELATION ' AND CLICK

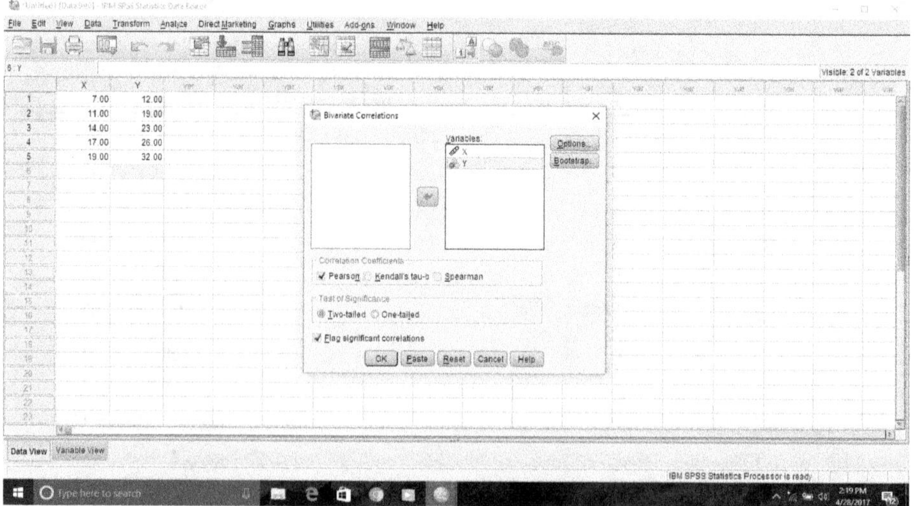

'BIVARIATE'. In the displayed TABLE take X and Y to the RHS as per indicator one after another. CLICK 'PEARSON' OPTION AND CLICK 'TWO-TAILED' OR 'ONE-TAILED'.OPTION as per your requirement. Then CLICK OK. option. We get the results displayed in a Table. The value of correlation coef and the probability value of its significance will be shown

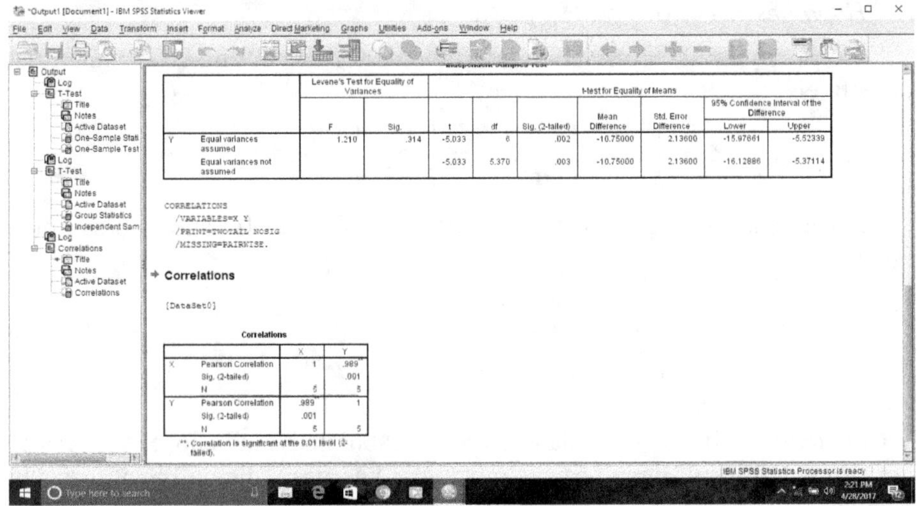

12.3 FISHER'S Z-TRANSFORMATION

When Correlation Coefficient (ρ) is not equal to zero in the population, the exact distribution of r is far from normal and the expression $\dfrac{r\sqrt{n-2}}{\sqrt{1-r^2}}$ does not follow student–t distribution. In order to overcome this difficulty, R.A. Fisher suggested Z-transformation where

$$Z = \frac{1}{2}\log_e\frac{1+r}{1-r}$$

It was found that Z follows normal distribution with standard error as $\dfrac{1}{\sqrt{(n-3)}}$.

12.3.1 One Sample Test

Null Hypothesis: H_o: $\rho = \rho_0$, H_1: $\rho \neq \rho_0$ here 'ρ_0' is the given value of ρ.

$$Z_s\frac{1}{2}\log_e\frac{1+r}{1-r}, \qquad Z_p\frac{1}{2}\log_e\frac{1+\rho_0}{1-\rho_0}$$

$$Z = \frac{\left|Z_s - Z_p\right|}{\sqrt{\dfrac{1}{n-3}}}$$

Conclusion: If Z (calculated) $>$ Z (tabulated) at chosen level of significance, the null hypothesis is rejected. Otherwise, the null hypothesis is accepted.

12.3.2   Two Sample Test

Null Hypothesis: H_o: $\rho_1 = \rho_2$, H_1: $\rho_1 \neq \rho_2$

Here ρ_1, ρ_2 are the correlation coefficients for the 1st and 2nd populations respectively.

$$Z_1 = \frac{1}{2}\mathrm{Log}_e\frac{1+r_1}{1-r_1}, \quad Z_2 = \frac{1}{2}\mathrm{log}_e\frac{1+r_2}{1-r_2}$$

r_1, r_2 are the sample correlation coefficients for the 1st and 2nd samples respectively.

$$Z = \frac{|Z_1 - Z_2|}{\sqrt{\dfrac{1}{n_1 - 3} + \dfrac{1}{n_2 - 3}}}$$

Conclusion: If Z (calculated) $>$ Z (tabulated) at chosen level of significance, the null hypothesis is rejected. Otherwise, the null hypothesis is accepted.

12.3.3   Test of the Homogeneity of the Set of Sample Correlation Coefficients

Let r_1, r_2,... r_k be k sample correlation coefficients based on sample sizes n_1, n_2 ..., n_k respectively and Z_1, Z_2, ..., Z_k be the corresponding Fisher's Transformation values with variances

$$\frac{1}{n_1 - 3}, \frac{1}{n_2 - 3}, ..., \frac{1}{n_k - 3} \text{ respectively.}$$

Null Hypothesis: All the sample correlation coefficients are homogeneous.

$$\chi^2 = \sum_{i=1}^{k}(n_i - 3)Z_i^2 - \frac{\left[\sum_{i=1}^{k}(n_i - 3)Z_i\right]^2}{\sum_{i=1}^{k}(n_i - 3)}$$

Conclusion: If χ^2 (calculated) $> \chi^2$ (tabulated) with (k–1) d.f. at chosen level of significance, the null hypothesis is rejected. Otherwise the null hypothesis is accepted.

12.3.4

If the correlation coefficients are found to be homogeneous, the best estimate of the correlation coefficients is given by

$$\hat{Z} = \dfrac{\displaystyle\sum_{i=1}^{k}(n_i - 3)Z_i}{\displaystyle\sum_{i=1}^{k}(n_i - 3)}$$

$\hat{Z}$ can be converted back to r using Z to r conversion table.

EXAMPLE: Test the homogeneity of the correlation coefficients 0.98, 0.99, 0.86, 0.88, 0.79, 0.72 and 0.83 based on sample sizes 5, 8, 9, 10, 10, 12 and 6 respectively.

Null Hypothesis: All the correlation coefficients are homogeneous.

TABLE 12.2

r_i	Z_i	$n_i{-}3$	$(n_i{-}3)\,Z_i$	$(n_i{-}3)\,Z_i^2$
0.98	2.298	2	4.596	10.562
0.99	2.646	5	13.230	35.007
0.86	1.293	6	7.758	10.031
0.88	1.376	7	9.632	13.254
0.79	1.071	7	7.497	8.029
0.72	0.908	9	8.172	7.420
0.83	1.188	3	3.564	4.234
		39	54.449	88.537

$$\chi^2 = \Sigma(n_i - 3)Z_i^2 - \dfrac{[\Sigma(n_i - 3)Z_i]^2}{\Sigma(n_i - 3)}$$

$$= 88.537 - \dfrac{(54.449)^2}{39} = 12.519$$

Conclusion: Here χ^2 (calculated) $< \chi^2$ (tabulated) i.e., 12.592. Hence the null hypothesis is accepted. Therefore all the correlation coefficients are found to be homogeneous.

12.4 RANK CORRELATION

Sometimes we may not be in a position to measure the variates but their ranks can be provided. In such cases the relationship between variates with the help of ranks can be obtained with the help of rank correlation coefficient and is denoted by (say)r_s. For example, in preparing a questionnaire (or schedule) in extension education, the questions are to be arranged in order of their importance. The questions would be sent to judges for their ranking. Rank correlation would measure the correlation between the ranks of the two judges in question. Since rank correlation coefficient is due to Spearman it is also known as Spearman's rank correlation coefficient. If there is a tie, the average of the ranks could be awarded to each of the tied observations by first giving ranks for all the tied observations as if they are not tied. Let X_i, Y_i be the ranks for the i-th object (individual), then 'r_s' is given by the formula.

$$r_s = 1 - \frac{6\Sigma(X_i - Y_i)^2}{n(n^2 - 1)} \quad \text{for i} = 1, 2..., n.$$

12.4.1   Test of Significance of r_s (n >10)

Null Hypothesis: $\rho_s = 0$ where 'ρ_s' is the rank correlation coefficient in the population.

$$t = \frac{r_s}{\sqrt{1 - r_s^2}} \sqrt{n - 2}$$

Conclusion: If t (calculated) > t (tabulated) with (n–2) d.f. at chosen level of significance, the null hypothesis is rejected. That is, there is significant correlation in the population. Otherwise, the null hypothesis is accepted.

EXAMPLE: In a certain crop production competition, two judges gave ranks for 10 entries as follows:

X	3	4	6	5	8	2	9	1	10	7
Y	2	7	3	4	9	1	10	5	8	6

Compute the correlation coefficient between the ranks of the two judges.

TABLE 12.3

S. No.	X_i	Y_i	$(X_i - Y_i)$	$(X_i - Y_i)^2$
1	3	2	1	1
2	4	7	–3	9
3	6	3	3	9
4	5	4	1	1
5	8	9	–1	1
6	2	1	1	1
7	9	10	–1	1
8	1	5	–4	16
9	10	8	2	4
10	7	6	1	1
				44

$$r_s = 1 - \frac{6\Sigma(X_i - Y_i)^2}{n(n^2 - 1)} = 1 - \frac{6 \times 44}{10 \times 99} = 0.73$$

12.5 COEFFICIENT OF CONTINGENCY

It is defined as the measure of the degree of association between the two attributes in a contingency table and is given by

$$C = \sqrt{\frac{\chi^2}{\chi^2 + N}}$$

where χ^2 is the calculated value of χ^2 in a contingency table for testing the independence of two attributes and N is the total frequency. The larger the value of C, the greater is the degree of association.

12.6 CORRELATION OF ATTRIBUTES

The correlation coefficient between two attributes in a r × r contingency table is

$$r = \sqrt{\frac{\chi^2}{N(r-1)}}$$

where χ^2 is the calculated value of χ^2 in r × r contingency table and N is the total frequency.

12.7 REGRESSION

The word 'regression' came into existence due to Galton while he was studying the heights of grand-fathers, fathers and sons. He observed that on the average the sons heights were in close agreement with grand-fathers rather than their fathers' heights. This type of backward relation between sons heights to grand-fathers heights, he termed it as "Regression". Son's height is a dependent variable (effect) and grand-father's height

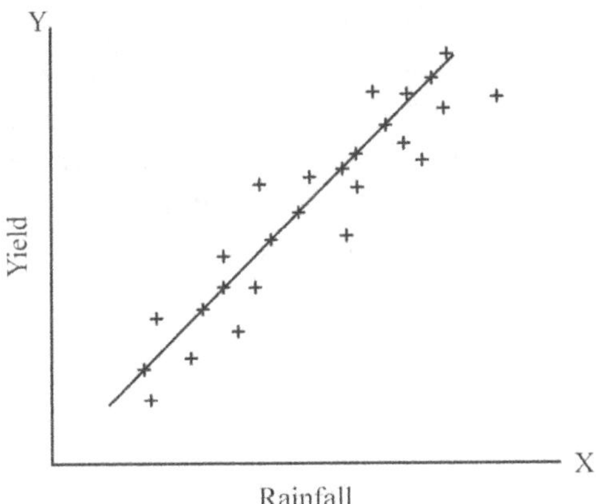

Fig. 12.3 Linear regression.

is an independent variable (cause). If the relation between sons heights and grand-father's heights shows a straight line on a scatter diagram then it can be said that there exists linear regression between the two variables. The straight line showing the relation is called 'regression line'. Consider two variates yield and rainfall. Where rainfall is cause and yield is effect. The scatter diagrams showing the paired observations of yield and rainfall are given in Fig. 12.3 and 12.4.

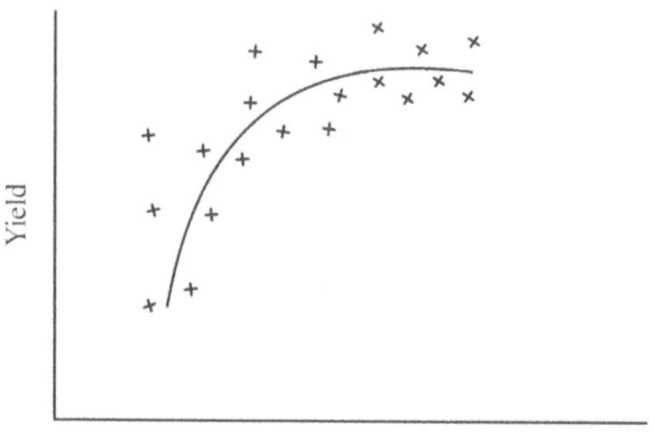

Fig. 12.4 Curve linear regression.

Fig.12.3 indicates linear regression and Fig. 12.4 shows curvilinear regression since if the rainfall goes on increasing yield also goes on increasing to a certain point and then declines due to excess rainfall. The equation of straight line is termed as 'regression equation' and is given by

$$Y = a + bX \qquad .....(12.3)$$

Where Y = dependent variable, X = independent variable; a = intercept and b = the regression coefficient (or slope) of the line a and b are also called as constants. The regression equation of Y on X is shown in Fig. 12.5.

$$b = \tan \theta = \frac{(Y_2 - Y_1)}{(X_2 - X_1)} \qquad .....(12.4)$$

The constants 'a and b' can be estimated with the help of 'least squares method' which states that the constants (a and b) should be so chosen that the sum of the squares of deviations of the observed values from the values obtained by suggested relationship will be minimum.

$$b = \frac{\Sigma X_i Y_i - \dfrac{(\Sigma X_i)(\Sigma Y_i)}{n}}{\Sigma X_i^2 - \dfrac{(\Sigma X_i)^2}{n}} = \frac{\text{Covariance (XY)}}{\text{Var (X)}} \qquad .....(12.5)$$

$$a = \overline{Y} - b\overline{X}$$

where b is called the estimate of regression coefficient of Y on X and it measures the change in Y for a unit change in X. Y and X are the means of Y and X respectively.

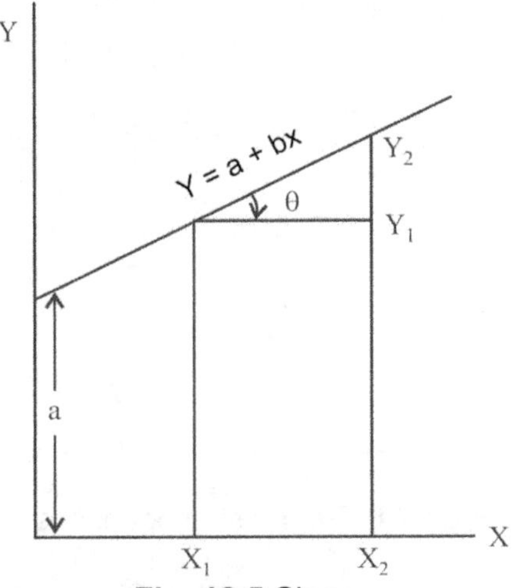

Fig. 12.5 Slope.

The regression equation of Y on X can also be written as

$$Y = \overline{Y} + b(X - \overline{X}) \qquad \qquad .....(12.6)$$

Similarly, if X is a dependent variate and Y is an independent variate, the regression equation of X on Y is given as

$$X = a' + b'Y \qquad \qquad .....(12.7)$$

where

$$b' = \frac{\Sigma X_i Y_i - \dfrac{(\Sigma X_i)(\Sigma Y_i)}{n}}{\Sigma Y_i^2 - \dfrac{(\Sigma Y_i)^2}{n}}$$

$$= \frac{\mathrm{Co\,var\,ieance\,}(XY)}{\mathrm{Var\,}(Y)} \qquad \qquad .....(12.8)$$

$$a' = \overline{X} - b'\overline{Y}$$

b' is known as the estimate of regression coefficient of X on Y and 'a' is an intercept. The regression of X on Y can also be written as

$$X = \overline{X} + b'(Y - \overline{Y}) \qquad \qquad .....(12.9)$$

The point of intersection of two regression lines is $(\overline{X}, \overline{Y})$ and is shown in Fig. 12.6.

The geometric mean of two regression coefficients is the simple.

Correlation coefficient. If r = –1, the two regression lines coincide on the negative side. If r = 1, the two regression lines coincide on the positive side. These are shown in the figures 12.7 (a), (b), and (c) respectively.

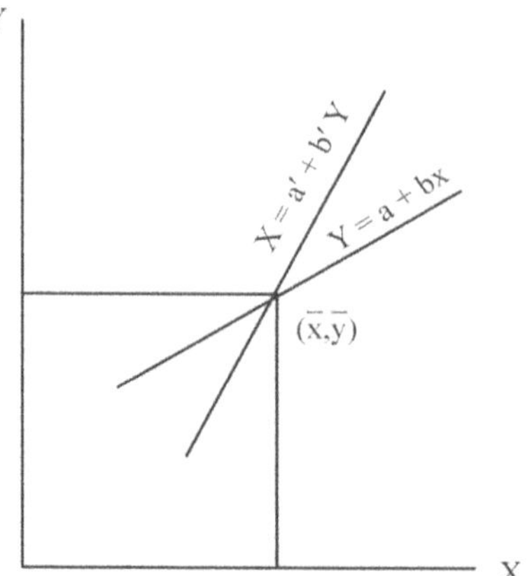

Fig.12.6 Regression equations.

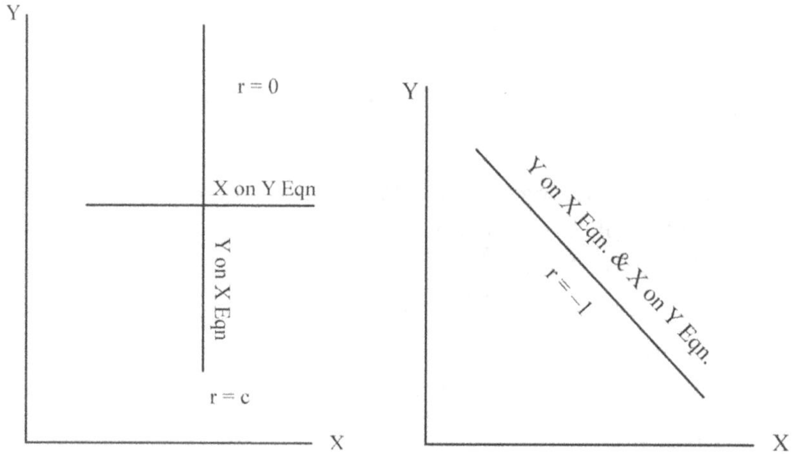

Fig. 12.7(a) Regression equations.   **Fig. 12.7(b)** Regression equations.

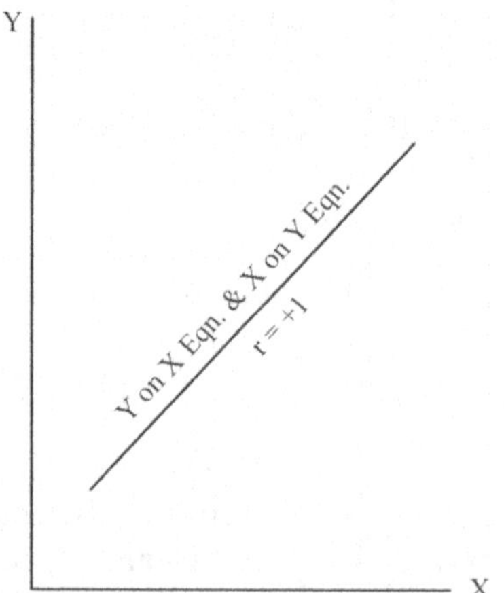

Fig. 12.7(c) Regression equations.

12.7.1   Test of Significance of Regression Coefficient

Null Hypothesis: $H_o: \beta = 0$, $H_1: \beta \neq 0$

where 'β' is called the regression coefficient of Y on X in the population.

$$t = \frac{|b - \beta|}{\text{SE of } b} = \frac{|b - 0|}{\sqrt{\dfrac{\Sigma(Y_i - \overline{Y})^2 - b^2\Sigma(X_i - \overline{X})^2}{(n-2)\,\Sigma(X_i - \overline{X})^2}}}$$

where n is the number of observations in the sample.

Conclusion: If t (calculated) > t (tabulated) with (n–2) d.f. at chosen level of significance, the null hypothesis is rejected. Otherwise, the null hypothesis is accepted.

12.7.2   Null Hypothesis

H_0: $\beta' = 0$, H_1: $\beta' \neq 0$ where 'β'' is the regression coefficient of X on Y in the population

$$t = \frac{|b' - \beta|}{\text{S.E. of } b'} = \frac{|b' - 0|}{\sqrt{\dfrac{\Sigma(X_i - \overline{X})^2 - b'^2\Sigma(Y_i - \overline{Y})^2}{(n-2)\,\Sigma(Y_i - \overline{Y})^2}}}$$

Conclusion: If t (calculated) > t (tabulated) with (n–2) d.f at chosen level of significance, the null hypothesis is rejected. Otherwise, the null hypothesis is accepted.

LINEAR REGRESSION WITH ONE INDEPENDENT VARIABLE
SPSS DATA ANALYSIS PROCEDURE

X	Y
(Indept)	(dept.)
16	9
21	13
26	16
29	20
33	22

Enter data under X (independent variable, ht.cm) and   Y(dependent variable, wt.kg) column wise. CLICK 'ANALYZE' from the TOOL BAR. Click OPTION ' REGRESSION' AND CLICK OPTION ' LINEAR'. We get DISPLAY TABLE. Take   X variable for independent variable and Y variable for dependent variable to RHS as per indicator. Then CLICK

'O.K'. option at the bottom of the TABLE. We get the analysis and results TABLE. In that we can find the values of R^2 and R (correlation coefficient) and next TABLE contains 'a' (constant) and 'b' (regression coefficient) values and test of significance of regression coefficient with F value and probability value indicating the level of significance.

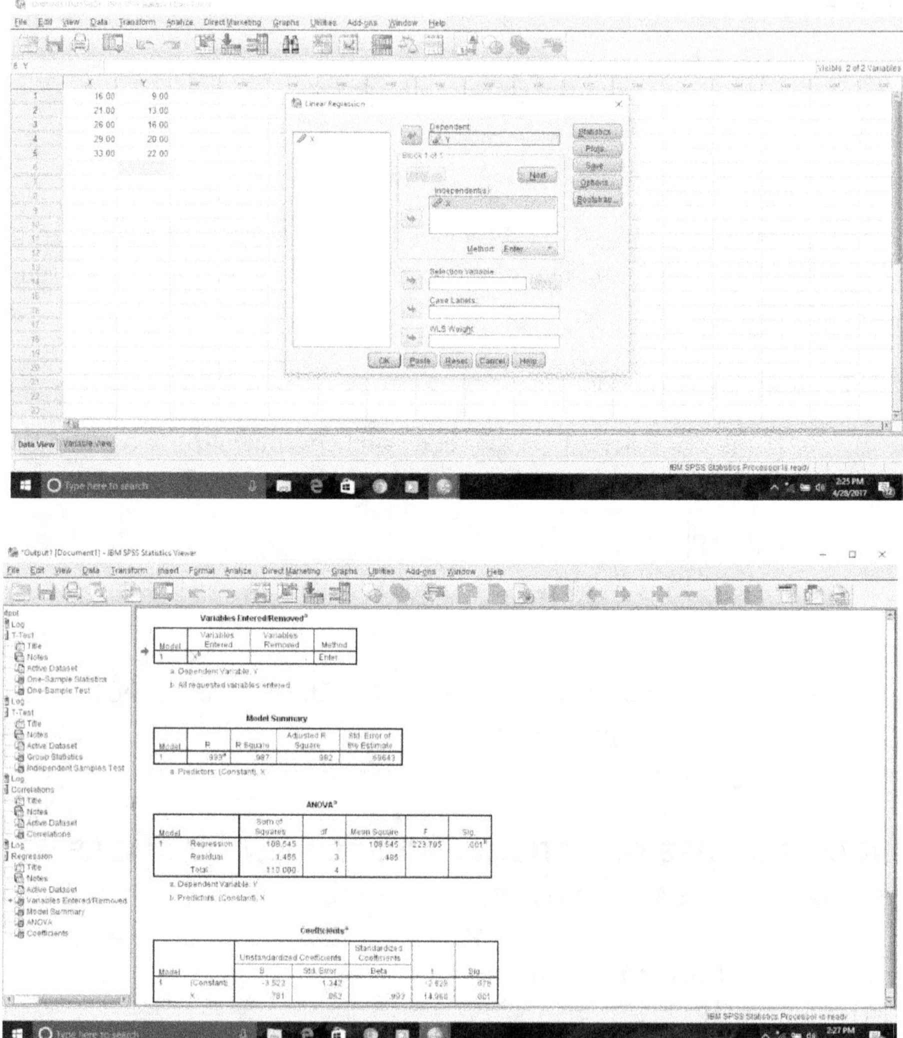

12.8   REGRESSION VS. CORRELATION

Correlation is a two-way relation whereas regression is a one-way relation. Correlation coefficient is independent of units whereas regression coefficient is in the units of dependent variate. Correlation coefficient always lies between -1 and $+1$ where as regression coefficient can lie between $-\infty$ and

$+\infty$. In the case of correlation one need not know, which is cause and which is effect whereas in regression, cause and effect are to be identified. For example, in the case of regression all the growth characters of plants can be taken as independent variates and yield as dependent variate. In the case of regression, one can predict the value of dependent variate for a given value of independent variate but that is not so in correlation. r^2 is known as coefficient of determination, $(1-r^2)$ as coefficient of non determination and $\sqrt{1-r^2}$ as coefficient of alienation. If the correlation coefficient is negative, both the regression coefficients are negative,

EXAMPLE: The following are the data based on average number of tillers and the corresponding yield for 10 samples each consisting of 5 plants of turmeric crop.

Average No. of tillers per sample	3.5	3.2	3.5	3.8	3.6	3.7	2.8	4.2	4.0	4.5
Yield (kgs)	2.0	1.8	1.9	2.1	2.0	2.3	1.7	2.5	2.6	3.0

Fit the linear regression equation of yield on number of tillers and test its regression coefficient. Also obtain the expected yield of turmeric given the average number of tillers is 6.0.

TABLE 12.4

S.NO.	Average No. of tillers(X)	Yield	d_{1i}	d_{2i}	$d_{1i}\,d_{2i}$	d^2_{1i}	d^2_{2i}	Expected yield (Y)
(1)	(2)	(3)	(4)	(5)	(6)	(7)	(8)	(9)
	(X)	(Y)						$(\hat{Y})$
1	3.5	2.0	0	−0.3	0	0	0.09	2.05
2	3.2	1.8	−0.3	−0.5	0.15	0.09	0.25	1.82
3	3.5	1.9	0	−0.4	0	0	0.16	2.05
4	3.8	2.1	0.3	−0.2	−0.06	0.09	0.04	2.28
5	3.6	2.0	0.1	−0.3	−0.03	0.01	0.09	2.13
6	3.7	2.3	0.2	0	0	0.04	0	2.21
7	2.8	1.7	−0.7	−0.6	0.42	0.49	0.36	1.50
8	4.2	2.5	0.7	0.2	0.14	0.49	0.04	2.60
9	4.0	2.6	0.5	0.3	0.15	0.25	0.09	2.44
10	4.5	3.0	1.0	0.7	0.70	1.00	0.49	2.83
			1.8	−1.1	1.47	2.46	1.61	

where $d_{1i} = (X_i - A)$ and $A = 3.5$, $d_{2i} = (Y_i - B)$ and $B = 2.3$

$$Y = a + bX$$

$$b = \frac{\Sigma d_{1i}d_{2i} - \dfrac{(\Sigma d_{1i})(\Sigma d_{2i})}{n}}{\Sigma d_{1i}^2 - \dfrac{(\Sigma d_{1i})^2}{n}}$$

$$b = \frac{1.47 - \dfrac{(1.8)\,(-1.1)}{10}}{2.46 - \dfrac{(1.8)^2}{10}} = 0.78$$

$$\overline{Y} = A + \frac{\Sigma d_{2i}}{n} = 2.3 + \frac{(-1.1)}{10} = 2.19$$

$$\overline{X} = A + \frac{\Sigma d_{1i}}{n} = 3.5 + 1.8/10 = 3.68$$

$$a = \overline{Y} - b\overline{X} = 2.19 - 0.78\,(3.68) = -0.68$$

The fitted regression equation is given by

$$Y = -0.68 + 0.78\,X$$

The expected yields are obtained by substituting the values of X given in column (2) and presented in column (9) of Table 12.4.

Test of Significance of Regression Coefficient

Null Hypothesis: H_0: $\beta = 0$, H_1 .$\beta \neq 0$

$$t = \frac{|0.78 - 0|}{\sqrt{\dfrac{\left[1.61 - \dfrac{(-1.1)^2}{10}\right] - (0.78)^2 \times 2.136}{(10-2)2.136}}} = 7.43$$

CONCLUSION: Here t (calculated), 7.43 > t (tabulated), 2.306 at 5 per cent level of significance, the null hypothesis is rejected. The regression coefficient of Y on X is significant.

EXERCISES

1. The following are the data on supply and price of rice in a certain market from 1960-69.

Year	1960	1961	1962	1963	1964	1965	1966	1967	1968	1969
Supply (Qtl)	210	214	217	215	230	239	205	198	225	232
Price (Rs./Qtl)	165	162	160	170	156	148	175	179	170	162

Fit the regression equation of price on supply of rice and also estimate the price when the supply is 240 quintals in the market. Compute also the correlation coefficient between supply and price of rice and test its significance.

2. The following results were recorded for the two variables X and Y
b = 0.4, variance (Y) = 25, variance (X) = 100

$$\overline{X} = 8.0, \ \overline{Y} = 12.5$$

Find the expected value of X when Y = 14 from the regression equation of X on Y where b is the regression coefficient of Y on X.

3. Plot the following data in a semi-logarithmic paper and fit a straight line by eye estimation. Hence find the relation between the number of bacteria and time.

Number of bacteria after time 't' hours they were first observed:

Time	0	1	2	3	4	5	6
Number	125	209	340	651	924	1525	1512

(B.A./B.Sc. Madras.1955)

4. Two judges in a beauty contest rank the competitors in the following order:

6	4	3	1	2	7	9	8	10	5
4	1	6	7	5	8	10	9	3	2

Do the judges appear to agree in their standard?

(B.A. Punjab, 1952)

5. The following data relate to the weight in pounds (X) and the height in inches (Y) of a sample of 1000 policemen.

$\Sigma X = 1,50,000, \ \Sigma(X)^2 = 22,725,000, \ \Sigma XY = 10,522,500$
$\Sigma Y = 70,000, \ \Sigma Y^2 = 42,36,000$

(i) Compute the two regression coefficients and correlation coefficient.

(ii) Write down the equations to the two lines of regression

(iii) Test whether the above sample could have been drawn from a bivariate population with correlation coefficient 0.52.

(iv) Estimate the height of a policeman with weight of 160 lbs and the weight of a policeman whose height is 6 feet.

6. (a) What is meant by Scatter diagram? Give its use,
Can the sample correlation coefficient be – 0.2 to 20 pairs of values if the true value of the correlation coefficient is 0.15 ?

7. What will be the value of the correlation coefficient if the two regression lines coincide?
The two regression line are given as
(i) 2X – 5Y + 10 = 0, and   (ii)   10X – 6Y – 20 = 0
and variance of Y = 16.

Find the mean and variance of X and also the correlation between X and Y.

8. The following are the 'chest' and 'Arm' circumference of 8 tribal children in the age group between 1-3 years.

Chest circumference (cm)	40	32	36	34	28	39	26	27
Arm circumference (cm)	10	6	7	8	5	9	7	4

Compute the correlation coefficient and also fit the regression equation of chest circumference on arm circumference and estimate the chest circumference given the arm circumference as 12 cms,

9. The following are the results on growth (gain in weight) of 10 pre-school going children at different levels of protein content in diet.

Protein content (per cent)	5	7	10	12	13	14	11	8	16	20
Gain in weight (10gm)	3	4	8	10	12	11	9	6	13	15

Compute the correlation coefficient and also test its significance. Fit also the regression equation of gain in weight on protein content in diet and test the significance of regression coefficient at 5 per cent level of significance.

10. Test the homogeneity of correlation coefficients obtained between iron content in the diet and the gain in weight 0.87, 0.76, 0.92, 0.64, 0.59, 0.85, 0.94, 0.67, 0.78 and 0.81 based on sample sizes 9, 12, 11, 13, 10, 16, 14, 10, 17, and 21 respectively.

11. Fit the regression equation of 'yield' on number of tiller's given the results obtained from 10 samples,

Yield (10 gm)	8	9	12	14	11	9	20	22	24	25
No. of tillers	3	3	5	7	6	5	8	7	10	11

Also estimate the yield when the number of tillers is 9.

12. The following table gives the age (X) and systolic blood pressure (Y) of 10 women.

Age	(X)	69	56	63	55	49	70	42	58	64	67
B.P	(Y)	154	147	149	150	142	156	135	151	150	153

Fit the linear regression equation of B.P on age and estimate B.P of a woman whose age is 60 years.

13. The following are the percentage of survival of Coconut seedlings and depth of plantation (cm) in 10 gardens.

	1	2	3	4	5	6	7	8	9	10
Depth (cm)	35	52	60	68	75	80	92	110	120	100
Survival (%)	40	48	50	52	60	64	75	78	80	85

Fit the regression equation of survival percentage of seedlings on depth of planting and estimate the survival percentage when the depth of plantation is 90 cm.

14. The following is the data on 'Glycimic' index and 'Fibre' content in different samples of Jowar millet.

Glycimic Index	6	5	12	14	6	8
Fibre content	10	18	9	7	13	16

Fit the regression equation of 'Glycimic Index', on Fibre content' in Jowar millet and test the regression coefficient at 5% level.

15. The following are the data on 'Total Protein' and "available protein" in 6 food products made from 'Ragi' Millet.

	1	2	3	4	5	6
Total Protein	16	22	14	9	13	11
Available Protein	8	13	9	5	6	7

Compute the 'Correlation Coefficient' between 'Total' and available protein, and test its significance at 1% level.

CHAPTER 13

MULTIPLE REGRESSION AND CORRELATION

13.1 INTRODUCTION

Multiple linear regression is an extension of simple linear regression by considering more than one independent variables. For example, in simple linear regression the relationship between yield on rainfall could be considered, whereas in multiple linear regression the relationship between yield on rainfall, fertilizer dose, height of the plant, number of gains per ear, ear, length, etc., for a paddy crop could be considered. Though with the help of simple linear regression the effect of these different characters on yield can be studied separately but with the help of multiple linear regression the influence of all the independent variables together as well as separately on dependent variable can be examined.

13.2 MULTIPLE LINEAR REGRESSION BASED ON TWO INDEPENDENT VARIABLES

Let $Y = a + b_1X_1 + b_2X_2$ be the multiple linear regression equation based on two independent variables X_1 and X_2 and dependent variable Y, b_1 and b_2 are the estimates of partial regression coefficients of Y on X_1 keeping X_2 constant and Y on X_2 keeping X_1 constant respectively, 'a' is the constant and is the distance between the base and the three dimensional plane of three variables Y, X_1 and X_2. In calculating simple regression coefficient of yield on level of irrigation, the effect of other characters like fertilizer dose would not be taken into account though they influence the yield indirectly. In fact these characters might influence the yield through irrigation level. In other words there might be interaction among these characters as well as with irrigation. The change in yield for a unit change in level of irrigation by removing the influence of other characters on yield is known as partial regression coefficient of yield on level of irrigation. In multiple linear regression equation $Y = a + b_1 X_1 + b_2 X_2$, b_1 and b_2 can also be denoted as $b_{Y1.2}$ and $b_{Y2.1}$ respectively since $b_{y1.2}$ refers to the partial regression coefficient of Y on X_1 keeping X_2 constant and $b_{Y2.1}$ refers to the partial regression coefficient of Y on X_2 keeping X_1 constant.

The least squares technique would be used to obtain the values of a, b_1 and b_2 which are the estimates of α, β_1 and β_2 respectively in the multiple linear regression equation $Y_J = \alpha + \beta_1 X_{1j} + \beta_2 X_{2j} + e_j$. The least squares technique states that the values α, β_1 and β_2 are estimated such that the sum of the   squares of the deviations of observed and expected values of

Y i.e., $\Sigma.e_J^2 = \Sigma\left(Y_J - \hat{Y}_j\right)^2$ should be minimum. For estimating the values

of α, β_1 and β_2 with least squares technique the   following normal equations would be obtained by partially differentiating the Sum, $\Sigma(Y - a - b_1 X_1 - b_2 X_2)^2$ with respect to a, b_1 and b_2 and equating to zero. The normal equations are

$$\Sigma Y = n\,a + b_1\Sigma X_1 + b_2\Sigma X_2$$
$$\Sigma X_1 Y = a\Sigma X_1 + b_1\Sigma X_1^2 + b_2\Sigma X_1 X_2 \qquad\qquad .....(13.1)$$
$$\Sigma X_2 Y = a\Sigma X_2 + b_1\Sigma X_2 X_1 + b_2\Sigma X_2^2$$

The   first   normal   equation   in   (13.1)   can   be   written   as $\overline{Y} = a + b_1\overline{X_1} + b_2\overline{X_2}$ and hence 'a' can be obtained from this equation as $a = \overline{Y} - b_1\overline{X_1} - b_2\overline{X_2}$. Instead of solving three normal equations in (13.1) for obtaining the values a, b_1 and b_2 they can be reduced to two by substituting the value of 'a' in the original multiple linear regression equation. Therefore, we have

$$Y = \overline{Y} + b_1(X_1 - \overline{X_1}) + b_2(X - \overline{X_2}) \qquad\qquad .....(13.2)$$

Let $y = (Y - \overline{Y})$, $x_1 = (X - \overline{X_1})$ and $x_2 = (X_2 - \overline{X_2})$; eqn.   .....(13.3)

reduces to $y = b_1 x_1 + b_2 x_2$ .....(13.4)

The normal equations for estimating the values of b_1 and b_2 are

$$\Sigma x_1 y = b_1\Sigma x_1^2 + b_2\Sigma x_1 x_2$$
$$\qquad\qquad .....(13.5)$$
$$\Sigma x_2 y = b_1\Sigma x_1 x_2 + b_2\Sigma x_2^2$$

The equations (13.4) can be written in matrix notation as

$$\underset{2\times 2}{\begin{bmatrix} \Sigma x_1^2 & \Sigma x_1 x_2 \\ \Sigma x_1 x_2 & \Sigma x_2^2 \end{bmatrix}} \underset{2\times 1}{\begin{bmatrix} b_1 \\ b_2 \end{bmatrix}} = \underset{2\times 1}{\begin{bmatrix} \Sigma x_1 y \\ \Sigma x_2 y \end{bmatrix}} \qquad .....(13.6)$$

where L.H.S. arrangement in eqn. (13.6) is called 2×2 matrix   and other arrangement in one column of two elements is called column vector and the R.H.S. arrangement is also called column vector. Now, we have

$$\begin{bmatrix} b_1 \\ b_2 \end{bmatrix} = \begin{bmatrix} \Sigma x_1^2 & \Sigma x_1 x_2 \\ \Sigma x_1 x_2 & \Sigma x_2^2 \end{bmatrix}^{-1} \begin{bmatrix} \Sigma x_1 y \\ \Sigma x_2 y \end{bmatrix} \qquad \dots(13.7)$$

The R.H.S. 2×2 matrix in eqn. (13.6) is called inverse matrix and is denoted by

$$\begin{bmatrix} \Sigma x_1^2 & \Sigma x_1 x_2 \\ \Sigma x_1 x_2 & \Sigma x_2^2 \end{bmatrix}^{-1} = \begin{bmatrix} C_{11} & C_{12} \\ C_{21} & C_{22} \end{bmatrix} \qquad \dots(13.8)$$

where C_{11}, C_{12}, C_{21} and C_{22} are called the elements of C–matrix and are also known as Gauss multipliers. Substituting the C–matrix in (13.7), we have

$$\begin{bmatrix} b_1 \\ b_2 \end{bmatrix} = \begin{bmatrix} C_{11} & C_{12} \\ C_{21} & C_{22} \end{bmatrix} \begin{bmatrix} \Sigma x_1 y \\ \Sigma x_2 y \end{bmatrix} \qquad \dots(13.9)$$

Since there are only two independent variables in the multiple linear regression equation the following procedure is adopted.

$$C_{11}\Sigma x_1^2 + C_{12} \Sigma x_1 x_2 = 1$$
$$C_{11}\Sigma x_1 x_2 + C_{22}\Sigma x_2^2 = 0 \qquad \dots(13.10)$$

Solving the equations in (13.10), C_{11} and C_{12} can be obtained. Similarly

$$C_{21}\Sigma x_1^2 + C_{22} \Sigma x_1 x_2 = 0$$
$$C_{21}\Sigma x_1 x_2 + C_{22}\Sigma x_2^2 = 1 \qquad \dots(13.11)$$

Solving two equations in (13.11), C_{21} and C_{22} can be obtained. In most of the situations $C_{12} = C_{21}$ because of symmetric variance matrix.

$$b_1 = C_{11}\Sigma x_1 y + C_{12} \Sigma x_2 y$$
$$b_2 = C_{21}\Sigma x_1 y + C_{22}\Sigma x_2 y \qquad \dots(13.12)$$

Also, we have $a = \overline{Y} - b_1 \overline{X}_1 - b_2 \overline{X}_2$. Therefore, the fitted multiple linear regression equation is obtained by substituting the values of a, b_1 and b_2 in $Y = a + b_1 X_1 + b_2 X_2$. The expected average values of Y could be computed by substituting the different pairs of values of X_1 and X_2 from this fitted equation.

13.2.1   Test of Significance of Overall Regression

Let β_1 and β_2 be the population partial regression coefficients of Y on X_1 and Y on X_2 respectively we have

Regression S.S. (R.S.S) $= b_1\Sigma x_1 y + b_2\Sigma x_2 y.$

Total S.S. (T.S.S.) $= \Sigma y^2$

Error S.S. (E.S.S.) = (Total S.S.–Regression S.S.)

Null Hypothesis: H_o: $\beta_1 = \beta_2 = 0$, H_1: $\beta_1 \neq \beta_2 \neq 0$

TABLE 13.1 ANOVA TABLE

Source	d.f	S.S	M.S.	F_{cal}
Regression	2	R.S.S	R.M.S.	$\dfrac{\text{R.M.S}}{\text{E.M.S}}$
Error	$\{(n{-}1){-}2\}$	E.S.S.	E.M.S.	
Total	$n{-}1$	T.S.S.		

where R.M.S. $= \dfrac{\text{R.S.S.}}{2}$ and E.M.S. $= \dfrac{\text{E.S.S.}}{(n-3)}$

CONCLUSION: If F (calculated) > F (tabulated) with 2, (n–3) d.f. at chosen level of significance, the null hypothesis is rejected. The two partial regression coefficients in the population are not equal and not equal to zero. In other words there is a significant overall regression of dependent variable on the two independent variables. Otherwise, the null hypothesis is accepted. In case F (calculated) is found significant the individual partial regression coefficients in the population are tested against zero.

13.2.2 Test of Significance of Partial Regression Coefficients

Case (i) *Null hypothesis:* H_o: $\beta_1 = 0$, H_1: $\beta_1 \neq 0$

$$t = \frac{|b_1|}{\sqrt{C_{11}(\text{E.M.S.})}}$$

CONCLUSION: If t (calculated) > t (tabulated) with error d.f. i.e., (n − 3) at chosen level of significance the null hypothesis is rejected. That is the partial regression coefficient in the population is different from zero. In other words there is significant influence of first independent variable on the dependent variable. Otherwise, the null hypothesis is accepted.

Case *(ii) Null hypothesis:* H_0: $\beta_2 = 0$, H_1: $\beta_2 \neq 0$

$$t = \frac{|b_2|}{\sqrt{C_{22}(\text{E.M.S.})}}$$

CONCLUSION: Same as in case (i)

It may be noted that (E.M.S.) is an unbiased estimate of variance of the dependent variable.

13.2.3 Multiple Correlation Coefficient

From Table 13.1, the coefficient of multiple determination (R^2) is obtained as $R^2 = \text{R.S.S./T.S.S.}$

The coefficient of multiple determination measures the extent of variation in dependent variable, which can be explained by the independent variables together.

$R=\sqrt{R^2}$ = Coefficent of multiple correlation where 'R' measures the goodness of fit of the regression plane when X's are fixed variables and Y is a stochastic variable and also when X's and Y are all stochastic variables. But when Y and X's are all fixed variables it measures the correlation between observed and expected values of Y in the multiple linear regression equation. R can also be written as $R_{Y.12}$ and can be considered as the correlation coefficient between Y and the group of variables (X_1, X_2) and can be expressed by writing as the correlation coefficient between Y and X_1, X_2. If Y is only stochastic variable R can be written as only $R_{Y.12}$ but If Y, x_1 and x_2 are all stochastic variables R can take as $R_{Y.12}$ $R_{1.Y2}$ and $R_{2.Y1}$. Further $0 \leq R \leq 1$.

For comparison among partial regression coefficients, the standard error of (b_1-b_2) is given as $\sqrt{(C_{11}+C_{22}-2C_{12})(E.M.S.)}$ and $t = |b_1 - b_2| / \sqrt{(C_{11}+C_{22}-2C_{12})(E.M.S.)}$. t (calculated) is compared with t (tabulated) with (n–3) d.f. The teast of the null hypothesis that the multiple correlation in the population is zero is identical to the F _ tests of the null hypothesis that

$$\beta_1 = \beta_2 = 0. \quad \text{i.e.} \qquad F = \frac{R^2(n-3)}{(1-R^2)2}$$

13.3 PARTIAL CORRELATION

Partial correlation between two variables when three variables are under question, is the correlation between two variables keeping third variable constant. Let $\rho_{12.3}$ be the population partial correlation coefficient between variables 1 and 2 keeping 3rd variable constant and $r_{12'.3}$ be the corresponding estimate based on the sample. Let $d_{1.3} = X_1 - b_{13}X_3$ and $d_{2.3} = X_2 - b_{23}X_3$ be the deviations of variables 1 and 2 from their regression on variable 3 respectively. The simple correlation coefficient between $d_{1.3}$ and $d_{2.3}$ is known as partial correlation coefficient $(r_{12.3})$, where

$$r_{12.3} = \frac{\Sigma d_{1.3}d_{2.3}}{\sqrt{\Sigma d_{1.3}^2 \ \Sigma d_{2.3}^2}} = \frac{r_{12} - r_{13}}{\sqrt{(1-r_{13}^2)(1-r_{23}^2)}} r_{23} \qquad \qquad .....(13.13)$$

where r_{12}, r_{13} and r_{23} in eqn. (13.12) are the simple correlation coefficients between variables 1 and 2, 1 and 3 and 2 and 3 respectively. Similarly $r_{13.2}$ and $r_{23.1}$ can be defined, where

$$r_{13\cdot2} = \frac{r_{13} - r_{12}\cdot r_{23}}{\sqrt{\left(1 - r_{12}^2\right)\left(1 - r_{23}^2\right)}} \quad\quad\quad .....(13.14)$$

$$r_{23\cdot1} = \frac{r_{23} - r_{12}\cdot r_{13}}{\sqrt{\left(1 - r_{12}^2\right)\left(1 - r_{13}^2\right)}} \quad\quad\quad .....(13.15)$$

13.3.1   Test of Significance of Partial Correlation Coefficient

Null hypothesis: H_o: $p_{12\cdot3} = 0$, H_1:$\rho_{12.3} \neq 0$

$$t = \frac{r_{12\cdot3}}{\sqrt{1 - r_{12\cdot3}^2}}\sqrt{n - 2 - 1}$$

CONCLUSION: If t (calculated) > t (tabulated) with (n–3) d.f. at chosen level of significance the null hypothesis is rejected. Otherwise the null hypothesis is accepted.

13.3.2 If four variables are considered, the partial correlation coefficient between two variables keeping other two variables is given by the formula.

$$r_{12\cdot34} = \frac{r_{12\cdot4} - r_{13\cdot4}r_{23\cdot4}}{\sqrt{\left(1 - r_{13\cdot4}^2\right)\left(1 - r_{23\cdot4}^2\right)}} \quad\quad\quad .....(13.16)$$

It can also be computed as

$$r_{12\cdot34} = \frac{r_{12\cdot3} - r_{14\cdot3}r_{24\cdot3}}{\sqrt{\left(1 - r_{14\cdot3}^2\right)\left(1 - r_{24\cdot3}^2\right)}} \quad\quad\quad .....(13.17)$$

Similarly other partial correlation coefficients $r_{13\cdot24}$, $r_{14\cdot23}$, $r_{23.14}$, $r_{24,.13}$, $r_{24,12}$ could be defined.

EXAMPLE: Compute the partial correlation coefficient between height of plant (X_1) and ear length (X_2) keeping number of panickels (X_3) constant for paddy crop and test its significance given the following.

Given $r_{12} = 0.92$, $r_{13} = 0.82$, $r_{23} = 0.93$

$$r_{12\cdot3} = \frac{r_{12} - r_{13}\,r_{23}}{\sqrt{\left(1 - r_{13}^3\right)\left(1 - r_{23}^2\right)}} = \frac{0.92 - (0.82)(0.93)}{\sqrt{\{1 - (0.82)^2\}\{1 - (0.93)^2\}}}$$

$$= 0.75$$

Test of Significance of $r_{12\cdot3}$

Null Hypothesis: H_o: $\rho_{12\cdot3} = 0$, H_1: $\rho_{12.3} \neq 0$

$$t = \frac{r_{12.3}}{\sqrt{1-r_{12.3}^2}}\sqrt{n-2-1} = \frac{0.75}{\sqrt{1-(0.75)^2}}\sqrt{10-2-1} = 3.00$$

CONCLUSION: Here t (calculated) value i.e. $3.00 > t$ (tabulated) 2.365 with 7 d.f. at 5 per cent level of significance the null hypothesis is rejected. Hence there is significant partial correlation between plant height and ear length removing the influence of number of panickles.

13.4   MULTIPLE LINEAR REGRESSION WITH MORE THAN TWO INDEPENDENT VARIABLES

If more than two independent variables are included in the multiple linear regression the equation becomes $Y = a + b_1X_1 + b_2X_2 + b_3X_3 + \ldots + b_kx_k$ where $X_1\, X_2, \ldots X_k$ are k independent variables and Y, dependent variable. The normal equations for fitting this equation are

$$\Sigma Y = na + b_1\, \Sigma X_1 + b_2\, \Sigma X_2 + \ldots + b_k\, \Sigma X_k$$

$$\Sigma X_1 Y = a\, \Sigma X_1 + b_1\, \Sigma X_1^2 + b_2\, \Sigma X_1 X_2 + \ldots + b_k\, \Sigma X_1 X_k$$

$$\Sigma X_2 Y = a\, \Sigma X_2 + b_1\, \Sigma X_2 X_1 + b_2\, \Sigma X_2^2 + \ldots = b_k\, \Sigma X_2 X_k \qquad \ldots\ldots(13.18)$$

$$\vdots \qquad \vdots \qquad \vdots \qquad \vdots \qquad \vdots \qquad \vdots$$

$$\Sigma X_k Y = a\, \Sigma X_k + b_1\, \Sigma X_k X_1 + b_2\, \Sigma X_k X_2 + \ldots + b_k\, \Sigma X_k^2$$

Using the first normal equation in (13.18) as $\overline{Y} = a + b_1\overline{X_1} + b_2\overline{X_2} + \ldots + b_k\overline{X_k}$ in the multiple linear regression equation for eliminating 'a', we have

$$Y = \overline{Y} + b_1(X_1 - \overline{X_1}) + b_2(X_2 - \overline{X_2}) + \ldots + b_k(X_k - \overline{X_k})$$

Let $(Y - \overline{Y}) = y, (X_1 - \overline{X_1}) = x_1, (X_2 - \overline{X_2}) = x_2 \ldots, (X_k - \overline{X_k}) = x_k$ in the above equation,

We have $y = b_1x_1 + b_2x_2 + \ldots + b_kx_k$ $\qquad\qquad \ldots\ldots(13.19)$

The normal equations (13.17) would become

$$\Sigma x_1 y = b_1\, \Sigma x_1^2 + b_2\, \Sigma x_1 x_2 \ldots + b_k\, \Sigma x_1 x_k$$

$$\Sigma x_2 y = b_1\, \Sigma x_2 x_1 + b_2\, \Sigma x_2^2 + \ldots + b_k\, \Sigma x_2 X_k \qquad\qquad \ldots\ldots(13.20)$$

$$\vdots \qquad \vdots \qquad \vdots \qquad \vdots$$

$$\Sigma x_k y = b_1\, \Sigma x_k x_1 + b_2\, \Sigma x_k x_2 + \ldots + b_k \Sigma x_k^2$$

Putting these equations in matrix notation, we have

$$
\underset{k \times k}{\begin{bmatrix} \Sigma x_1^2 & \Sigma x_1 x_1 & \cdots & \Sigma x_1 x_k \\ \Sigma x_2 x_1 & \Sigma x_2^2 & \cdots & \Sigma x_2 x_k \\ \cdots & & & \vdots \\ \Sigma x_k x_1 & \Sigma x_k x_2 & \vdots & \Sigma x_k^2 \end{bmatrix}} \underset{k \times 1}{\begin{bmatrix} b_1 \\ b_2 \\ \vdots \\ b_k \end{bmatrix}} = \underset{k \times 1}{\begin{bmatrix} \Sigma x_1 y \\ \Sigma x_2 y \\ \vdots \\ \Sigma x_k y \end{bmatrix}} \qquad \ldots\ldots(13.21)
$$

The k x k matrix in eqn. (13.21) is called the variance–covariance matrix. The solution of the $k \times 1$ column vector of regression coefficients is

$$
\begin{bmatrix} b_1 \\ b_2 \\ \vdots \\ b_k \end{bmatrix} = \begin{bmatrix} \Sigma x_1^2 & \Sigma x_1 x_2 & \cdots & \Sigma x_1 x_k \\ \Sigma x_2 x_1 & \Sigma x_2^2 & \cdots & \Sigma x_2 x_k \\ \vdots & \vdots & \vdots & \vdots \\ \Sigma x_k x_1 & \Sigma x_k x_2 & \cdots & \Sigma x_k^2 \end{bmatrix}^{-1} \begin{bmatrix} \Sigma x_1 y \\ \Sigma x_2 y \\ \vdots \\ \Sigma x_k y \end{bmatrix} \qquad \ldots\ldots(13.22)
$$

The inversion of the matrix will be done with the help of Doolittle method given in Chapter 14.

$$
\begin{bmatrix} \Sigma x_1^2 & \Sigma x_1 x_2 & \cdots & \Sigma x_1 x_k \\ \Sigma x_2 x_1 & \Sigma x_2^2 & \cdots & \Sigma x_2 x_k \\ \vdots & \vdots & & \vdots \\ \Sigma x_k x_1 & \Sigma x_k x_2 & \cdots & \Sigma x_k^2 \end{bmatrix}^{-1} = \begin{bmatrix} C_{11} & C_{12} & \cdots & C_{1k} \\ C_{21} & C_{22} & \cdots & C_{2k} \\ \vdots & \vdots & & \vdots \\ C_{k1} & C_{k2} & \cdots & C_{kk} \end{bmatrix} \qquad \ldots\ldots(13.23)
$$

$$
\begin{aligned}
b_1 &= C_{11}\,\Sigma x_1 y + C_{12}\,\Sigma x_2 y + \ldots + C_{1k}\,\Sigma x_k y \\
b_2 &= C_{21}\,\Sigma x_1 y + C_{22}\,\Sigma x_2 y + \ldots + C_{2k}\,\Sigma x_k y \\
\vdots & \qquad\qquad \vdots \qquad\qquad \vdots \qquad\qquad \vdots \\
b_k &= C_{k1}\,\Sigma x_1 y + C_{k2}\,\Sigma x_2 y + \ldots + C_{kk}\,\Sigma x_k y
\end{aligned}
\qquad \ldots\ldots(13.24)
$$

13.4.1   Test of Significance of Overall Regression

Regression S.S $= b_1\,\Sigma x_1 y + b_2\,\Sigma x_2 y + \ldots + b_k\,\Sigma x_K y$

Total S.S $= \Sigma y^2$

Error S.S. $=$ Total S.S. $-$ Regression S.S.

Null hypothesi: H_o: $\beta_1 = \beta_2 \ldots = \beta_K = 0$, H_1:$\beta_1 \neq \beta_2 \neq \ldots \neq \beta_K \neq 0$

TABLE 13.2 ANOVA TABLE

Source	d.f.	S.S	M.S	F_{cal}
Regression	k	R.S.S	R.M.S = R.S.S./k	R.M.S./E.M.S
Error	n −1− k	E.M.S	E.M.S. = E.S.S./ (n−1−k)	
Total	n−1	T.S.S.		

CONCLUSION: If F (calculated) $>$ F (tabulated) with (k, n–1–k) d.f. at chosen level of significance, the null hypothesis is rejected. Otherwise, it is accepted.

13.4.2   Test of Significance of Partial Regression Coefficients

Null hypothesis: $H_o: \beta_i = 0, \quad H_1: \beta_i \neq 0$

Where β_i is the population partial regression coefficient of i-th variable.

$$t = \frac{|b_i|}{\sqrt{C_{1i}(\text{E.M.S.})}}$$

Where bi is the i-th partial regression coefficient in the sample. C_{ii} is the i-th diagonal element in the C-matrix.

CONCLUSION: if t (calculated) $>$ t (tabulated) with (n – I – k) d.f. at chosen level of significance, the null hypothesis is rejected. Otherwise, it is accepted.

EXAMPLE: Fit the multiple regression equation and test its overall regression for the following data of yield (Y) on height of plant (X_1), ear length (X_2) , number of panickles (X_3) and test weight (X_4) obtained from samples of 15 plots in an experimental field of wheat.

S. No.	Yield (kg) (Y)	Height of plant (cms) (X_1)	Ear length (cms) (X_2)	Number of panickles (X_3)	Test weight (gms) (X_4)
1	2.50	100	14	23	50
2	3.15	105	16	29	62
3	2.80	104	13	25	55
4	3.20	107	17	31	64
5	2.95	106	16	28	47
6	3.17	103	18	30	58
7	3.35	107	19	35	67
8	3.42	108	21	28	72
9	3.05	105	16	28	45
10	3.28	102	19	30	60
11	3.40	106	18	32	71
12	3.50	108	22	37	80
13	3.55	107	20	38	75
14	3.60	106	21	40	68
15	3.10	104	19	28	61
16	3.72	109	23	39	69
17	3.25	95	17	36	59
18	3.10	86	16	37	64
19	3.65	89	18	29	63
20	2.95	91	15	31	58

$$\Sigma Y = 64.69, \ \Sigma X_1 = 2048, \ \Sigma X_2 = 358, \ \Sigma X_3 = 644, \ \Sigma X_4 = 1248$$

The variance-covariance matrix is

$$\begin{bmatrix} 866.8 & 151.8 & 84.4 & 286.8 \\ 151.8 & 133.8 & 189.4 & 339.8 \\ 84.4 & 189.4 & 465.2 & 636.4 \\ 286.8 & 339.8 & 636.4 & 1542.8 \end{bmatrix}$$

The inverted or C-matrix is

$$\begin{bmatrix} .001644 & -.003389 & -.001098 & -.000012 \\ -.003389 & .028261 & -.007432 & -.002529 \\ .001098 & -.007432 & .006923 & -.001423 \\ .000012 & -.002529 & -.001423 & .001794 \end{bmatrix}$$

The values of a, b_1, b_2, b_3 -and b_4 and the values of R, R^2 are presented in Table 13.3.

TABLE 13.3 MULTIPLE LINEAR REGRESSION

a	b_1	b_2	b_3	b_4	R	R^2
1.942154	-.007455	.082629**	.006722	.005773	.8907	.7932**
	(.006320)	(.026199)	(.012967)	(.006602)	...	

**Significant at 1 per cent level.

The values in the parentheses are the standard errors of the corresponding partial regression coefficients.

MULTIPLE REGRESSION WITH MORE THAN ONE INDEPENDENT VARIABLE

SPSS DATA ANALYSIS PROCEDURE

X_1	X_2	Y
21	13	9
24	14	11
29	16	15
32	19	18
26	17	14
35	21	19

Enter X_1 ,(independent, Protein content)X_2 (independent, calcium) and Y (dependent variable, gain in wt.kg.) column wise as shown above. CLICK ANALYZE FROM THE TOOL BAR. CLICK OPTION 'REGRESSION' AND THEN OPT FOR 'LINEAR'. Take Y to the dependent variable slot and X_1 and X_2 to the independents slot to the RHS as per indicator. Then 'CLICK' O.K. ' at the bottom of the TABLE. Then we get the DISPLAY

TABLE containing results of the analysis. The table contains 'R^2 and 'R' values and 'a' and b_1 and b_2 values and their test of significance and probability of significance. Further test of overall significance will be tested and 'F' (calculated) value will also be given.

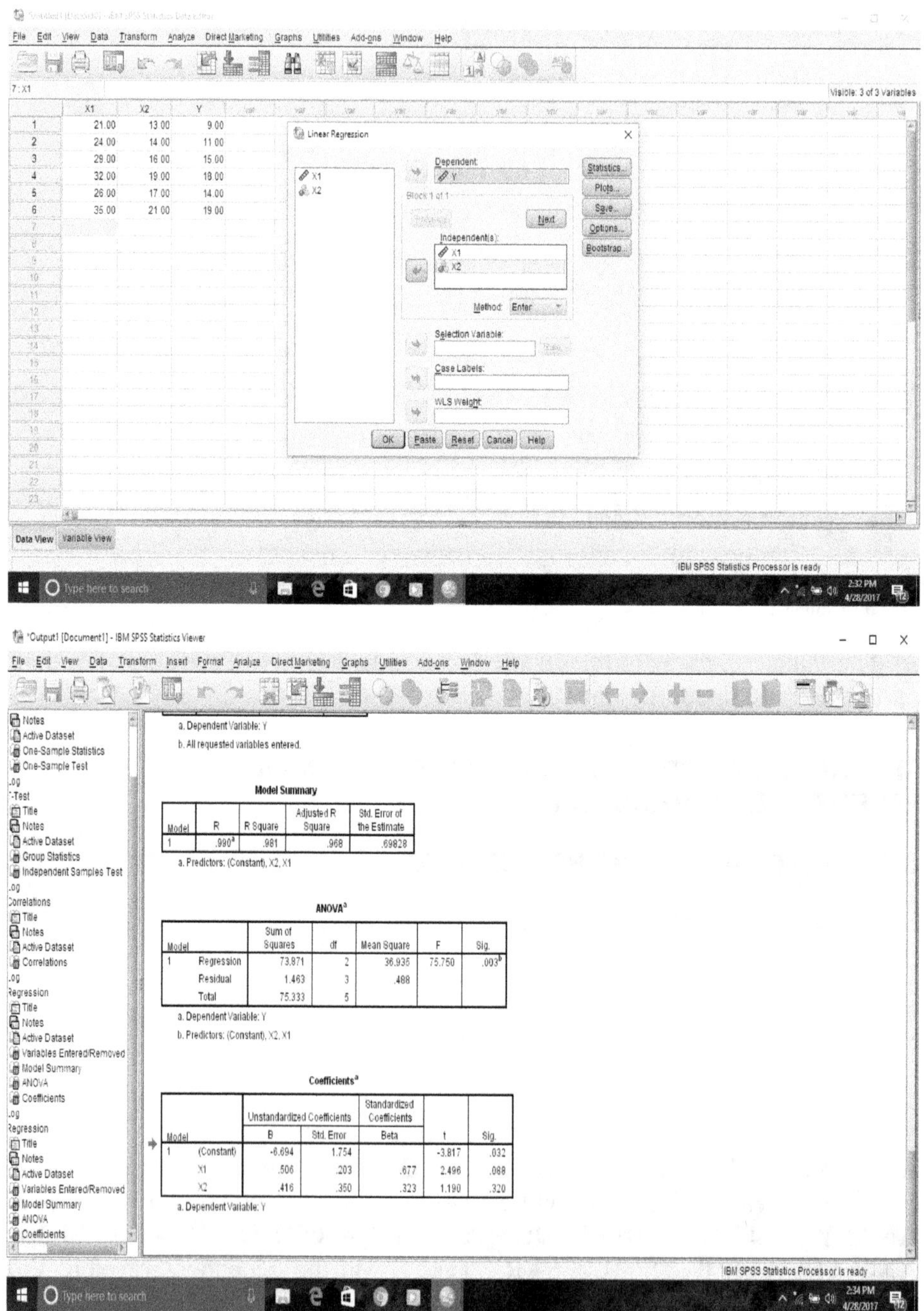

13.4.3

It may be noted that at certain stage in the multiple linear regression a particular independent variable has to be dropped because of its less utility. Therefore, one need not compute afresh the inverse of variance -covariance matrix and analysis of variance table. Let X_i be the variable to be deleted out of $X_1, X_2, \ldots X_k$, then we have

$$b_r^1 = b_r - C_{r1}b_1 / C_{ii} \qquad C_{rr}^1 = C_{rr} - \frac{C_{ri}^2}{C_{ii}}$$

$$C_{rs}^1 = C_{rs} - \frac{C_{ri}\, r_{si}}{C_{ii}}$$

$$\text{and Cor. (E.S.S.)} = \frac{\left(ESS + \dfrac{b_i^2}{c_{ii}}\right)}{(n-k)}$$

where Cor. (E.S.S.) is the corrected error Sum of squares in the analysis of variance table for testing the overall regression.

Applications

1. Prasad (1987) investigated the influence of factors determining the assets of beneficiaries in an evaluative study of Integrated rural development in Ananthapur district (Andhra Pradesh) using the multiple regression analysis by Assets (Y) as dependent variable; Familiy size (X_1) Non-productive assets (Rs) (X_2) ; Consumption expenditure on health and recreation (Rs/capita) (X_3) ; Expenditure on other items (Rs/capita) (X_4) ; per capita Total Income (X_5); Number of man days of employment (man days/man unit) (X_6) ; Total productive expenditure (Rs/house hold) (X_7) and Land (X_8) as dummy (X_8) as independent variables.

2. Kumari (1993) carried out multiple regression analysis with height-for-age percent standard as dependent variable (Y) and per cent Calorie adequacy (X_1) per cent protein adequacy (X_2), per cent calorie through supplement (X_3), Initial body weight (X_4), Total episodes of morbidity (X_5), episodes of URI (X_6), episodes of fever (X_7), episodes diarrhea (X_8) and number of packets purchased (X_9) as independent variables.

    She also worked out path coefficient analysis for the factors associated with weight gain and found X_9 (No. of packets) had highest direct effect and X_1 (Calorie adequacy) has highest indirect effect through X_4 (Initial body weight).

3. Raju (1993) carried out multiple regression analysis by taking cane yield (Y) as dependent variable and Number of Canes (X_1)' Length (X_2) , weight (X_3) and Diameter (X_4) on plant and Ratoon crops of sugarcane with different levels of N and K as independent variables. He found that contributions of stalk characters to cane yield (R^2) in

plant crop was as high as 88 per cent. These values were higher for potassium fertilizer than for nitrogen. Among stalk characters, diameter followed by weight of cane and to some extent, Stalk population appeared to be important for cane yield.

4. Laxmi Devi (1983) carried out path coefficient analysis with Total role performance of rural woman as dependent variable (Y) and seven significant independent variables were selected on the basis of their zero-order correlation coefficients with the dependent variable. These independent variables are Education (X_1), caste (X_2), family size (X_3), Leadership status (X_4), Family norms $(X_5)'$ Cosmopolite value (X_6) and liberalism value (X_7). She found that highest direct effect was recorded by X_4 (Leadership status) followed by X_1 (Education). Further highest indirect effect was recorded by X_2 (caste) followed by X_7 (Liberalism value). It was found that X_2 (caste) had highest indirect effect through X_1 (Education).

5. Rani (1987) used path coefficient analysis to explain direct and indirect effects of exogenous (independent) variables on the functional literacy status of adult learners as endogenous (dependent) variable. She found that Age (X_1), Occupation (X_6), Extension activity participation (X_{11}), Achievement motivation (X_{13}), farm size (X_{17}), experience in farming (X_{16}) had the first six maximum direct effects on the functional literacy status of adult learners in descending order. Age (X_1) had highest indirect effect followed by Mass media participations (X_{12}), X_1 variable had highest indirect effect through X_{16}(Experience in farming). It was interesting to note that X_1 (Age) had both highest direct and indirect effect on functional literacy status of adult learners

EXERCISES

1. Fit a multiple linear regression with height of plant (X_1) and number of panickles (X_2) as independent variables and yield (Y) as dependent variable and test the overall regression given the following data based on 10 plants of paddy in an experiment.

	Yield (kg) *(Y)*	*Height of plant* *(X₁)*	*No. of panickles* *(X₂)*
1	0.30	10	22
2	0.25	9	18
3	0.31	13	21
4	0.32	12	24
5	0.35	13	23
6	0.41	11	25
7	0.44	14	26
8	0.43	15	25
9	0.45	13	26
10	0.42	14	25

2. The following are the data on gain in weight (y), protein content in diet (X_1), Fat (X_2) and calcium (X_3) of children. Fit a multiple regression equation and test the overall regression and partial regression coefficients

S.No.	Wt(y) gain	Proteins (X_1)	Fat (X_2)	Calcium (X_3)
1	3	15	6	4
2	8	21	8	6
3	13	22	7	5
4	10	18	9	7
5	15	23	10	8
6	9	13	6	4
7	17	21	11	9
8	11	13	7	5
9	16	15	8	7
10	18	14	9	8

CHAPTER 14

D^2-STATISTICS AND DISCRIMINANT FUNCTIONS

14.1 D^2-STATISTICS

Some times there is need to know whether there is any significant difference between two populations and if there is significant difference how much distance exists between them based on several characters. For example, in examining two species of plants based on several phynotypical characters whether these two species differ significantly and if they differ significantly how much is the distance between them are the questions to be answered. Mahalanobis (1936) developed a statistic known as D^2 – statistic to measure the distance between two populations where each population was described by 'p' characters where $p > 1$. Let n_1, n_2 be two sizes of samples drawn at random from the two normal populations with means $\mu^{(1)}$, $\mu^{(2)}$ respectively with common variance-covariance matrix, Σ based on p characters. That is on each of the n_1 and n_2 units, p characters would be observed and let W_{1j} be the estimate of covariance between i-th and j-th variates,

$$W_{ij} = \frac{1}{n_1 + n_2 - 2}\left[\sum_{k=1}^{n_1}\left(X_{ik}^{(1)} - \overline{X}_i^{(1)}\right)\left(X_{jk}^{(1)} - \overline{X}_j^{(1)}\right) + \sum_{k=1}^{n_2}\left(X_{ik}^{(2)} - \overline{X}_i^{(2)}\right)\left(X_{jk}^{(2)} - \overline{X}_i^{(2)}\right)\right]$$

where $\overline{X}_i^{(1)}$ $\overline{X}_i^{(2)}$ be the means of i-th variate in the first and second samples respectively for $i = 1, ..., p$. Let D_p^2 be the distance between two normal populations estimated from the samples,

$$D_p^2 = \sum_{i=1}^{p}\sum_{i=1}^{p}W^{ij}\left(\overline{X}_i^{(1)} - \overline{X}_i^{(2)}\right)\left(\overline{X}_j^{(1)} - \overline{X}_j^{(2)}\right)$$

where (W^{iJ}) is the inverse matrix of the variance covariance matrix (W_{1j}).

14.1.1 Test of Significance

The test of significance is given to test the significant difference between means of the two samples for all the p-characters based on D^2 –statistic.

Null hypothesis: $\mu_i^{(1)} = \mu_i^{(2)}$ for $i = 1, 2, ..., P$ where $\mu_i^{(1)} = \mu_i^{(2)}$ are the means of the i-th variate for the first and second populations respectively.

$$F \frac{n_1 n_2 (n + n_2 - p - 1)}{p(n_1 + n_2)(n_1 + n_2 - 2)} D_p^2$$

Conclusion: If F (calculated) $>$ F (tabulated) with p and $(n_1 + n_2 - 1 - p)$ d.f at chosen level of significance, the null hypothesis is rejected. Otherwise, it is accepted.

EXAMPLE: Rao (1978) conducted a study to know whether different classification is necessary for awarding letter grades like, A, B, C and D in different branches (subjects) of study in the internal assessment system of examination by taking individual marks in subjects Statistics, Agronomy and Animal Science for about 75 students who were in B.Sc. (Ag.) second year of S.K.N. College of Agriculture, University of Udaipur, Jobner (Rajasthan). These were randomly selected from the students of first semester for the three years from 1965-66 to 1967-68. The marks of four interim one hour examinations and the final examination were recorded in the said subjects. The procedure of obtaining D^2 –statistics among the three subjects is given as follows.

These three subjects were considered since they represent different lines of study. The marks for the three subjects in the different one hour examinations and final examination were tested to ascertain whether the marks follow multivariate normal distribution or not. Five concomitant variates here are the marks in different one hour examinations and final examination.

14.1.2 D^2 – Statistic Between Agronomy and Animal Science

A sample of 70 student marks from each of Agronomy and Animal Science were selected at random from the first semester of the years from 1965-66 to 1967-68. The pooled variance – covariance matrix is given in Table 14.1 along with d_i values where $d_i - \overline{X}_i^{(1)} - \overline{X}_i^{(2)}$ be the difference between two sample means belonging to two populations for the i-th variate for i = 1, 2, ...,5.

TABLE 14.1 VARIANCE COVARIANCE MATRIX

X_1	X_2	X_3	X_4	X_5	d_i
1.35	1.04	0.75	0.45	1.18	2.38
	3.15	0.66	0.68	1.01	2.08
		2.29	1.00	1.01	0.31
			3.42	0.61	3.65
				1.82	0.25

The pivotal condensation method given by Rao (1962) for obtaining the values of D_i^2 for $i = 1, 2, \ldots, 5$ is presented here.

TABLE 14.2

Row No.	X_1	X_2	X_3	X_4	X_5	d_i
1	1.35	1.04	0.75	0.45	1.18	2.38
2		3.15	0.66	0.68	1.01	−2.08
3			2.29	1.00	1.01	0.31
4				3.42	0.61	3.65
5					1.82	−0.25
6						0.00
10	1	.770370	.555556	.333333	.874074	1.762962
11	.770370	2.348815	.082222	.333334	.100963	−3.913480
12	.555556		1.873333	.750000	.354445	−1.012222
13	.333333			3.270000	.216667	2.856667
14	.874074				.788593	−2.330295
					$-D_1^2 =$	−4.195850
15	$L_1(X) = 1.762962$					
20	.327982	1	0.35006	.141916	.042984	−1.666150
21	.528589	.035006	1.870455	.738331	.350911	−8.75228
22	.224005	.141916		3.222695	.202339	3.412051
23	.840960	.042984			.784253	−2.162075
24	$L_2(X) =$ 3.046513	−1.666150			$-D_2^2 =$	−10.716295
30	.282599	0.18715	1	.394733	.187607	−4.467922
31	.015353	.128098	.394733	2.931252	.063823	3.757532
32	.741793	.036417	.187607		.718420	−1.997876
33	$L_3(X) =$ 3.293852	−1.649770	−.457922			−11.125833
40	.005238	.043701	.134663	1	.021773	1.281886
41	.741459	.033628	.179012	.021773	.717030	−2.079690
42	$L_4(X) =$ 3.274170	−1.813978	−.973923	1.281866	$-D_4^2 =$	−15.942561
50	1.034069	.046899	.249657	.030365	1	−2.900422
51	$L_5(X) =$ 5.424713	−1.716443	−.454714	1.345036	− 2.900422	−21.974540 $= -D_5^2$

Steps: Row 10 was obtained by sweeping out Row 1, i.e., by dividing the elements in Row 1 by leading element 1.35. Row 11 was obtained by multiplying each element in Row.10 by 1.04 and subtracting from corresponding elements of Row 2. In the first column of Rows 11,12,13,14 and 15 rewrite the elements of Row 10 in the same order. Rows 12,13,14 and 15 were obtained by multiplying the elements in Row 10 with the elements in Row 1 from 3rd to 6th element i.e., 0.75, 0.45, ... , 2.38 respectively and subtracting from the corresponding elements in Rows 3, 4, 5 and 6 respectively. For example, 0.75 was multiplied with 0.555556, 0.333333, ..., 1.762962 in Row 10 and subtracted from 2.29, 1.00, ..., 0.31 respectively of Row 3 for obtaining Row 12.

The same procedure would be followed for .the Rows starting from 20, 30, 40 and 50. For example, Row 20 was obtained -by sweeping out Row 11 by the leading element 2.348815. Row 21 was obtained by multiplying the elements in Row 20 by 3.15 and substracting from corresponding elements in Row 2. Rewrite the elements in Row 20 from 3rd element to 6th element in the 2nd column of Rows 21 to 24.

The Rows 10, 20, 30, 40 and 50 are called pivotal rows and the elements in the last column of the Rows 15, 24, 33, 42 and 51 are the values of D_i^2 for $i = 1, 2,..., 5$. These were computed for Agronomy Vs Statistics and Animal Science Vs Statistics and are presented in Table 14.3.

TABLE 14.3

D_i^2	Agronomy Vs Animal Science	Agronomy Vs Statistics	Animal Science Vs Statistics
1	4.1985	3.8926	0.1852
2	10.7163	6.2394	0.2570
3	11.1258	6.2432	0.9674
4	15.9426	11.9048	1.8860
5	21.9745	12.1361	4.7714

F-test was computed to test the significant difference between mean values of two populations based on 5 concomitant variables as given in Section 14.1.1. In this case there would be three F values to be computed for testing the difference between mean values based on 5 examinations marks for Agronomy Vs Animal Science, Agronomy Vs Statistics and Animal Science Vs Statistics. For example, the calculating value of F was obtained for Agronomy Vs Animal Science as follows.

$$F = \frac{n_1 n_2 (n_1 + n_2 - p - 1)}{p(n_1 + n_2)(n_1 + n_2 - 2)} D_p^2$$

where $n_1 = n_2 = 70$; $p = 5$ and $D_5^2 = 21.9745$, then we have

$$F = \frac{70 \times 70 \, (70 + 70 - 5 - 1)}{5(70 + 70) \, (70 + 70 - 2)} \times 21.9745 = 149.3629$$

Here the F (Calculated), 149.3629 > F (Tabulated), 3.16 with 5 and 134 d.f.at 1 per cent level of significance. Hence the null hypothesis is rejected. In other words, the difference between mean values of marks of Agronomy and Animal Science was significant. Similarly the F values were computed for the differences between mean values .of Agronomy Vs Statistics and Animal Science Vs Statistics and are presented in the Table 14.4.

TABLE 14.4

Pair	*F*
Agronomy Vs Animal Science	149.3629**
Agronomy Vs Statistics	21.7122**
Animal Science Vs Statistics	8.5383**
**Significant at 1 per cent level.	

From the Table 14.4 it can be noted that all the F-values were found to be significant at 1 per cent level there by indicating that the true mean marks between the three subjects were significantly different from each other.

14.2 DISCRIMINANT FUNCTIONS

To classify an individual or an object into two populations (or groups) the discriminant functions are being used. If an individual or object was characterised by a single character, then the individuals having the character value greater than a predetermined value would be classified in one group and the rest in another group. For example, the per acre yield of a particular variety of a paddy crop exceeded, say, 25 quintals then it could be considered belonging to high yielding variety group otherwise, not. If an individual (or object) was characterised by more than one character or multiple characters, a suitable linear function would be used for classification into one of the two groups by taking into consideration of the measurements of all the characters. Let X_1, X_2,... , X_p be the measurements on p characters of an individual (or an object) then the coefficients .a_1, a_2, ... , a_p in the linear function $a_1X_1 + a_2X_2 + ..., + a_pX_p$ would be determined in such a way that the linear function would be able to discriminate between the two groups. That is, the coefficients (a_i's) were so chosen that the difference between mean values of the p characters of the two groups is maximum subject to the condition that variance of the linear function is unity.

Let $(a_1d_1 + a_2d_2 + ... + a_pd_p)^2$ be the square of the linear function of the differences of mean values of two groups for the p characters, where $d_i = \mu_i^{(1)} - \mu_i^{(2)}$ be the difference between mean values for the two groups for i = 1,2, ... , p and this function was maximised with respect to variance of the linear function,

$$\sum_{i=1}^{p} \sum_{j=1}^{p} W_{ij} \, a_i \, a_j$$ as unity where (W_{1j}) is the variance covariance matrix of the p characters based on sample size, say, n. Here the two groups assumed to follow multivariate normal distribution with means $\mu_i^{(1)}$ and $\mu_i^{(2)}$ respectively and same variance convariance matrix $\left(W_{ij}\right)$. After maximizing, the values of coefficients are given as

$$a_1 = W^{11}d_1 + W^{21}d_2 + \ldots + W^{p1}d_p$$
$$a_2 = W^{12}d_1 + W^{22}d_2 + \ldots + W^{p2}d_p$$
$$\vdots \qquad \vdots \qquad \qquad \vdots \qquad \vdots$$
$$a_p = W^{1p}d_1 + W^{2p}d_2 + \ldots + W^{pp}d_p$$

The discriminant function using this solution was the best one to discriminate (or classify) between two groups when the groups follow multivariate normal distribution with same variance covariance matrix. The discriminant function using this solution is given as:

$$(W^{11}d_1 + W^{21}d_2 + \ldots + W^{p1}d_1)X_1 + (W^{12}d_1 + W^{22}d_2 + \ldots + W^{p2}d_p) \, X_2$$
$$+ \ldots + (W^{1p}d_1 + W^{2p}d_2 + \ldots + W^{pp}d_p) \, X_p$$

EXAMPLE: The discriminant functions for discriminating between marks of Agronomy, Animal Science and Statistics for the example given in Section 14.1 are obtained in this section.

The coefficients for the discriminant function could be simultaneously obtained along with D^2-values from the pivotal condensation method given in Section 14.1. The coefficients for the discriminant function separating Agronomy and Animal Science were successively obtained from the Table 14.2 in column 1 of the Rows 15, 24, 33, 42 and 51. For example, the discriminant function separating Agronomy and Animal Science with one character is given as 1.762962 X_1 where 1.762962 is the value of $L_1(X)$ in column 1 of Row 15 in Table 14.2. If three characters are under consideration, the discriminant function is given as 3.293852 X_1 – 1.649770 X_2 – 0.467922 X_3 where 3.293852, – 1.649770 and – 0.467922 are the coefficients given in Columns 1, 2 and 3 of Row 33 in Table 14.2.

If all the five characters are under consideration, the discriminant function is given by

$$5.4247 \, X_1 - 1.7164 \, X_2 - 0.4547 \, X_3 + 1.3450 \, X_4 - 2.9004 \, X_5$$

where the coefficients in this function were obtained from columns 1, 2, 3, 4 and 5 of Row 51 in Table 14.2. The mean values of Agronomy and Animal Science are given in the following Table 14.5.

TABLE 14.5

Character	Agronomy	Animal Science
1	6.22	3.84
2	4.26	6.34
3	6.30	5.99
4	8.55	4.90
5	5.42	5.67

Substituting the mean values for Agronomy in Table 14.5 into the discriminant function with five characters the mean value for the function was obtained as 19.3448, i.e., 5.4247 (6.22) − 1.7164 (4.26) − 0.4547 (6.30) + 1.3450 (8.55) − 2.9004 (5.42) =19.3448 whereas the mean value for the function in the case of Animal Science was obtained as − 2.6296 and middle value for the two subjects is given as:

$$\frac{19.34448 + (-2.6296)}{2} = 8.3576$$

All the students marks with the values of the discriminant function above 8.3576 would be assigned to Agronomy (or considered as Agronomy marks) and all others to Animal Science.

The error of wrong classification would be computed as the area to the right of standard normal deviate. For example, the error of wrong classification for a student mark of Agronomy to be classified as Animal Science is obtained for the area above 8.3576 and the errors would follow normal distribution with mean as − 2.6296 and variance as $D_5^a = 21.9745$ with a standard normal deviate as,

$$\frac{8.3576 - (-2.6296)}{\sqrt{21.9745}} = 2.3438.$$

After entering into the tables of standard normal distribution, the area to the right of 2.3438 was obtained as 0.0096' which is the error of wrong classification.

The discriminant function for separating the marks of Agronomy and Statistics was similarly obtained as $0.9671 X_1 − 1.0371 X_2 + 0.4059 X_3 + 1.6866 X_4 − 0.4301 X_5$ and the discriminant function for separating the marks of Animal Science and Statistics was obtained as $− 1.3479 X_1 − 0.4363 X_2 + 0.1871 X_3 + 0.6048 X_4 + 1.7734 X_5$. The errors of wrong classification could also be calculated for Agronomy Vs Statistics and Animal Science Vs Statistics as on the same lines of Agronomy Vs Animal Science.

Application: The discriminant functions are also quite useful in studies like discriminating small farmers from others in Agricultural Economics by taking several ancillary variates such as area under holding, net income, area under high yielding varieties, size of the family, area under irrigation, etc.

CHAPTER 15

INTRODUCTION TO PROBIT ANALYSIS

15.1 INTRODUCTION

If X is a normal variate with mean μ and standard deviation σ then $\dfrac{X-\mu}{\sigma}$ follows normal distribution with mean zero and standard deviation unity. The normal distribution with mean zero and standard deviation unity is known as standard normal distribution and its curve is shown in Fig. 15.1 given below.

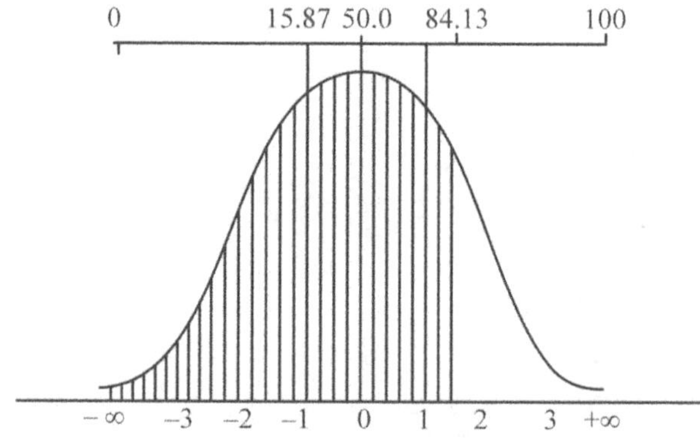

Fig. 15.1 Standard normal curve.

In Fig. 15.1 the total area under the standard normal curve was assumed as 100 and the areas to the left of the values given on X-axis are given on the top line. For example the area between $-\infty$ to $+1$ is given as 84.13 per cent $-\infty$ to 0 as 50 per cent and $-\infty$ to -1 as 15.87 per cent and so on. The values $\dfrac{X-\mu}{\sigma}$ on the X-axis are called normal equivalent deviates or standard normal variates or deviates. The percentages given on the top line were worked out on the basis of standard normal distribution for each of the standard normal deviate and are available in statistical tables. On the other hand if the percentages of areas are available the corresponding standard normal deviates could be worked out and are available in statistical tables (Fisher and Yates tables). The standard normal deviates were transformed to 'probits' by adding 5 to each of them to avoid negative values. The analysis based on these 'probits' is called 'probit analysis'.

The curve drawn between X-values and the corresponding area percentages resembles 'Sigmoid curve' assuming that X follows normal distribution and is shown in Fig. 15.2. Instead of percentages if probits are considered on Y-axis the relationship between X-values and probits would turn out to be a straight line as shown in Fig. 15.2 in place of 'Sigmoid curve'. The closer the points towards the straight line the closer would be the approximation of X-values to the normal distribution. The transformation of percentages to probits was found useful in the analysis of biological, economic and psychological data.

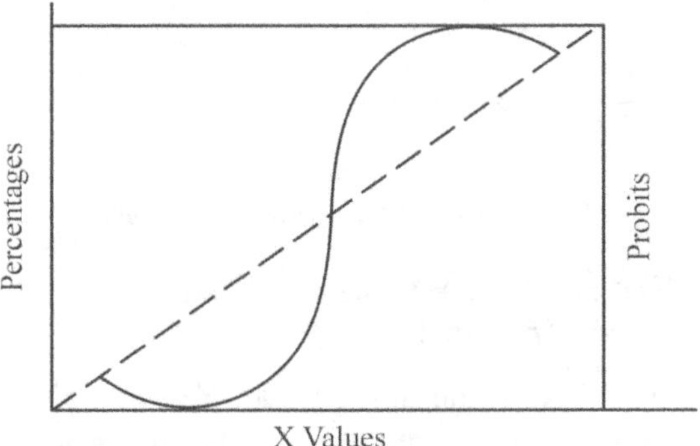

Fig. 15.2 Sigmoid curve.

15.2 ANALYSIS OF BIOLOGICAL DATA

To test the chemical preparation (stimulus) for its toxicity it would be applied at different levels of intensity to different batches of subjects at random. The batches of subjects should be of homogeneous nature and the environment under which they were to be maintained also should be uniform. For any subject there will be level of concentration of the chemical preparation below which response does not occur and above which response occurs. This level of concentration is known as 'tolerance' of the subject. In biological populations, the distribution of tolerances does not in general follow normal distribution. However if 'd' be the tolerance of the subject than 'log d' follows approximately normal distribution. Here 'd' is called dose in terms of actual concentration and 'log d' is called dosage or dose metameter. The curve drawn between 'log d' on X-axis and percentage of the subjects killed (or effected) on Y-axis will be of sigmoid type as shown in Fig. 15.3.

Therefore the relationship between log d and the probits corresponding to percentage killed would be of straight line nature as shown in Fig. 15.3 itself. The probit transformation is effected so that the statistical inferences could be drawn from the straight line with the help of linear regression analysis instead of Sigmoid curve. The dose giving a 50 per cent kill is referred to' as Lethal dose (L D 50) and 90 per cent kill as L D 90. Similarly the dose giving 50 per cent effected 90 per cent effected are denoted by E D 5O and E D 90 respectively.

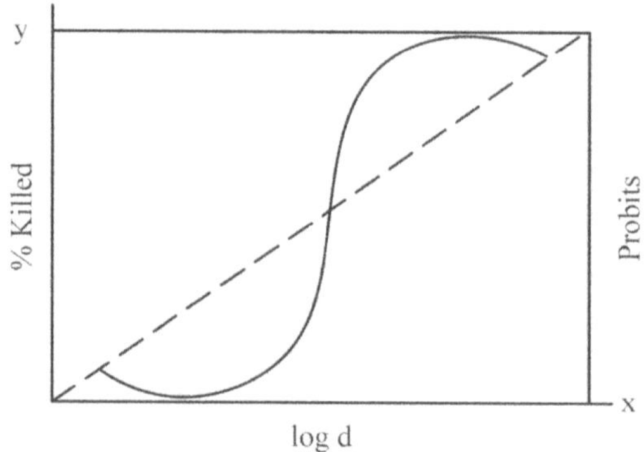

Fig. 15.3 Sigmoid curve.

15.2.1 Fitting a Probit Regression Line Through Least Squares Method

Let Y be the probit and X be the dosage than Y = a + bX be the regression equation for the straight line connecting dosage and probits. However there is an essential difference between fitting a usual regression line and the probit regression line. In the usual regression line the var (Y) at all values of X is small whereas in the case of probit regression line the var (Y) will be minimum at LD 50 and increases to infinity at 0 per cent kill at one end and at 100 per cent kill on the other end. In order to minimize the effect of fluctuations of Var (Y) the values at each point are weighted with the inverse of variance. In the case of determining LD 100, the levels beyond this dosage will not contribute anything to the result. Similarly levels below zero per cent kill will not contribute but levels above and below LD 50 would contribute to the precision of the result. Hence more precision could be achieved for LD 50 and therefore it *is* taken as 'potency' of the chemical preparation and is generally considered as the important statistic. The change in probit value for a unit change in dosage is termed as 'sensitivity' of the chemical preparation. The 'potency' and 'sensitivity' of the chemical preparation are measured with the help of regression equation fitted to the straight line relationship given in Fig. 15.3.

If Y = a + bX be the equation of a straight line for the relationship between 'dosage' and probits, 'b' represents sensitivity and 'm' represents potency of the chemical preparation. The value of m can be obtained from the fitted equation Y = a + bX by substituting 5 for 'Y' and obtaining the value of X. If 'P' be the probability of kill and 'Q' be the survival the correct weighing coefficient is given by w= Z^2/PQ where Z is the ordinate of the normal distribution at P. The different values of 'w' for different probit values of Y are tabulated in 'statistical tables' by Fisher & Yates (1948).

Since P and Q are the parameters it is difficult to obtain the values of w unless the equation of the straight line is known prior which is not possible. One solution to overcome this difficulty is to fit a straight line drawn close to the points of a scatter diagram and obtain values of m and b and provisional probits Yp and hence the confidence limits for m and b.

Another method is to draw approximate line to the scatter diagram and obtain provisional probits Yp and w. Another line would be drawn based on provisional probits and w, which is better than the first one. Provisional probits and w would be obtained using second line. A third line would be drawn using provisional probits and w. This process would be continued till no further change in line is noticed. This method of fitting is known as 'Iterative process'. In practice no more than two lines need to be drawn.

EXAMPLE: The following Table gives the procedure for fitting the dosage mortality curve as well as to obtain the confidence limits for potency and sensitivity of a stimulant by taking the data based on the mortality of one day old male adults of Drosophila with phosphamidon in cabbage wrapper leaves stripping solution. In col. (2), of Table 15.1 d was multiplied with 10^3 for taking logarithm in order to avoid negative values in logarithms. In col. (4), arithmetic mean of 3 replications was taken. In col. (5) the percentages were corrected up to first decimal place. The empirical probits in col. (6) were obtained by entering into Statistical tables by Fisher & Yates (1948) for different values in col. (5). The linear regression equation would be fitted for the values given in col. (2) as. independent variate and the values in col. (6) as dependent variate and this is given as $\hat{Y}$ = 1.81+1.74 X. Using this regression equation provisional probits would be obtained by substituting different values of X of col. (2) in the fitted regression equation. Provisional probits could also be obtained directly from the graph itself. The values of w in col. (8) were obtained by entering into 'Statistical tables' for different values of provisional probits given in col. (7) The cols. (9) to (11) could be filled in easily. The values of P_1 were obtained from 'Statistical tables' by reading backwards for the different values of provisional probits given in col. (7).The cols. (12) to (14) were furnished accordingly without any difficulty. The total of col. (14) is a χ^2 and tested against tabulated value of χ^2 with (k–2) d.f. where 'k' refers to the number of batches of subjects and '2' refers to the number of constraints for constants 'm' and 'b' in the regression equation. Here χ^2 measures the heterogeneity between the observed and expected number of subjects killed (or effected) and the heterogeneity factor is given by $\chi^2/(k–2)$ which is denoted by 'C'. Here in this example χ^2 was found to be not significant at (5–2) d.f. The variance of b is given by

$$\text{Var (b)} \; \frac{1}{\left\{ \Sigma nwX^2 - \dfrac{(\Sigma nwX)^2}{\Sigma nw} \right\}} = \frac{1}{S.S(X)}$$

TABLE 15.1

Amount of phosphamidon in I ml of stripping un μg (d)	Log (d ×10³) =X	No. of insects repli-cations) (n)	No. of insects killed (average of 3 replications) (r)	Correct ed percent age mortali ty P	Emperi cal prodit y_e	Provisi onal probit y_p	Weight w	nw	nwX	nwX²	$P_1 / r - np^1$		$P_1 = \dfrac{\left(r - np^1\right)^2}{np^1 q^1}$ 100
(1)	(2)	(3)	(4)	(5)	(6)	(7)	(8)	(9)	(10)	(11)	(12)	(13)	(14)
0.20	2.30	20	16.0	80.0	5.84	5.81	.4964	9.928	22.8344	52.5191	.792	0.16	.0078
0.16	2.20	20	14.7	73.5	5.63	5.64	.5474	10.948	24.0856	52.9883	.739	−0.08	.0016
0.08	1.90	20	10.7	53.5	5.09	5.12	.6329	12.658	54.0502	45.6954	.548	−0.26	0.136
0.04	1.60	20	6.7	33.5	4.57	4.59	.5986	11.972	19.1552	30.6483	.341	−0.12	0.112
0.02	1.30	20	37	18.5	4.10	4.07	.4616	9.232	12.0016	15.6021	.176	0.18	.0112
Control	-	20	0	0									
								54.738 102.1270		197.4532			0.0374 = X²

Source: Ramasubbaiah K. (1971), Ph.D. Thesis approved by i.A.R.I., New Delhi

The fitted regression equation is $\hat{Y} = 1.81 + 1.74\,X$

$$\text{Var (b)} = \cfrac{1}{\left[197.4532 - \cfrac{(102.1270)^2}{54.738}\right]} = 0.1447$$

and the confidence limits for 'β' the population regression coefficient (or sensitivity) are given by

$$b \pm 1.96\sqrt{\text{Var(b)}} = 1.74 \pm 1.96\sqrt{0.1447} = (2.4856,\ 0.9944).$$

Assuming χ^2 found to be significant, the corrected variance of b is

$$V_1(b) = \frac{C}{\text{S.S(X)}} \text{ where } \quad C = \frac{\chi^2}{(k-2)} = \frac{.0374}{3} = 0.125$$

$$= 0.125/6.9105 = .0018$$

The confidence limits for 'β' the sensitivity in the population are

$$b \pm t\sqrt{V_1(b)} = 1.74 \pm 3.182\sqrt{.0018} = (1.875,\ 1.605)$$

where t is the tabulated value of t with (k–2) d.f. at 5 per cent level of significance.

The confidence limits of 'potency 'in the population 'M' are $m \pm 1.96\sqrt{\text{Var (m)}}$. The value of m is obtained by substituting $Y = 5$ in the regression equation $\hat{Y} = 1.81 + 1.74\,X$, $m = 1.8333$

where $\text{Var (m)} = \dfrac{1}{b^2}\left\{\dfrac{1}{\Sigma nw} + \dfrac{(m - \overline{X})^2}{\text{S.S(X)}}\right\}$

and $\overline{X} = \dfrac{\Sigma nwx}{\Sigma nw}$,

$$\text{Var (m)} = \frac{1}{(1.74)^2}\left\{\frac{1}{54.738} + \frac{(1.8333 - 2.0577)^2}{6.9105}\right\}$$

$$= 0.008441$$

Therefore the confidence limits for potency are

$$1.8333 \pm 1.96\sqrt{.008441} = (2.0134,\ 1.6532)$$

The correction for heterogeneity would be used for Var (m) in case χ^2 found to be significant. The procedure for this is given in the following section 15.3.

TABLE 15.2

Amount of phosphamidon in I ml of stripping un μg (d)	Log (d × 10^3) =X	No. of insects per rep-tication (n)	No. of insects killed (average of 3 repli cations 1ml)(r)	Corrected percentage mortality p	Emperi cal probit Y_e	Provisional probit Y_p	Weight w	nw	nwX	Working Probit Yw	nw Yw	nw X Yw
(1)	(2)	(3)	(4)	(5)	(6)	(7)	(8)	(9)	(10)	(11)	(12)	(13)
0.16	2.20	20	17.0	85.0	6.04	6.08	.4115	8.230	18.106	6.27	51.6021	113.5246
0.08	1.90	20	15.0	75.0	5.67	5.59	.5602	11.204	21.2876	5.67	63.5267	12037007
0.04	1.60	20	10.6	53.0	5.08	5.11	.6336	12.672	20.2752	5.09	64.5005	10.2008
0.02	1.30	20	7.0	35.0	4.61	4.63	.6052	12.104	15.7352	4.62	55.9205	72.6966
0.01	1.0	20	4.0	20.0	4.16	4.15	.4870	9.740	9.7400	4.16	40.5184	40.5184
Control	-	20	0									
								53.950	85.1440		276.0682	450.6411

Source: Ramasubbaiah K. (1971), Ph.D. thesis approved by i.A.R.I., New Delhi, $\hat{Y} = 2.54 + 1.61$

The working probit is obtained as $\hat{Y} (-p/Z) + p/Z$.

15.3 MAXIMUM LIKELIHOOD METHOD

If ϕ is the probability density of the observations then the likelihood of the parameters occurring in ϕ is defined to be any function proportional to ϕ the constant of proportionality being independent of the parameters. The principle of maximum likelihood consists in accepting as the best estimate of parameters those values of the parameters which maximize the likelihood for a given set of observations. The maximum likelihood estimates satisfy such important properties as consistency efficiency and normality when the size of sample is large. Therefore this method is preferred though it involves more computations. The procedure for fitting the dosage mortality curve and the confidence limits for 'potency' and 'sensitivity' are given in the Table 15.2 .by taking the data based on the mortality of one day old male adults of Drosophila with Phosphamidon in bhindi leaf stripping solution. The cols. (1) to (6) would be furnished in a manner as described in Table 15.2. The provisional probits are obtained from the graph by fitting an eye fit of the straight line in col. (7). The values of w would be obtained by entering to 'Statistical tables' for different values of provisional probits. The cols. (9) and (10) would be furnished without any difficulty. The working probits in col. (11) could be read from the 'Statistical tables' for different values of provisional probits given in col. (7) as follows:

$$\text{Minimum working probit} + \frac{P}{100} \times (\text{range})$$

The 'minimum working probit' and 'range' values are obtained by entering into "Fisher and Yates tables'.

The 'potency' and 'sensitivity' and their confidence limits are computed for a chemical preparation as follows:

$$\text{Sensitivity, } b = \frac{\Sigma nwXY - \dfrac{(\Sigma nwX)(\Sigma nwYw)}{\Sigma nw}}{\Sigma nwX^2 - \dfrac{(\Sigma nwX)^2}{\Sigma nw}} = \frac{\text{S.P}(XYw)}{\text{S.S}(X)}$$

$$= \frac{\left[450.6411 - \dfrac{(85.144)(276.0682)}{53.950}\right]}{142.9157 - \dfrac{(85.144)}{53.950}} = 1.7503$$

$$\text{Var (b)} = \frac{1}{\Sigma nwX^2 - \dfrac{(\Sigma nwX)^2}{\Sigma nw}} = \frac{1}{\text{S.S}(X)}$$

$$\text{Var (b)} = \frac{1}{8.5413} = 0.1171$$

The confidence limits for 'β' the sensitivity in the population are $b \pm 1.96$ $\sqrt{\text{Var}(b)}$ $=1.7503 + 1.96 \sqrt{.1171} = (2.4210, 1.0796)$

The heterogeneity between the observed and expected number of subjects killed (or effected) will be measured by χ^2 which is

$$\chi^2 = \left\{ \Sigma nwY_w^2 - \frac{(\Sigma nwY_w)^2}{\Sigma nw} \right\} - b \left\{ \Sigma nwXY_w - \frac{(\Sigma nwX)(\Sigma nwYw)}{\Sigma nw} \right\}$$

$$= S.S.(Y) - b\,[S.P.(XY_w)]$$

$$= [1438.9584 - (276.0682)^2/53.950] = 1.7503\,(14.9497) = 0.12$$

This χ^2 would be tested with tabulated χ^2 with (k–2) d.f at 5 per cent level. Here χ^2 is found to be not significant at 5 per cent level. If χ^2 is found to be significant, a correction would be applied to V (b) i.e

$$V_1(b) = \frac{C}{S.S.(X)}$$

where $C = \dfrac{\chi^2}{(k-2)}$ .The confidence limits for 'β' are $b \pm t\sqrt{V_1(b)}$

where 't' is the tabulated value of t with (k–2) d.f at 5 per cent level of significance. From the regression equation we have

$$Y = \overline{Y}_w + b(m - \overline{X})$$

Therefore, $m = \dfrac{Y - \overline{Y}_w}{b} + \overline{X}$

where 'm' is the 'potency' of a chemical preparation corresponding to LD 50 and

$$\overline{Y}_w = \frac{\Sigma nwYw}{\Sigma nw} = \frac{276.0082}{53.950} = 5.1171$$

$$\overline{X} = \frac{85.144}{53.950} = 1.5782$$

$b = 1.7503,\ m = 1.6451$

The potency of a chemical preparation for LD 50

$$= \frac{\text{Anti log } 1.6451}{10^3} = .04417$$

Similarly the concentrations for LD 90 and other levels could be computed from the regression equation by substituting LD 90 in place of m, where LD 50 and LD 90 are the doses giving a per cent killed of 50 and 90 respectively. The confidence limits for m are given in section 15.2 when χ^2

is not found significant. If χ^2 is found significant the exact confidence limits are

$$m + \frac{h(m - \overline{X})}{1 - h} \pm \frac{t}{b(1 - h)} \sqrt{\left[\frac{1 - h}{\Sigma nw} + \frac{(m - \overline{X})^2}{S.S.(X)}\right]} \times C$$

where $h = \dfrac{t^2 C}{b^2 [SS(X)]}$, $C = \dfrac{\chi^2}{(k - 2)}$ and 't' is the tabulated value of t with

(k–2) d.f. at 5 per cent level of significance. Since χ^2 is not found to be significant, the confidence limits for 'potency' are m $m \pm 1.96\sqrt{Var(m)}$

where Var $(m) = \dfrac{1}{b^2}\left[\dfrac{1}{\Sigma nW} + \dfrac{(m - \overline{X})^2}{S.S.(X)}\right]$

$$= \frac{1}{(1.7503)^2}\left[\frac{1}{53.90} + \frac{(1.6451 - 1.5782)^2}{8.5413}\right]$$

$$= .0062$$

The confidence limits are $1.6451 \pm 1.96\sqrt{.0062}$

$$= (1.7994, 1.4908)$$

Similarly the potency of a chemical preparation corresponding to LD 90 is obtained by substituting for Y = 9 in the relation.

15.4 APPLICATION TO ECONOMIC DATA

Bal and Bal (1973) used probit analysis in forecasting the demand for fertilizers for wheat crop in Punjab. Let D_{ti} be demand of i-th fertilizer in the t-th year, A_t be the area under wheat (hectares) in the t-th year and F_{ti} be the amount (kgs) of i-th fertilizer to be used in one hectare in the t-th year then we have

$$D_{ti} = A_t . F_{ti}$$

Let A be the total cultivated area and is assumed to be constant.

To estimate the total demand for fertilizers it is required to estimate the total area under wheat and the amount of fertilizers to be used per hectare for future years.

Let $F_{ti}{}^r$, $F_{ti}{}^0$, $F_{ti}{}^a$ be the amounts of i-th fertilizer dose to be used in one hectare in the t-th year according to recommended dose, optimum dose and average dose respectively. For definitions of recommended dose, optimum dose and average dose lease refer to Bal and Bal (1973). If the points are

plotted with each of the percentages as $\dfrac{(At.100)}{A}, \dfrac{(F_u^0.100)}{F_{ti^r}}, \dfrac{\left(F_{ti}^a 100\right)}{F_{ti^r}}$ on the

Y-axis and time, t on the X-axis the figures obtained by joining these points assumed to follow 'Sigmoid curve' in case. Then transformation of 'percentages' to 'probits' make 'sigmoid curve' to straight line relationship in each situation.

Therefore, the probit regression lines are

$$\text{Probit} \quad \frac{(At.100)}{A} = a_o + a_{it}$$

$$\text{Probit} \quad \frac{\left(F_{ti}{}^{o}.100\right)}{F_{tl}{}^{r}} = b_o + b_{it}$$

$$\text{Probit} \quad \frac{\left(F_{ti}{}^{a}100\right)}{F_{tl}{}^{r}} = c_o + c_{it}$$

Each of the above equations can be fitted by choosing two points on the time scale in the usual process. With the help of these equations for amount of fertilizers per hectare as well as area under wheat can be made using different values of t. Hence the total demand for each type of fertilizer could be worked out with the help of the relation $D_{ti} = A_t \, F_{ti}$.

APPLICATIONS

1. Subbaratnam (1979) computed LT 50 and LT 90, values by taking "Time' as a factor in place of dose of insecticide in dosage mortality curve. LT (50) was defined as the lethal time required to arrive at 50 percent mortality and LT 90 was defined as lethal time required to arrive at 90 per cent mortality on the number lapsed days (X).

2. Rao (1982) also worked out the percentage reduction of residue and half life value. Half life value was defined as the "time required for half of a given quantity of the material to dissipate". This was obtained with the help of regression equation

$$Y = a + bX$$

where
$Y = \log$ of residue (ppm)

$X = $ elasped days after treatment

$a = \log$ of initial deposit

$b = $ slope of the line

$T_{1/2} = \log (e)/k$

where
$k = b \times 2.303$ (Gunthar and Blinn (1955)

$T_{1/2} = $ Time required for half of a given quantity of the material to dissipate (Half-life).

3. Rao (1982) prepared standard dosage mortality curves with the test organism. Drosophilas rnalanogaster Meig. Insecticidal solutions of monocrotophos (µg/ml) (0.2, 0.3, 0.4, 0.6, 0.8, 1.0 and 1.2),

quinalphos (μg/ml) (0.005, 0.01, 0.015, 0.020, 0.025, 0.030 and 0.035), phosalone (μg/ml) (0.2, 0.3, 0.4, 0.6, 0.8, 1.0, 1.2, and 1.4) and carbaryl (μg/ml) (0.4, 0.6, 0.8, 1.0, 1.2, 1.4 and 1.6) were used. For each concentration, three replications were maintained. For preparing the standard dosage mortality curve one day old male vinagar flies were utilised. Mortality counts were made 18 hours after exposure in all the concentrations and control. The resulting mortality data were subjected to probit analysis.

He also worked out the safety interval for the permissible consumptions of crops sprayed with insecticide using the formula.

$$t.\,tol = \frac{(\log k_2 - \log tol)}{K_1}$$

where

t.tol = minimum number of days to lapse before the insecticide reaches the tolerance limit.

K_2 = initial deposit (ppm)

tol = tolerance limit of the insecticide

K_1 = regression coefficient of the equation of RL 50 of log ppm residue (Y).

PART II

EXPERIMENTAL DESIGNS

CHAPTER 16

EXPERIMENTAL DESIGNS

16.1 INTRODUCTION

When more than two treatments are under study t-test has to be performed for every pair of treatments for testing the null hypothesis. Sir R.A. Fisher developed a technique called Analysis of variance with which the simultaneous comparison of any number of treatments is possible. Here treatments are known as the objects of comparison. In the Analysis of variance technique, the variances due to different sources of variation were analysed and tested with F-test for the null hypothesis of equality of treatment means in the population. Before proceeding to discuss the analysis of variance technique in detail some of the concepts used are studied.

(i) *Randomization:* Random allocation of treatments to different experimental units is known as Randomization. All the experimental units will be listed and a number from random number tables will be taken and the first treatment will be allotted to the experimental unit having the serial number equal to random number. This is further explained in Section 16.4. Randomization ensures the validity of statistical tests like F-test, t-test, etc. It may be recalled that one of the assumptions in two-sample t-test is that the samples are drawn independently and at random. Similarly F-test is not applicable unless treatments are not allotted at random to different experimental units.

(ii) *Replication:* Repetition of treatment to different experimental units is known as Replication. Replication of treatment reduces experimental error. The standard error of treatment mean is $\dfrac{\sigma}{\sqrt{r}}$ where 'σ' is standard deviation of treatment in the population and r is number of replications. As r, number of replications, increases the standard error of mean decreases. Also in the analysis of variance the replication of treatments provides estimate of experimental error which is essential for the application of F-test.

(iii) *Local control:* Grouping of homogeneous experimental units is known. as 'local control'. .The 'local control' helps in reduction of experimental error. In agricultural field experiments, the neighbouring plots expected to have homogeneous environmental conditions such as soil fertility, depth of soil, etc. and hence the neighbouring plots could be grouped into blocks. In animal experiments animals having same age, litter, body weight, sex, lactation, etc. could be grouped. This grouping of homogeneous experimental units will reduce the experimental error as the differences between blocks can be removed from the experimental error in the analysis of variance table.

16.2 UNIFORMITY TRIAL

This is conducted in a field to know the nature of the soil fertility gradient. In fact the uniformity trial is conducted on a site where an actual experiment is contemplated for minimizing experimental error using the concept of 'local control'. In this method the experimental field was divided into small plots of equal size, shape and uniform cultural operations including treatments would be given to each of the plots in the field. The yields of all the plots would be recorded. The differences in yields for the different plots were attributed to the differences Inherent in soil fertility of the field. The coefficients of variations could be obtained by taking different combinations of neighbouring small plots along and across row. For example, along and across row the number of plots would be taken as (1, 1), (1, 2), (1, 4), etc. Similarly plots across and along row can be taken as the Combinations (1,2), (1, 4):etc. The coefficients of variations could be arranged in a two-way table with number of plots along row on one side and number of plots across row on the other side. It is generally observed that the more the number of smaller plots the less is the coefficient of variation. .In this experiment, the average yield per plot for the entire experiment is considered as zero and the above average yields of plots would be considered as positive values and the below average yields would be taken as negative values. Let $\overline{Y}$ be the average yield of all plots considered in the experiment and Y_1 be the yield of

1st plot then the fertility gradient can be worked out as $\dfrac{(Y_1 - \overline{Y})}{\overline{Y}} \times 100$.

Similarly the fertility gradient for the second plot would be $\dfrac{(Y_2 - \overline{Y})}{\overline{Y}} \times 100$

where Y_2 be the yield for the 2nd plot in the experiment, etc. These fertility

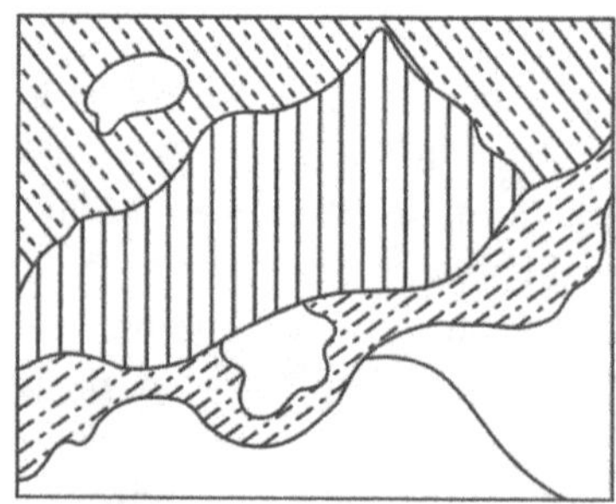

Fig. 16.1 Fertility contour map.

gradient percentages would further be divided into groups of equal size with say, 10 per cent as the size of each group. The diagram showing these fertility gradient percentages is known as 'fertility contour map' and is shown in Fig.16.1.

16.3 ANALYSIS OF VARTANCE

The observational value in an experiment can be assumed to be the sum of the components such as *(i)* general mean, *(ii)* the effect of the treatment applied, *(iii)* the environmental effect and *(iv) the* residual effect. The residual effects might be due to extraneous causes which are not foreseen before hand. This 'residual effect' is also known as 'experimental error'. The relation between observational value and the different components of variation is denoted by the following 'mathematical model'

$$Y_{1JP} = \mu + \alpha_i + \beta_j + E_{1jp} \qquad\qquad .....(16.1)$$

where Y_{ljp} represents the observational value on the, p-th experimental unit in the j-th block of i-th treatment for $I = 1, 2, ... , t, j = 1, 2, ... r$, and $p = 1, 2, ... , k$ where p is the total number of experimental units in each block and "μ" refers to the general mean,α_i the effect of i-th treatment, β_J the effect of j-th block and E_{ijp} the residual effect (the experimental error) for the (i, i, p)-th unit. In the model μ, α_1, β_j are unknown constants (parameters) based on population values, These are estimated with the help of least squares method. Since the model (16,.1) is a general one for most of the designs except in the case of 'split plot' and 'incomplete block' designs where in some additional assumptions are to be made. For procedure of estimating the different parameters in the model and their test of significance by 'least squares method' the reader is advised to refer to Cochran and Cox (1957).

16.3.1   Assumptions underlying in the Analysis of Variance model

(i)   The treatment and block effects in model 16.1 are additive. That is, the difference between true effects of any two treatments is same for all the blocks. In other words the comparison between any two treatments can be made irrespective of whether they are in the same block or in different blocks.

(ii)   The residual effects, (experimental errors) E_{ijp}'s are independent from observation to observation and are assumed to follow normal distribution with zero mean and common variance σ^2. Additional assumptions are also made for the designs 'Split plot' and 'incomplete block' but they are not discussed here.

16.4 COMPLETELY RANDOMIZED DESIGN

This design is useful when the experimental material (or experimental field) is of uniform nature. Since for the pot culture experiments and laboratory experiments, the experimental material is expected to be uniform in nature, the completely randomized design is often used. The completely randomized design is not much used in field experiments because the experimental field may not be uniform in soil fertility, etc. Though the method of analysis described here is for field experiment it is also identical for potculture, laboratory and animal experiments.

The experimental field is divided into plots of equal size either of rectangular or square shape. The net plot size may be taken as 5×4, 6×4 sq. metres, etc. according to the situation. Let i-th treatment be repeated r_i times for $\Sigma r_i = n$. Thus there are in all n experimental units. The field layout is shown in Fig. 16.2. Each treatment would be allotted to the plot in the field at random by listing all the plots serially from 1 to n with the help of random number tables.

The procedure of consulting random number tables is described in Chapter 17. If a particular random number comes in the selection, the first treatment would be allotted to the plot having serial number equal to that random number. Similarly the second treatment would be allotted to the plot corresponding to the second random number. This procedure would be continued till all the plots in the field are exhausted. In other words any treatment may occur any where in the experimental field. The additive model of Analysis of variance is $Y_{ij} = \mu + \alpha_i + E_{ij}$ where μ be the general mean, α_i be the effect of i-th treatment, E_{ij} are experimental errors which are independent and normally distributed with mean zero and common variance σ^2 and Y_{ij} be the observational value on the j-th unit of the i-th treatment for j= 1,2,…,t.

The method of analysis is given as follows:

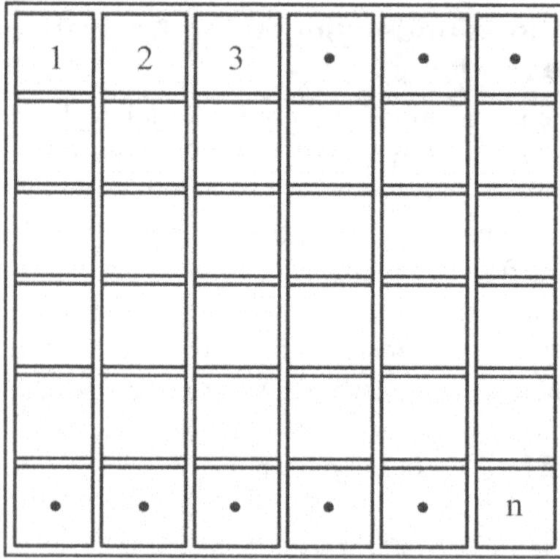

Fig. 16.2 Completely randomized design.

$$\text{Correction factor (C.F.)} = \frac{G^2}{n}$$

where G is the total of the observational values for all the n plots. Total sum of squares (T.S.S) = $\Sigma_{i,j} Y^2_{ij}$ - C.F.

$$\text{Treatment sum of squares (Tr.S.S)} = \sum_{i=1}^{t} \frac{T_i^2}{r_i} - \text{C.F.}$$

where T_i is the total of i-th treatment and is based on r_i experimental units (i = 1, 2, ... , t), Error Sum of' squares (E.S.S) = (T.S.S.)-(Tr.S.S). Further analysis would be carried out by furnishing the above sum of squares in the following analysis of variance table - *Null Hypothesis:* $\alpha_1 = \alpha_2 = ... = \alpha_t$

TABLE 16.1 ANOVA TABLE

Source	*d.f*	*Fcal*	*Ftab d,f*
Treatment	t-1	Tr.M.S/E.M.S	(t-1),(n-1)
Error	n-t		
Total	n-1		

where Tr. M.S. = Tr. S.S/(t -1), E.M.S = E.S.S./(n-t)

CONCLUSION: If F (calculated) > F (tabulated) with *(t-1)*. (n-t) d.f, at chosen level of significance, the null hypothesis is rejected. That is, there is significant difference between treatment effects. Otherwise, the null hypothesis is accepted. If the null hypothesis is rejected, the pairwise comparison of treatment means will be done by computing critical difference (C.D.) when the number of replications for each treatment is same and is

given by t (tab. at error d.f) $\times \dfrac{\sqrt{2(\text{E.M.S})}}{r}$ where $r_1 = r_2 ... = r_t = r.$

The treatment means are arranged first in descending order of magnitude. If the difference between the two treatment means is less than the C·.D. value, then this is indicated by drawing underline connecting those two treatments; otherwise not. This is shown in the following Fig. 16.3.

$$\overline{T_3} \qquad \overline{T_2} \qquad - - - \qquad \overline{T_t} \qquad - - - \qquad \overline{T_7} \qquad \overline{T_4}$$

Fig. 16.3

When the number of replications are unequal for the treatments the usual student's t-test would be conducted for testing the significant difference between each pair of treatment means by taking standard error for the difference between 'u-th and v-th treatment means $\sqrt{\text{E.M.S.}\left(\dfrac{1}{r_u} + \dfrac{1}{r_v}\right)}$ where r_u and r_v are replications for u-th and v-th treatments respectively.

The standard error (S.E) of treatment means is $\sqrt{\dfrac{\overline{(E.M.S.)}}{r}}$ when the number of replications are equal for each treatment and C.V. $= \dfrac{\sqrt{E.M.S.}}{G/n} \times 100$ where (C.V.) is the coefficient of variation in the experiment.

It was observed in some cases that even if F-test shows no significant difference between all the treatment means student's t-test indicates significant difference between some pairs of treatment means. This might be due to the fact that some non- independent comparisons were made at the same chosen level of significance. The independent comparisons are those where there is no common treatment mean between the different comparisons. For example, if there are four treatment means such as T_1, T_2, T_3 and T_4 the independent comparisons are (T_1–T_2 and' (T_3–T_4) or (T_1–T_3) and (T_2 – T_4) since there is no common treatment mean either in the first pair or in the second pair. However, (T_1–T_2) and (T_2–T_3) are non-independent comparisons. Whenever non-independent comparisons are involved the level of significance would become more than the chosen level of significance. Cochran & Cox (1957) pointed out that in such cases with three treatments the observed value of t is greater than the tabulated in 13 per cent of the cases, with 6 treatments it would be about 40 per cent, with 10 treatments it would be about 60 per cent and for 20 treatments it would be as large as 90 per cent. In' other words if three treatments are compared at 5 per cent level of significance the actual level of significance would be 13 per cent. To avoid this difficulty Duncan suggested multiple range test and is given in the next section.

SPSS DATA ANALYSIS PROCEDURE

X	Y
1	20
1	22
1	24
1	28
2	6
2	9
2	13
2	15
3	17
3	14
3	16
3	19

Enter X (Group/levels/millets) and Y(Iron content) variables with data column wise as given above with X as Fixed Factor and Y as dependent variable. CLICK ANALYZE   FROM THE TOOL BAR AND TAKE OPTION 'General Linear Model' and opt for 'UNIVARIATE'. We get display table where you take X variable to FIXED FACTOR ENCLOSURE AND 'Y ' to DEPENDENT VARIABLE enclosure. Then click on O.K. Now we get display table containing results of one-way Analysis of variance with F-values and probabilities. Here the difference between levels of X factor is tested. Conclusions can be drawn based on the results obtained.

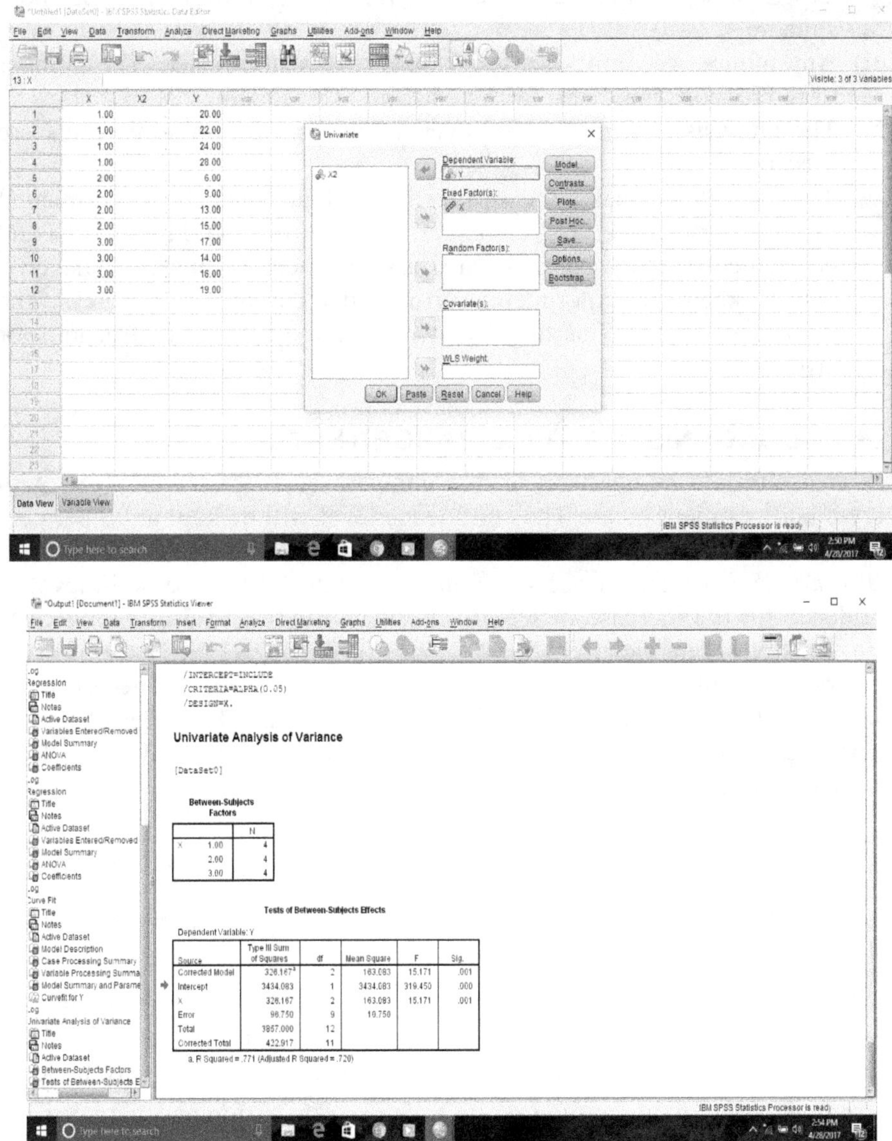

16.4.1 Salient features of completely randomized design

(i)   This design is most commonly used in laboratory experiments such as in Ag Chemistry, Plant Pathology, Animal experiments and Food science and nutrition experiments, where the experimental material is expected to be homogeneous.

(ii) This design is useful in Pot culture experiments where the same type of soil is usually used. However, in Greenhouse experiments care has to be taken with regard to sunshade, accessibility of air along and across the bench before proceeding to analyse the experiment. In these experiments randomized block or latin square design may be more suitable

(iii) Any number of replications and treatments can be used. The number of replications may vary from treatment to treatment.

(iv) The analysis remains simple even if information on some units are missing.

(v) The only draw back with this design is that when the experimental material is heterogenous, the experimental error would be inflated and consequently the treatments are less precisely compared. The only way to keep the experimental error under control is to increase the number of replications thereby increasing the degrees of freedom for error.

16.5   DUNCAN'S MULTIPLE RANGE TEST

This is a multiple comparison test to compare each treatment mean with every other treatment mean. The confidence level will be kept at 95 per cent for the sets of two means, $(0.95)^2$ i.e., 90.25 per cent for sets of three means, $(0.95)^3$ *i.e. 85.7* per cent for sets of four means, etc. The decrease in level of confidence for the increase in sets of means makes the 'Multiple range test' more sensitive than the usual least square difference (L.S.D) test for the detection of real differences between treatment means. The different 'significant studentized ranges (S.S.R) at 5 per cent and 1 per cent levels at different error d.f 's were given by Duncan (1955). Let R_p be the 'least significant range' for the 'p' treatments for p = 2, 3 ... then

$$R_p = \sqrt{\frac{(E.M.S.)}{r} \times (S.S.R)} \qquad \dots\dots(16.2)$$

where (S.S.R) is the 'significant studentized range' and r is the number of replications for each treatment.

The treatment means are arranged in descending order of magnitude and the differences between every pair of treatment means are compared with R_p values. For example, there are 4 treatment means which are arranged in discending order of magnitude as $\overline{T}_2\overline{T}_4\overline{T}_1\overline{T}_3$. The difference between $\overline{T}_2$ and $\overline{T}_3$ would be compared with R_4 since between $\overline{T}_2$ and $\overline{T}_3$ there are 4 treatment means involved as a closed interval. Similarly R_3 would be used comparing $\overline{T}_2$ with $\overline{T}_1$ and R_2 for $\overline{T}_2$ and $\overline{T}_4$. If there is no significant difference between any two treatment means it would be indicated by underline otherwise not. This test can be applied without resorting to F-test. For further reading on this topic, please refer to Federer (1967).

(i) *Confidence intervals:* Let $\overline{T_i}$, $\overline{T_c}$ be the treatment means for i-th treatment and control respectively, then the confidence intervals for the difference between them is

$$\left(\overline{T_i} - \overline{T_c}\right) \pm t_D \; X \sqrt{\frac{2(E.M.S.)}{r}} \qquad\qquad .....(16.3)$$

For the procedure of partition of treatment Sum of squares, the reader is advised to refer to Cochran and Cox (1957).

16.6   TRANSFORMATIONS

The important assumption underlying in the analysis of variance model is that experimental errors are independently distributed with constant variance σ^2. To ensure that experimental errors are independent from unit to unit, randomization is done in the allocation of treatments in the experiment. However, some times the experimental errors may not have common variance and may not follow normal distribution due to several reasons. These variations of experimental errors are of two types. One is called 'irregular' and the other as 'regular' type.

(i) *Irregular variations*: The irregular variation of experimental errors may occur due to certain irratic variation by one or two treatments from others. For example, the control plots may give more variation than the other plots in the insecticide spraying experiment. The error mean square may be inflated due to more variation of the control plots. Consequently the real differences between the different insecticides may not be property detected. This sort of irregular variation in the experimental errors can be tackled by subdividing the sum of squares' for error. The student's t-test could be used with different standard errors for different pairs of treatment means,

(ii) *Regular variations*: Here the treatment means are related with their variances in some order. If the distribution in the parent population is known then the relation between treatment means and variances could be known. In order to make the experimental errors follow normal distribution some transformation could be applied to the original data so that the transformed data follow at least approximately normal distribution. The following are some of the widely used transformations.

16.6.1   Square Root Transformation

If the original data follow approximately poisson distribution wherein the mean and variance are equal, the square root transformation is useful in bringing the original distribution to normal distribution. For example, the number of plants infected with a particular disease in a given area, the number of insects of a particular species in a given area, the number of noxious (or weed seeds) in a sample of seeds, the bacterial colonies on a plate count, etc. follow poisson distribution. In these cases $\sqrt{Y}$ transformation is recommended for Y, (the original data) when the original values lie between '0' and '20'. However, when the original data consist of zeros, the $\sqrt{\left(Y+\dfrac{1}{2}\right)}$ or $\sqrt{(Y+1)}$ transformation is used. After the transformation is effected, the usual analysis of variance would be carried out.

16.6.2   Angular Transformation

When the original data follow Binomial distribution wherein mean is related with variance , the experimental error variance is bound to change from one treatment mean to another, In such, case Angular transformation of $\mathrm{Sin}^{-1}\sqrt{Y}$ would be used for Y to stabilize variance so that the transformed data follow approximately normal distribution. This transformation requires the original data in percentages since the original data assumed to follow Binomial distribution.

16.6.3   Logarithmic Transformation

Here log Y or log (Y+1) is used for Y the observational value in the original data. If the original· data are in large numbers where the variances of treatment means are proportional to the square of the treatment means the "logarithmic transformation' is used to stabilize variances. If 'zero' values, present in data log (Y+1) transformation is used instead of log Y. This transformation is useful for Economic data like income. log (Y+1) transformation behaves like $\sqrt{\left(Y+\dfrac{1}{2}\right)}$ transformation for small values up to 10 and differs little from log Y there onwards.

16.6.4   Reciprocal transformation

This is to use 1/Y for 'Y, the observed value in the original data. The variances would differ from treatment to treatment if 'time' is a characteristic under study. In this situation reciprocal transformation converting the original values to reciprocals would bring stabilization to the variances of the treatments.

16.7 RANDOMIZED BLOCK DESIGN

In this design the whole experimental field is divided into blocks of homogeneous units based on soil fertility, etc. This grouping is done in such a way that there would be more homogeneity within blocks with respect to soil fertility so as to reduce the error sum of squares in the analysis of variance. The knowledge of the fertility gradient in the experimental field can be ascertained either through uniformity trial or through the results of previous experiments conducted on the same field... Since the neighbouring plots are expected to be having uniform fertility gradient they would be grouped in blocks in such a way that if the fertility gradient of the experimental field is in the direction of North to South the blocks would be formed in the direction of East to West as shown in Fig, 16.4. Similarly the fertility gradient is in the direction of East to West the blocks would be formed in the direction of North to South. However, in Animal Husbandry experiments, the grouping of animals can be done on the basis of litter, sex, initial body weight, etc., This design is also called as one-way elimination of heterogeneity design or two-way classification of analysis of variance in laboratory and animal experiments. The additive model of Analysis of variance is

$$Y_{ijp} = \mu + \alpha_i + \beta_j + E_{ijp} \qquad\qquad .....(16.4)$$

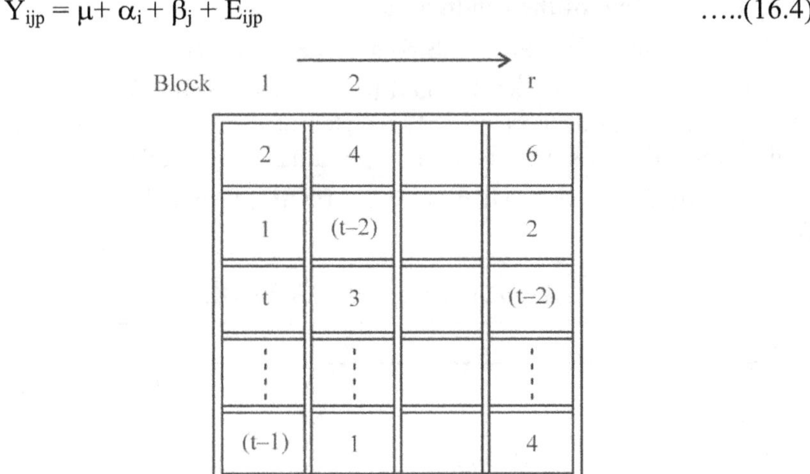

Fig. 16.4 Randomized block design.

where Y_{ijp} be the observational value on the p-th unit of the j-th block in which i-th treatment occurs for i = 1, 2 ... t; j = 1, 2, ... r; p = 1, 2, ... , k; μ be the general mean, α_i be the i-th treatment effect, β_j be the j-th block effect and E_{ijp} are experimental errors which are independently and normally distributed with zero mean and common variance σ^2. Here the randomization is restricted in one-way in the sense that all the treatments are applied to each block at random. Since all the treatments occur once in each block making block as a complete replication. The differences in blocks are eliminated from the experimental error for 'effective comparison of treatment means. The procedure of computing F-test is as follows:

$$C.F. = \frac{G^2}{rt}$$

where G is the total of the observational values of all the rt experimental units assuming that k=t in the model (16.4)

$$\text{Total sum of squares (T.S.S)} = \sum_{i,j} Y_{ij}^2 - C.F$$

where Y_{ij} by the observational value on the experimental unit having i-th treatment in the j-th block.

$$\text{Block sum of squares (B.S.S)} = \sum_{j} \frac{B_j^2}{t} - C.F.$$

where B_j is the total of the j-th block.

$$\text{Treatment Sum of squares (T}_r\text{.S.S)} = \sum_{j} \frac{T_i^2}{r} - C.F.$$

where T_1 is-the total of the i-th treatment.

Error sum of squares (E,S.S) = T,S.S – (B.S.S + Tr.S.S).

The above sum of squares would be furnished in the following ANOVA Table for carrying out the F-test. The different mean squares are obtained by dividing the sums of squares by corresponding degrees of freedom.

Null Hypothesis: (i) H_0: All treatment effects arc equal, H_1: All treatment effects are not equal.

TABLE 16.2 ANOYA TABLE

Source	*d.f.*	*fcal*	*ftab*	*d.f*
Blocks (replications)	r–1	B.M.S/E.M.S	(r–1)	(r–1) (t–1)
Treatments	t–1	T$_r$.M.S/E.M.S	(t–1)	(r–1) (t–1)
Error	(r–1) (t–1)			
Total	rt–1			

Conclusion: If F (calculated) > F (tabulated) with corresponding d.f. at chosen level of significance, the null hypothesis is rejected. If F (calculated) is found significant, the critical difference (C.D.) would be computed for the pair-wise comparison of treatment means. If L.S.D. test (or C.D.) is not suitable Duncan's multiple range test may be applied. The C.D. for treatment means is given as $t_{(\text{tab. error d.'.})} \times \sqrt{\dfrac{2(\text{E.M.S.})}{r}}$.

The C.D. is compared with the difference of treatment mean by arranging them in discending order of magnitude.

16.7.1 Some Salient Features of R.B. Design

(i) This design is widely used in Agricultural field experiments. (ii) Any number of replications can be used in the experiment. (iii) This design is more efficient, in general, compared to C.R. design since the replication sum of squares would be eliminated from error sum of squares which result in efficient comparison of the treatments. When the experimental unit is one for each treatment in a block, the error sum of squares is nothing but the interaction sum of squares of treatments and replications. (iv) If any treatment shows irratic variation (or irregular type of variation), the error sum of square can further be sub-divided into single d.f and the comparison of treatments can be done. (v) Even if some treatments are missing in one or more blocks, the statistical analysis can be performed with the help of 'missing plot technique'. If the number of missing units are more in any particular block, the statistical analysis can still be done by removing the entire block from the analysis. (vi) If certain treatments are to be repeated in a block each of those would be applied to two units in every block. (vii) In the Greenhouse experiments if the access of air current and sunlight are across the bench then the block would be formed along the bench and vice-versa. (viii) In animal husbandry experiments the characteristics like age, sex, vigour, litter, breed, initial body weight etc., could be considered for forming blocks. Milks of the same fat content would be considered as blocks in dairy experiments. Shelves of the germinator could be blocks in Horticulture experiments. Soils of same type would farm blocks in soil science experiments.

16.7.2 Missing Plot Technique

Certain observations will be missing in the experimental data due to floods, rodents, failure to record, etc., in field experiments; death of an animal due to sickness in the animal husbandry experiment; breakdown of an equipment in the laboratory experiment, etc. In such situations parameters can be estimated by the 'least squares' procedure based on the observations present. Since this method is complicated' Yates suggested a procedure to insert a value for each missing observation. The total and error d.f. would be reduced to the extent of number of missing units. However, the analysis of variance would remain simple as before.

If a single value is missing in randomized block design it can be substituted by the value

$$Y = \frac{rR + rT - G}{(r-1)(t-1)}. \qquad \ldots\ldots(16.5)$$

where r, t be the number of blocks and treatments respectively. R be the total of the units in the block in which missing unit occurs, T be the total of the remaining units of the treatment' which is missing and G be the grand total. In the analysis of variance, 1 d.f. would be subtracted from the error and total d. f. The inflation in the treatment sum of squares is to the extent of

$$\frac{\left[R-(t-1)Y\right]^2}{t(t-1)} \qquad .....(16.6)$$

where Y is the value obtained for the missing unit obtained from (16.5). The standard error of the difference between the treatment mean with a missing unit and other treatment-mean is given as

$$\sqrt{E.M.S.\left[\frac{2}{r}+\frac{t}{r(r-1)(t-1)}\right]} \qquad .....(16.7)$$

16.7.3   Estimation of Gain in Efficiency Over Completely Randomised Design

The error mean square expected in case when the one-way elimination of heterogeneity is not used is

$$E.M.S.\ (C.R) = \frac{n_r(B.M.S.)+(n_t+n_e).(E.M.S)}{n_r+n_t+n_e} \qquad .....(16.8)$$

where n_r, n_t, and n_e are the d.f. for blocks, treatments and error respectively in R.B design.

Efficiency of R.B. design over C.R. design =

$$100 \times \frac{E.M.S\ (C.R.)}{E.M.S\ (R.B.)} \quad \frac{(n_1+1)\ (n_2+3)}{(n_2+1)\ (n_1+3)}$$

where $\dfrac{(n_1+1)\ (n_2+3)}{(n_2+1)\ (n_1+3)}$ is called the precision factor and n_i,

n_2 are the. error d.f. for R.B and C.R designs respectively.

TWO-WAY ANALYSIS OF VARIANCE WITHOUT INTERACTION (ONE UNIT PER CELL)

SPSS DATA ANALYSIS PROCEDURE

Suppose the two-way Anova Table is given as follows.

Y

X	1	2	3
1	6	8	13
2	12	16	18
3	20	24	27

The data will be entered in the SPSS Excel sheet as follows

X	Y	R
1	1	6
1	2	8
1	3	13
2	1	12
2	2	16
2	3	18
3	1	20
3	2	24
3	3	27

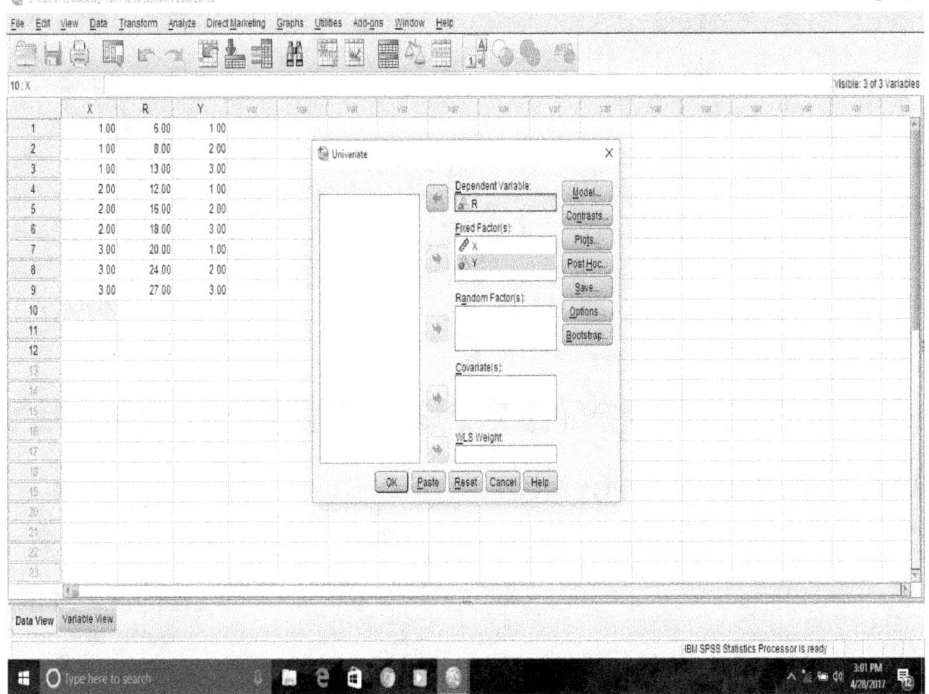

Enter the variables X, Y, and R and then enter data column wise as in the FORMAT above manner. CLICK 'ANALYZE 'from TOOL BAR AND TAKE option 'GENERAL LINEAR MODEL' and take OPTION again 'UNIVARIATE' .Take 'X' and 'Y' as FIXED FACTORS AND 'R' AS DEPENDENT VARIABLE in the display table. CLICK 'MODEL' and take option" CUSTOM'. Take 'X' and 'y' factors to the Right Hand Side without interaction XY and CLICK 'CONTINUE'. We can use DISCRIPTIVE

STATISTICS AND CONTRASTS OF MEANS FOR X AND Y as and when required. Then CLICK O.K. OPTION FINALLY. We get Table of results of Two-way Analysis of Variance with two-factors as 'X' and 'Y' and F-values and corresponding probabilities. Here Error term is nothing but interaction of X and Y.

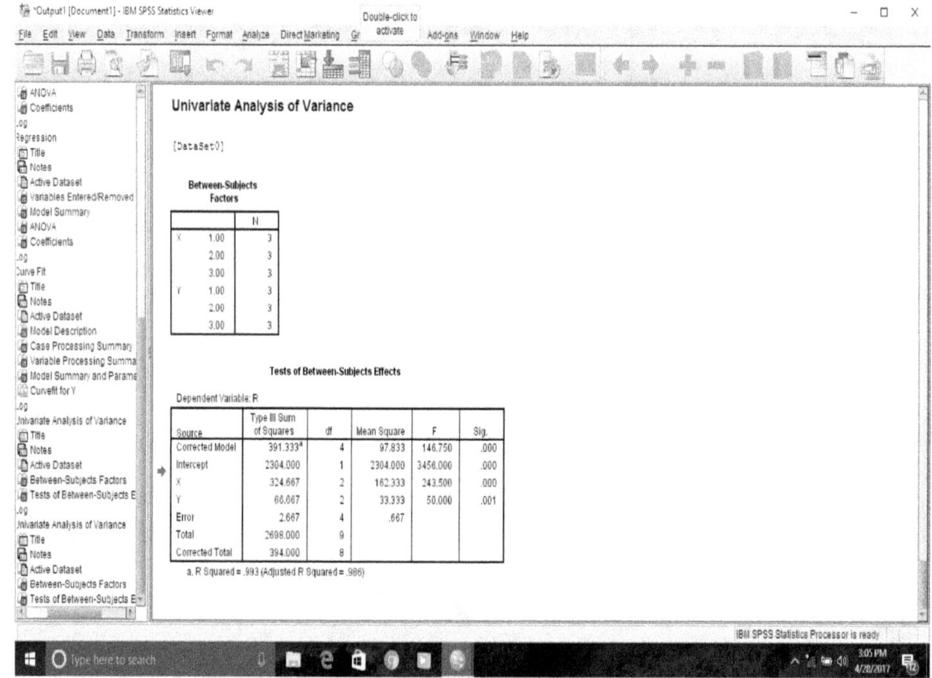

16.8   LATIN SQUARE DESIGN

Since Latin letters like A, B, C, etc., are arranged in a square, this design is known as 'Latin square'. In the experimental field, the soil heterogeneity was eliminated in two ways by grouping the units into rows and columns. If the fertility gradient is in the direction of East to West then the grouping would be done in the direction of North to South as columns and also if the fertility gradient is in the direction of North to South then the grouping would be done in the direction of East to West as rows, This design is also known as two-way elimination of heterogeneity design.. For laboratory experiments this is known as three-way classification of analysis of variance, since there are three sources of variation due to rows, columns and treatments. In this design the treatment should appear once and only once in each row and column. The number' of treatments, rows and columns are all equal. Each row and column is a complete replication. The lay out of 4 x 4 Latin square design is shown in Fig. 16.5.

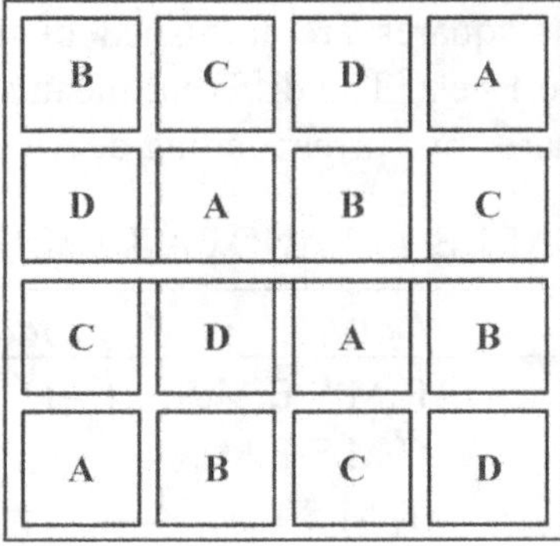

Fig. 16.5 Latin square.

Here the total number of experimental units are 16 and are arranged in 4 rows' and 4 columns. The 'Latin square' may also be arranged in long strip in case the experimental' field is a slope on hilly tract where the uniform blocks as rows 'and the order within each block as columns. It may be noted that If a significant interaction was suspected between any two of the factors like treatments, rows and columns then 'Latin square' would not be a suitable design.

The additive model of analysis of variance for this design is

$$Y_{ijk} = \mu + \alpha_i + \beta_j + \gamma_k + E_{ijk} \qquad \qquad .....(16.9)$$

where Y_{1Jk} be the observational value on the k-th column of the j-th row having i-th treatment for i, j, k, = 1, 2, ... r. The computational procedure is given as follows:

$$C.F. = \frac{G^2}{r^2}$$ where G be the grand total and r be the number of rows,

columns or treatments.

$$\text{Total sum' of squares (T.S.S.)} = \sum_{(i)j,k} Y_{ijk}^2 - C.F.$$

$$\text{Row sum of squares (R.S.S.)} = \sum_{j} \frac{R_i^2}{r} - C.F.$$

$$\text{Column sum of squares (C.S.S.)} = \sum_{k} \frac{C_k^2}{r} - C.F.$$

$$\text{Treatment sum of squares (Tr. S.S.)} = \sum_{i} \frac{T_i^2}{r} - C.F.$$

Error sum of squares (E.S.S.) = T.S.S. - (R.S.S. + C.S.S. + Tr.S.S.).

The different sum of squares are furnished in the following ANOVA table for carrying out the F-test. The different mean squares are obtained by dividing the sums of squares by corresponding degrees of freedom.

TABLE 16.3 ANOVA TABLE

Source	*d.f*	*fcal*	*Ftabd.f*	
Rows	r-1	R.M.S/E.M.S	(r–1)	(r–1) (r–2)
Columns	r-1	C.M.S/E.M.S		-do-
Treatments	r-1	T_r.M.S/E.M.S		-do-
Error	(r-1) (r-2)			
Total	r^2-1			

Conclusion: If F (calculated) > F (tabulated) with corresponding d.f. at chosen level of significance, the null hypothesis is rejected. Otherwise, it is accepted. In case the null hypothesis is rejected,

$$C.D. = t_{(tab\ with\ error\ d.f.)} \times \sqrt{\frac{2(E.M.S)}{r}}.$$

16.8.1   Some Salient Features of Latin Square Design

(i) This design is found useful wherever two-way heterogeneity is present in the experimental material such as in animal husbandry, dairy, psychology and industrial experiments, wherein the number of treatments experimented are limited. (ii) It is having limited application in agricultural field experiments since the number of treatments and replications is the same.

R.B. design is more practicable in field experiments than Latin square unless the field is known to possess two-way soil fertility gradient. However, three-way analysis of variance can be done in situations like testing the quality of tobacco by .different buyers at different locations and timings. (iii) In general, Latin squares 4 × 4 to 12 × 12 would be used in practice. For the Latin squares less than 4 × 4 order, the error d.f. would be too small and the experimental units would be too large for the designs beyond 12 × 12 size. The E.M.S. per -unit would increase with the increase in the size of latin square as in the case of size of block in R.B. design. Several Latin squares of the size 2 × 2 or 3×3 may be found useful in certain situations to keep error d.f. reasonably adequate. (iv)For this design, randomization would be carried out in three stages by taking first Latin square and then randomizing rows, Columns and treatments. The initial arrangement of symbols (treatments) would be taken as in Fig. 16.6 for 4 × 4 Latin square.

A   B   C   D

B   C   D   A

C   D   A   B

D   A   B   C

Fig. 16.6

(i) *Randomization of rows:* .To randomize rows in Fig. 16.6 random numbers (3, 1 ,4, 2) would be selected out of 4 from the 'table of random numbers. The rows would be rearranged accordingly in the order of random numbers as given in Fig. ·16.7.

C   D   A   B

A   B   C   D

D   A   B   C

B   C   D   A

Fig. 16.7

(ii) *Randomization of columns:* The columns in Fig. 16.6 would be randomized by selecting random numbers (2, 4, 3, 1). The rearrangement of columns according to random numbers is

D   B   A   C

B   D   C   A

A   C   B   D

C   A   D   B

Fig. 16.8

(iii) *Randomization of treatments:* The symbols (treatments) in Fig. 16.8 would be randomized denoting A,B,C, and D by numbers 1, 2, 3 and 4 respectively and selecting random numbers (3,1,4,2) out of 4. The arrangement of symbols would be done by replacing the symbols A, B,C and D *with* C, *A*, D and B respectively in the arrangement given - in Fig. 16.8 and the final arrangement is in Fig. 16.9

B	A	C	D
A	B	D	C
C	D	A	B
D	C	B	A

Fig. 16.9

16.8.2   Missing Plot Technique

If an observation in single unit is missing then it is estimated by

$$Y = \frac{r(R + C + T) - 2G}{(r-1)(r-2)} \qquad .....(16.10)$$

where R, C and T are the totals of the remaining units in. the row, column and the treatment respectively in which the missing unit occurs, G be the grand total and r be the number of replications for each treatment. The total and error d.f.'s would be reduced to the 'extent of number of missing observations.. For example, error d.f. and total d.f, would be reduced by two if two observations are missing. The amount of upward bias in the treatment sum of squares for a single missing unit is

$$\frac{\left[G - R - C - (r-1)T\right]^2}{(r-1)^2 (r-2)^2} \qquad .....(16.11)$$

The standard error of the difference between the mean of the treatment having missing observation and the mean of any other treatment is given as

$$\sqrt{\text{E.M.S}\left[\frac{2}{r} + \frac{1}{(r-1)(r-2)}\right]} \qquad .....(16.12)$$

The effective number of replications for each of the treatment means in obtaining standard error is computed as follows.

Each observation on one treatment is credited 1 replication if the other treatment is present in both the row and column, 2/3 if the other treatment is present in either row or column; 1/3 if the other treatment is not present in both the row and column and zero if the treatment itself is absent. For example, consider the following Latin square design given in Fig. 16.10 with missing units in parentheses. The effective number of replications for A starting from 1st row=0+2/3+2/3+1 =7/3. The effective number of replications for D starting from Ist row=2/3+1+0+2/3=7/3. The S.E. for (A–D) = $\sqrt{\text{E.M.S.}(3/7 + 3/7)}$.

D	B	(A)	C
C	A	B	D
A	(D)	C	B
B	C	D	A

Fig. 16.10

16.8.3   Estimation of Efficiency of L.S. Design over R.B. Design

If the experiment is laid out as R.B. design with rows as blocks, the error mean square is

$$\text{E.M.S (R.B.)} = \frac{n_r(\text{R.M.S}) + (n_t + n_e)\,(\text{E.M.S})}{n_r + n_i + ne} \qquad \dots\dots(16.13)$$

where R.M.S. and E.M.S. be the mean squares for rows and error respectively and n_r, n_t and n_e be the d.f. for rows, treatments and error respectively in the L.S. design.

$$\text{Efficiency of L.S. design} = \frac{\text{E.M.S. (R.B.)}\;(n_1 + 1)\,(n_2 + 3)}{\text{E.M.S. (L.S.)}\,(n_2 + 1)\,(n_1 + 3)} \times 100 \qquad \dots\dots(16.14)$$

where n_1, n_2· be the d.f. of error for L.S. and R.B. designs respectively. Similarly the efficiency of L.S. design over R.B. design if columns are taken as blocks can be obtained.

SPSS DATA ANALYSIS PROCEDURE

The data for LSD will be entered column wise as follows in the SPSS Excel sheet with R(rows), C(columns0,T(treatments),Y(observational data).

C	A	D	B	E
6	4	10	15	20
E	C	A	D	B
22	8	3	14	18
B	E	C	A	D
17	25	5	2	13
A	D	B	E	C
6	13	19	27	4
D	B	E	C	A
15	18	26	7	5

R	C	T	Y
1	1	3	6
1	2	1	4
1	3	4	10
1	4	2	15
1	5	5	20
2	1	5	22
2	2	3	8
2	3	1	3
2	4	4	14
2	5	2	18
3	1	2	17
3	2	5	25

3	3	3	5
3	4	1	2
3	5	4	13
4	1	1	6
4	2	4	13
4	3	2	19
4	4	5	27
4	5	3	4
5	1	4	15
5	2	2	18
5	3	5	26
5	4	3	7
5	5	1	5

Here A = 1, B = 2, C = 3, D = 4, E = 5 are taken for treatments.

After entering the data in SPSS Excel sheet, CLICK 'ANALYZE 'from the TOOL BAR and opt for 'GENERAL LINEAR MODEL' and 'UNIVARIATE' IN THE LEFT SIDE OPTIONS IN THE TABLE. Enter R.C,T as 'FIXED FACTORS ' and Y as "DEPENDENT VARIABLE' in the RHS as per indicator. Press 'MODEL' OPTION. Click CUSTOM' OPTION & 'INTERACTION'. In the displayed table take all 'Main effects 'or take Individually RCT to the RHS. Click' CONTINUE 'OPTION 'IN THE BOTTOM OF THE TABLE. In the displayed table click 'O.K." then we get table with ANOVA table containing all results required including F-Values, probability values etc.

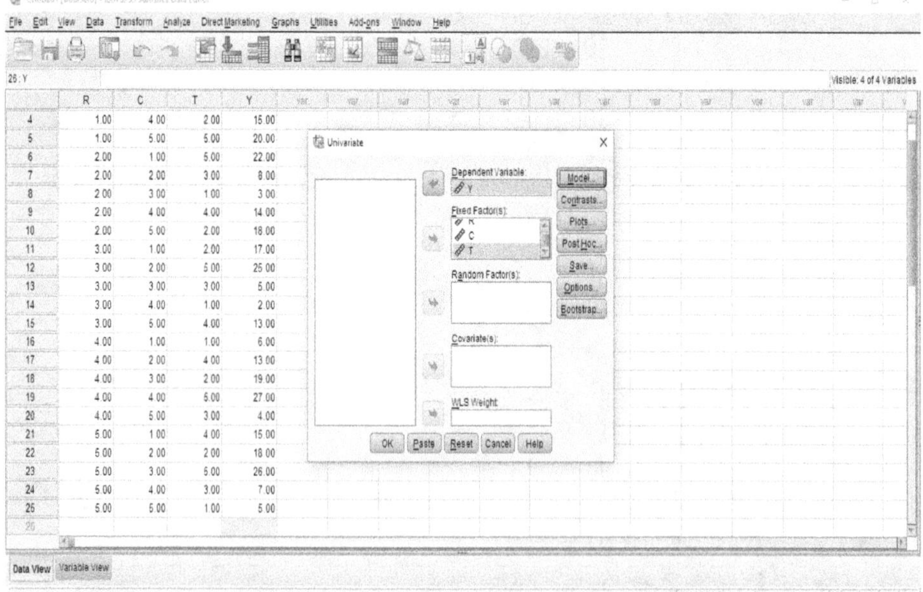

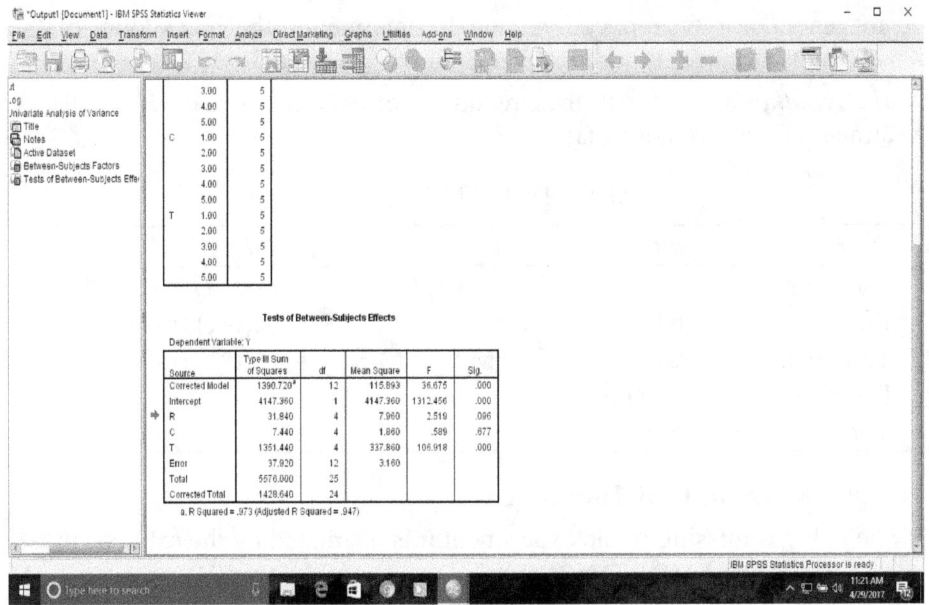

Tests of Between-Subjects Effects

Dependent Variable: Y

Source	Type III Sum of Squares	df	Mean Square	F	Sig.
Corrected Model	1390.720[a]	12	115.893	36.675	.000
Intercept	4147.360	1	4147.360	1312.456	.000
R	31.840	4	7.960	2.519	.096
C	7.440	4	1.860	.589	.677
T	1351.440	4	337.860	106.918	.000
Error	37.920	12	3.160		
Total	5576.000	25			
Corrected Total	1428.640	24			

a. R Squared = .973 (Adjusted R Squared = .947)

16.9 CROSS OVER DESIGN

This design is used when certain experimental units usually give a higher performance than other units in a replication. For example, in a dairy experiment on cows two rations are to be compared with respect to milk yield. The two periods of lactation (morning and evening) constitute one replication. The morning lactation would always be higher than the evening with respect to milk yield. In order to nullify the effect of inherent variation in the two periods of lactation of the cows, which are even in number, would be divided into two groups at random and the first treatment would be given in the morning lactation for one group and the second treatment would be given at the time of morning lactation for the second group. Let P, Q be two rations which are given to ten cows as shown in Fig. 16.11.

P	Q	P	Q	Q	P	Q	P	Q	P
Q	P	Q	P	P	Q	P	Q	P	Q

Fig. 16.11 Cross over design.

If there are three experimental units in each replication then the number of replications should be multiple of three. In this experiment if the rate of decline of milk yield is not constant then a group of Latin squares of size 2×2 may be more appropriate than the cross over design.

In general for 'r' replications and 't' treatments, the breakdown of the Analysis of variance for a cross over design is given in Table 16.4.

Null Hypothesis: H_0: All the treatment effects are equal, H_1: All the treatment effects are not equal.

TABLE 16.4 ANOVA TABLE

Source	*d.f*	*fcal*	*Ftabd.f*
Columns	r-1	C.M.S/E.M.S	(r–1),(t–1) (r–2)
Rows	t-1	R.M.S/E.M.S	(t–1),(t–1) (r–2)
Treatments	t-1	Tr.M.S./E.M.S	(t–1),(t–1) (r–2)
Error	(t–1) (r–2)		
Total	tr-1		

16.9.1 Missing Plot Technique

If one value is missing in the experiment it is estimated by the expression

$$Y = \frac{rC + t + (R + T) - 2G}{(t - 1)(r - 2)} \quad\quad \text{.... (16.15)}$$

Where C, R be the totals of remaining units in column, row respectively where the missing unit occurs and T be the total of the remaining units for the treatment which is missing.

16.9.2 Group of Latin Squares

In the experiment given in 16.9 if the superiority of one experimental unit over the other is not constant for all replications then the cross over design may not be applicable since the real difference between rows cannot be estimated through cross over design. In such case several Latin squares may be preferable. In the dairy experiment the two cows having same rate of decline in the milk yield for the two lactation periods may be .grouped into Latin squares of size 2×2 as in Fig. 16.12.

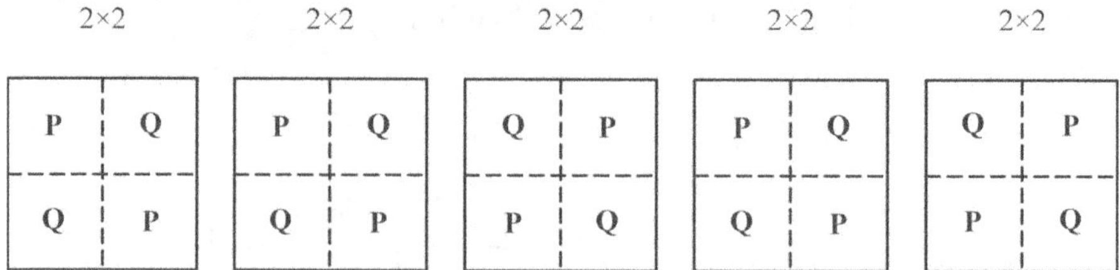

Fig. 16.12 Group of Latin squares.

In general, with 's' Latin squares each of size r x r, the analysis of variance is given in the Table 16:5.

Null Hypothesis: All the treatment effects are equal.

TABLE 16.5 ANOVA TABLE

Source	d.f	fcal	Ftabd.f.
Squares	s-1	S_q.M.S./E.M.S	
Columns within squares	s (r–1)	C_W.M.S./E.M.S	
Rows within squares	s(r-1)	Rw M.S/E.M.S	
Treatments	r–1	T_r.M.S./E.M.S	$(r-1), [s(r-1)^2-(r-1)]$
Error	$s(r-1)^2-(r-1)$		
Total	$s.r^2-1$		

It may be noted that C_W.S.S. and R_W.S.S. can be obtained by adding up of the C.S.S. and .R.S.S for each Latin square respectively. Alternatively these are obtained by subtracting the S_q.S.S. from the column and Row Sum of squares respectively, when all Latin squares taken together. The rest of the sum of squares in Table 16.5 would be obtained as usual.

16.10   FACTORIAL EXPERIMENTS

In this experiment several factors each at different levels are tested for comparison in the layouts of R.B. design, L.S. design, etc.

16.10.1 2^2 experiment

If there are two factors each at two levels then there would be 4 treatment combinations in all and is denoted by 2^2 where the index indicate the number of factors and the number of levels by base. Suppose that the experiment was laid out in R.B. design with 6 replications and the hypothetical mean yields are presented in the following Table 16.6.

TABLE 16.6

		n_o	n_i	Response to n_1
	p_o	10	18	8
	p_1	14	26	12
Response to	p_1	4	8	

From Table 16.6, 8 and 12 are the simple effects of nitrogen at Po and P_1 levels of phosphorus respectively, $\dfrac{8+12}{2}=10$ is the main effect of nitrogen and $\dfrac{12-8}{2}=2$ is kown as is interaction of nitrogen with phosphorus.

and $\dfrac{12-8}{2}=2$ is kown as is interaction of nitrogen with phosphorus.

Similarly 4 and 8 are the simple effects of phosphorus at n_o and n_1 levels of nitrogen respectively. $\dfrac{4+8}{2}=6$ is the main effect of phosphorus and

$\dfrac{8-4}{2}=2$ is the interaction of phosphorus with nitrogen which is same as interaction of nitrogen with phosphorus. In general, $(n_1p_o) - (n_0p_0)$ and $(n_1P_1-n_0P_1)$ are the simple effects of nitrogen. $\dfrac{1}{2}\left[(n_1p_0)-(n_0p_0)+(n_1p_1)-(n_0p_1)\right]$ is

the main effect of nitrogen and $\dfrac{1}{2}\left[\{(n_1p_1)-(n_0p_1)\}-\{(n_1p_0)-(n_0p_0)\}\right]$ is

the interaction of nitrogen with phosphorus. Similarly simple effects and main effect of phosphorus can be defined.

The simple effects of nitrogen and interaction effects are shown in Fig. 16.13 and Fig. 16.14 respectively.

The distances shown in Figs. 16.13 (a) 16.13 (b) and 16.14 with broken lines with proper divisors are the simple effects of nitrogen at P_o, P_1 and interaction effects respectively. Fig.16.14. Interaction of 'N' with 'P'.

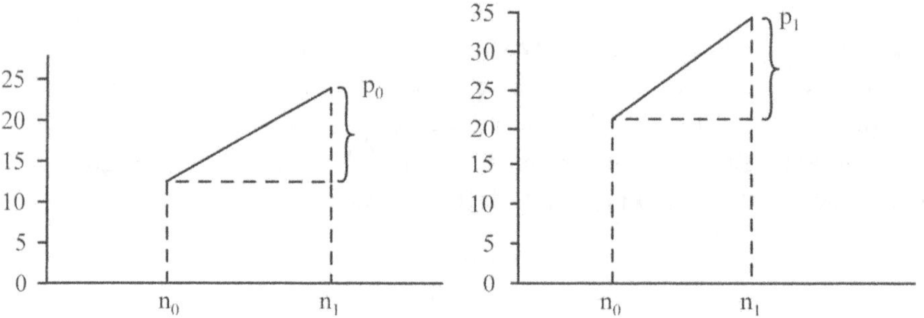

Fig. 16.13(a) Simple effect of 'N' at P_o.   **Fig. 16.13(b)** Simple effect of 'N' at P_1.

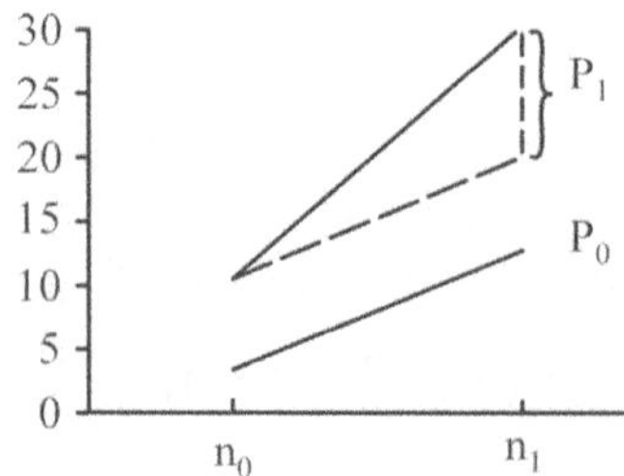

Fig. 16.14 Interaction of 'N' with 'P'.

The partition of d.f. in analysis of variance table for this experiments given in the following Table 16.7.

TABLE 16.7 ANOVA TABLE

source	d.f	Fcal	$F_{tab}(d.f)$
Replications	5		
Treatments	3		
N	1	N.M.S./E.M.S	1,15
P	1	P.M.S./E.M.S	1,15
NP	1	NP.M.S./E.M.S	1,15
Error	15		
Total	23		

The T.S.S. and R.S.S. can be computed in the usual way. The main effects of N and P and interaction NP S.S's are' obtained from the following two-way table 16.8.

TABLE 16.8

	n_o	n_1	Total
p_0	10	18	28
p_1	14	26	40
Total	24	44	68

The data in different cells of the Table 16.8 are the totals of the yields over 6 replications in R,B, design. The sum of squares for main effects of N and P are calculated from the marginal totals and 'the Table S.S. from the cells values is obtained as follows.

$$\text{N.S.S.} = \frac{1}{2 \times 6} \ [(24)^2 + (44)^2] - \frac{(68)^2}{24}$$

$$\text{P.S.S} = \frac{1}{2 \times 6} \ [(28)^2 + (40)^2] - \frac{(68)^2}{24}$$

$$\text{Table S.S.} = \frac{1}{2 \times 6} \ [(10)^2 + (18)^2 + (14)^2 + (26)^2] - \frac{(68)^2}{24}$$

$$\text{NP.S.S.} = \text{Table S.S.} - (\text{N.S.S.} + \text{P.S.S.})$$

Conclusion: If F(calculated) > F(tabulated) with (1,15) d.f. at 'chosen level of significance, the null hypothesis is rejected. Otherwise the null hypothesis is accepted. If the null hypothesis is accepted for NP interaction then it can be assumed that the two factors are independent with respect to yield. Otherwise, the two factors are dependent on each other.

16.10.2 Even Vs odd rule

TABLE 16.9

Factriol effect	Treat.combination				*Divisor*
	(1)	(n)	(p)	(np)	
M	+	+	+	+	4
N	–	+	–	+	2
P	–	–	+	+	2
N P	+	–	–	+	2

In Table 16.9 if the number of common letters between factorial effect and treatment combination and the number of letters in factorial effect are both even or both odd then +; sign, otherwise '-' sign' is written. The sum of different factorial effects are

$$N.S.S. = \frac{1}{2^2 \times 6}\{[np]+[n]-[p]-[1]\}^2$$

$$P.S.S. = \frac{1}{2^2 \times 6}\{[np]-[n]+[p]-[1]\}^2$$

$$NP.S.S. = \frac{1}{2^2 \times 6}\{[np]-[n]-[p]+[1]\}^2$$

Where [np], [n], [p] and [1] are the totals over all replications

Which receive the treatment combination np, n, p and 1 respectively and '6' is the number of replications used in this experiment and (1) is the treatment combination which receive first levels of both the factors.

2^2 FACTORIAL R B D LAY OUT

SPSS DATA ANALYSIS PROCEDURE

The data entry will be done as follows by taking N and P two factors , R as replication and Y as observational factor.

N	P	R	Y
0	0	1	5
1	0	1	8
0	1	1	6
1	1	1	12
0	0	2	7
1	0	2	13
0	1	2	9
1	1	2	17
0	0	3	8
1	0	3	10
0	1	3	9
1	1	3	16

Here the data will be entered column wise as mentioned above. CLICK 'ANALYZE' from the 'TOOL BAR'. Choose 'GENERAL LINEAR MODEL' AND OPT FOR 'UNIVARIATE '. We get a display table. Take 'N',P' 'R' as 'FIXE D FACTORS' AND 'Y' AS DEPENDENT FACTOR' to the Right Hand Side as per indicator. Further CLICK 'MODEL' and select 'CUSTOM' and 'INTERACTION' take 'NP' as interaction to RHS by pressing 'CTRL' KEY ALONG WITH 'N' & 'P' for making 'NP'..Then press 'OPTION 'for getting means of 'N','P' and 'NP'. Finally CLICK 'O.K.' OPTION. We get a table with ANOVA TABLE CONTAINING F-values of 'N','P', NP 'and 'R' and means of them. We can also OPT FOR RESIDUAL PLOTS ETC.,

16.10.3 2^3 Experiment

There will be 3 factors each at 2 levels. So there will be 3 main effects, 3 two factor interactions (first order interactions) and one three factor interaction (second order interaction). The method of analysis of this experiment is similar to 2^2 experiment. However, we shall present here the 'Yates algorithm' by illustration wherein the sum of squares of 'Main effects' and 'interactions' could be obtained readily without resorting to different two-way tables and three-way table as required in the previous method. Yates' method is, however, applicable only to 2^n series of experiments for n $\geq$ 2 .

EXAMPLE: This example represents part of a' Laboratory investigation of the yield of a nitration process, the resulting product forming the base material for a wide range of dyestuffs and medicinal products. This part of the investigation dealt with the effect of the factors, A (time of addition of nitric acid), B(time of stirring), C (effect of heel).

The heel is the residium left behind on the pan by the previous batch. The levels of the factors are A: 3 hours and 8 hours, B : 1 hour and 5 hours and C: absence of heel and presence of heel. Observations were recorded for each combination and presented in the Table 16.10.

TABLE 16.10 PER CENT YIELD FROM NITRATION EXPERIMENT

Time of stirring	No heel time of addition of HNO_3		With heel Time of addition of HNO_3	
	3 hours	8hours	3 hours	8hours
1hour	86.5	88.5	86.5	89.0
	85.0	89.2	83.4	87.2
5hour	82.5	83.2	83.0	83.5
	84.0	85.0	83.8	82.0

Computational procedure: The data are analysed by 'Yates algorithm', The treatment combinations are written in the table systematically by introducing each factor and the combination of that factor with the prevoius combinations as given in Table 16.11.

TABLE 16.11 YATESALGORITHM

Treatments	Treatment total	(1)	(2)	(3)	
(1)	171.5	349.2	683.9	1362.3	G
a	177.7	334.7	678.4	12.9	A
b	166.5	346.1	7.9	−28.3	B
ab	168.2	332.3	5.0	−12.1	C
c	169.9	6.2	−14.5	−5.5	AB
ac	176.2	1.7	−13.8	−2.9	AC
bc	166.8	6.3	−4.5	0.7	BC
abc	165.5	−1.3	−7.6	−3.1	ABC

The column (1) in Table 16.11 is obtanied from the column of treatment totals by adding them successively taking two at a time and subtracting values from the succeeding values taking two at a time. For example,171.5 + 177.7 = 349.2, 166.5 + 168.2 = 334.7 177.7 – 171.5 = 6.2 168.2 = 166.5 = 1.7 etc

The columns (2) and (3) are obtained from columns (1) and (2) respectively on similar lines as in Column (1). For example, the column (2) is obtanied from column (1) as 349.2 + 334.7 == 683.9, 346.1 + 332.3 =: 678.4 etc. 334.7 - 349.2= 14.5 332.3 -346.1 3.8_ etc. The number of column should be as the number of factors in the experiment. In 2^3 experiment three columns are maintained since there are three factors. 'The column (3) gives the Grand total and factorial effect totals in the same order as treatment combinations as indicated in the table. For example, 1362.3, 12.9... 28.3 etc., are the values of G, A, B etc., respectively. The factorial effect means are obtanied by dividing the factorial effect totals with 2^{n-1}. r when n, r are the number of factors and replications respectively. For example, the factorial effect mean of B is

$$\frac{-28.3}{2^{3-1} \times 2} = -3.54, \text{ for } AB = \frac{-12.1}{4 \times 2} = -1.51, \text{ etc.}$$

From Table 16.11 the Sums of squares for 'main effects' and 'interactions' are obtained as follows.

$$\text{A.S.S.} = \frac{(A)^2}{2^3 \times 2} = \frac{(12.9)^2}{16} = 10.40$$

$$\text{B.S.S.} = \frac{(B)^2}{2^3 \times 2} = \frac{(-28.3)^2}{16} = 50.06$$

$$\text{AB.S.S.} = \frac{(AB)^2}{2^3 \times 2} = \frac{(-12.1)^2}{16} = 9.15$$

$$\text{C.S.S} = \frac{(C)^2}{2^3 \times 2} = \frac{(-15.5)^2}{16} = 1.89$$

$$\text{AC.S.S.} = \frac{(AC)^2}{2^3 \times 2} = \frac{(-2.9)^2}{16} = 0.53$$

$$\text{BC S.S.} = \frac{(BC)^2}{2^3 \times 2} = \frac{(0.7)^2}{16} = 0.0306$$

$$\text{ABC.S.S.} = \frac{(ABC)^2}{2^3 \times 2} = \frac{(-3.1)^2}{16} = 0.60$$

$$\text{Replication S.S} = \frac{1}{8}\left[(682.7)^2 + (679.6)^2\right] - \frac{(1362.3)^2}{16} = 0.60$$

$$\text{Total S.S} = 116075.97 - \frac{(1362.3)^2}{16} = 84.64$$

Error S.S. Total S.S – (Repl. S.S + A.SS + B.SS + AB.SS + C.SS + AC.SS + BC.SS + ABC.SS)

TABLE 16.12 ANOVA TABLE

Source	d.f	S.S	M.S.	Fcal
Replications	1	0.60	0.60	0.37
Treatments	7			
A	1	10.40	10.40	6.38*
B	1	50.06	50.06	30.71*
AB	1	9.15	9.15	5.61*
C	1	1.89	1.89	1.16
AC	1	0.53	0.53	0.33
BC	1	0.03	0.03	0.02
ABC	1	0.60	0.60	0.37
Error	7	11.38	1.63	
Total	15	84.64		

*Significant at 5 per cent level.

It may be noted that the same procedure of analysis would be followed for any 2^n experiment where n is the number of factors involved in the experiment. In 2^n experiment there would be n main effects, $\binom{n}{2}$ two factor interactions, $\binom{n}{3}$ three factor interactions, etc. and one 'n' factor interaction.

SPSS DATA ANALYSIS PROCEDURE

2^3 FACTORIAL EXPERIMENT IN RBD

Here N,P,K factors each at 2 levels and replications, R =3.

The data will be entered column wise as follows.

N	P	K	R	Y
0	0	0	1	5
1	0	0	1	8
0	1	0	1	6
1	1	0	1	12
0	0	1	1	9
1	0	1	1	14
0	1	1	1	11

1	1	1	1	15
0	0	0	2	6
1	0	0	2	11
0	1	0	2	10
1	1	0	2	15
0	0	1	2	7
1	0	1	2	13
0	1	1	2	12
1	1	1	2	14
0	0	0	3	9
1	0	0	3	13
0	1	0	3	17
1	1	0	3	20
0	0	1	3	8
1	0	1	3	15
0	1	1	3	14
1	1	1	3	19

Here the data will be entered column wise as mentioned above. CLICK 'ANALYZE' from the 'TOOL BAR'. Choose 'GENERAL LINEAR MODEL' AND OPT FOR 'UNIVARIATE '. We get a display table .Take 'N',P,'K'' 'R'   as 'FIXE D FACTORS' AND 'Y' AS DEPENDENT FACTOR' to the Right Hand Side as per indicator. Further Press 'MODEL' AND OPT FOR 'CUSTOM' and 'INTERACTION' take 'NP,NK, PK, and 'NPK ' as interaction to RHS by pressing 'CTRL' KEY ALONG WITH 'N' OR 'P',K, for making 'NP, NK, PK, and NPK 'interactions..Then press 'OPTION 'for getting means of 'N','P' and 'NP. NK, PK, and NPK'. Finally CLICK 'O.K.' OPTION. We get a table with ANOVA TABLE CONTAINING F-values of 'N','P', NP', NK, PK, and NPK and 'R' and means of them. We can also OPT FOR RESIDUAL PLOTS ETC.,

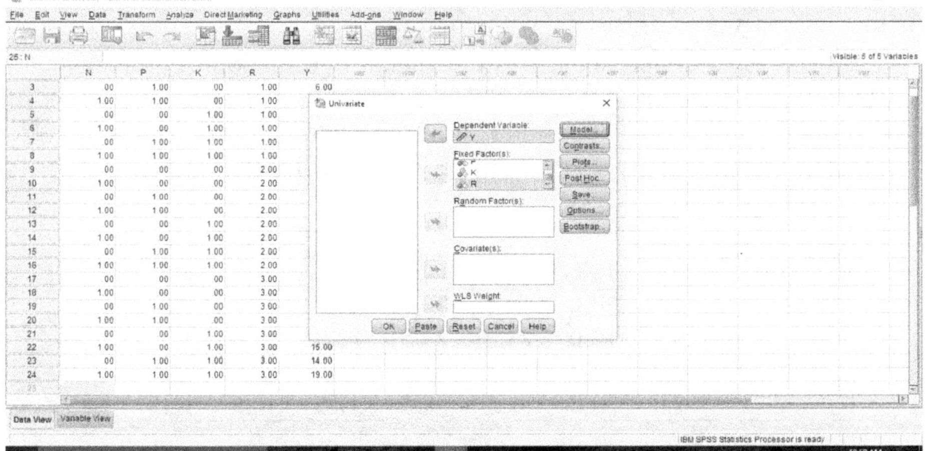

Between-Subjects Factors

		N
N	.00	12
	1.00	12
P	.00	12
	1.00	12
K	.00	12
	1.00	12
R	1.00	8
	2.00	8
	3.00	8

Tests of Between-Subjects Effects

Dependent Variable: Y

Source	Type III Sum of Squares	df	Mean Square	F	Sig.
Corrected Model	325.375[a]	9	36.153	9.273	.000
Intercept	3337.042	1	3337.042	855.913	.000
N	126.042	1	126.042	32.328	.000
P	92.042	1	92.042	23.608	.000
K	15.042	1	15.042	3.858	.070
R	84.083	2	42.042	10.783	.001
N * P	1.042	1	1.042	.267	.613
N * K	.375	1	.375	.096	.761
P * K	3.375	1	3.375	.866	.368
N * P * K	3.375	1	3.375	.866	.368
Error	54.583	14	3.899		
Total	3717.000	24			

16.10.4 3^n Factorial Experiment

In this experiment there are n factors each at 3 levels. In all there will be n main effects, $\binom{n}{2}$ two factor interactions, $\binom{n}{3}$ three factor interactions, etc.

Each main effect will have 2 d.f., each two factor interaction will have $2^2=4$ d.f. each three factor interaction will have $2^3=8$ d.f., etc. Sum of squares of two factor interactions, three factor interactions are obtained from two-way tables, three-way tables respectively in the usual process.

16.10.5 Mixed Factorial Experiment

If the levels of factors are not equal in a factorial experiment then it is known as mixed factorial experiment (asymmetrical factorial experiment). The method of analysis is illustrated with an example.

EXAMPLE: An experiment was conducted in 3 randomized blocks with 3 varieties of sugarcane (V) and 4 levels of Nitrogen (N). The levels of nitrogen are given as 40, 70, 100 and 130 kgs, per acre. The yields (-50) in tons of cane per acre are presented in the following Table 16.13.

TABLE 16.13

Replications	V_1N_1	V_1N_2	V_1N_3	V_1N_4	V_2N_1	V_2N_2	V_2N_3	V_2N_4
1	18.5	20.0	17.5	30.0	6.5	9.0	14.0	15.2
2	16.5	18.0	25.0	23.5	13.0	16.0	3.0	16.5
3	12.0	14.0	11.0	35.0	7.5	5.6	14.5	23.2
Total	47.0	52.0	53.5	88.5	27.0	30.6	31.5	54.9

Table 16.13 Contd...

TABLE 16.13 Contd...

Replications	V_3N_1	V_3N_2	V_3N_3	V_3N_4	Total
1	10.1	16.5	13.6	6.0	176.9
2	12.4	18.0	15.0	7.1	184.0
3	14.0	18.2	7.4	4.4	166.8
Total	36.5	52.7	36.0	17.5	527.7

The sum of squares of 'main effects', 'interactions', etc. are computed as follows:

$$C.F. = \frac{(527.7)^2}{36} = 7735.20$$

Total S.S .(T.S.S) = $[(18.5)^2 + (20.0)^2 + \ldots + (4.4)^2] - 7735.2 = 1715.19$

Replication S.S. (R.S.S)

$$= \frac{1}{12}\left[(176.9)^2 + (184.0)^2 + (166.8)^2\right] - 772.84 = 12.45$$

The two-way table for finding the 'main effects' and 'interaction' sum of squares is

TABLE 16.14

	N_1	N_2	N_3	N_4	Total
V_1	47.0	52.0	53.5	88.5	241.0
V_2	27.0	30.6	31.5	54.9	144.0
V_3	36.5	52.7	36.0	17.5	142.7
	110.5	135.3	121.0	160.9	527.7

$$V.S.S. = \frac{1}{12}[(241.0)^2 + (144.0)^2 + 142.7)^2] - 7735.20 = 529.82$$

$$N.S.S. = \frac{1}{9}[110.5)^2 + (135.3)^2 + \ldots + (160.9)^2] - 7735.20 = 15.8.82$$

Table 16.14 S.S = $\frac{1}{3}(V_1N_1)^2 + (V_1N_2)^2 + \ldots + (V_3N_4)^2.] - C.F.$

where V_iN_j) is the total of V_iN_j treatment combination over all replications.

Table 16.14 S.S·= $\frac{1}{3}[(47.0)^2 + (52.0)^2 + \ldots + (17.5)^2] - 7735.20 = 1261.77$.

$V \times N.S.S.$ = Table 16.14 S.S.– (V.S.S. + N.S.S·) = 573.13

Error S.S. = T.S.S. – [R.S.S + Table (16.14) S .S] = 440.97

TABLE 16.15 ANOVA TABLE

Source	d.f.	S.S	M.S.	Fcal
Replications	2	12.45	6.23	0.31
Treatments	11			
V	2 ⎤	529.82	264.91	13.22**
N	3 ⎬	158.82	52.94	2.64
VN	6 ⎦	573.13	95.52	4.77**
Error	22	440.97	20.04	
Total	35	1715.19		

**Significant at 1 percent level

$$\text{C.D. (varieties)} = t_{tab} \text{ (Crror d.r.)} \times \sqrt{\frac{2(E.M.S.)}{4 \times 3}}$$

$$\text{C.D (V)} = 2.074 \times \sqrt{\frac{2(20.04)}{12}} = 3.80$$

$$\text{C.D. (VN)} = t_{(tab\ error\ d.f.)} \times \sqrt{\frac{2(E.M.S.)}{3}}$$

$$= 2.074 \times \sqrt{\frac{2(20.04)}{3}} = 7.59$$

The means of V as well as VN interaction are given in Table 16.16

TABLE 16.16

	N_1	N_2	N_3	N_4	Mean
V_1	15.67	17.33	17.83	29.50	20.08
V_2	9.00	10.20	10.50	18.30	12.00
V_3	12.17	17.57	12.00	5.83	11.89

C.D (V) = 3.80 and C.D (VN) = 7.59 at 5 per cent level.

Since this experiment was conducted with, nitrogenous fertilizer at equidistant levels, the main effect of nitrogen can, further be divided into linear, quadratic and cubic components each with 1 d.f. and the interaction d.f. can be split up into VN_l, VN_q and VN_c each with 2 d.f. Though the main effect of N is not significant, the procedure of computing linear, quadratic and cubic components for main effect of N is presented here for the sake of illustration, though it is not necessary in the present case. The coefficients for the linear;' quadratic, cubic etc. compents are, obtained from' statistical Tables of Fisher and Yates (1948). In the present example, the coefficients for linear, quadratic and cubic components for nitrogen levels respectively are (–3, –1, +1, +3); (+1, –1, –1+1), and (–1, +3, –3, + 1). Multiplying the values in the first row of Table 16.16 by the coefficients (– 3, –1 +1, – 3) in that order and adding up, the linear component of nitrogen (N_l) for the variety V_1 would be obtained, Similarly N_1 for varieties V_2 and V_3 could be obtained. Multiplying the values in the first row by coefficients (+1–.1, –1,+1) in that order and adding up, the quadratic component of nitrogen (Nq) for the variety V_1 Would be obtained. Similarly N_q for V_2 and V_3 could be obtained, Multiply the values in the first row by the coefficients (–1, +3, – 3, + 1) in that order and adding up, the cubic component (N_c) for V_1 would be obtained. Similarly N_e for V_2 and V_3 could be obtained. The resulting values are presented in the following Table 16.17.

TABLE 16.17

	N_l	Nq	Nc
V_1	126.0	30.0	37.0
V_2	84.6	19.8	25.2
V_3	–73.7	–34.7	31.1
Total	136.9	15.1	93.3

$$N_1.\text{S.S.} = \frac{(136.9)^2}{3 \times 3\left[(-3)^2 + (-1)^2 + (1)^2 + (+3)^2\right]} = 104.12$$

$$N_q.\text{S.S.} = \frac{(15.1)^2}{3 \times 3\left[(+1)^2 + (-1)^2 + (-1)^2 + (+1)^2\right]} = 6.33$$

$$N_c.\text{S.S.} = \frac{(93.3)^2}{3 \times 3\left[(-1)^2 + (+3)^2 + (-3)^2 + (+1)^2\right]} = 48.36$$

$$VN_1.\text{S.S.} = \frac{1}{3\left[(-3)^2 + (-1)^2 + (+1)^2 + (+3)^2\right]}\left[(126.0)^2 + (84.6)^2 + (-73.7)^2\right]$$

$$-\frac{(136.9)^2}{180} = 370.29$$

$$VN_q.\text{S.S.} = \frac{1}{3\left[(+1)^2 + (-1)^2 + (-1)^2 + (1)^2\right]}\left[(30.0)^2 + (19.8)^2 + (-34.7)^2\right]$$

$$-\frac{(15.1)^2}{36} = 201.68$$

$$VN_c.S.S. = \frac{1}{3\left[(-1)^2 + (+3)^2 + (-3)^2 + (+1)^2\right]}\left[(37.0)^2 + (25.2)^2 + (31.1)^2\right]$$

$$-\frac{(93.3)^2}{180} = 1.16$$

The ANOVA Table 16.15 may be rewritten as in Table 16.18

TABLE 16.18 ANOVA TABLE

Source	d.f.	S.s	M.s.	Fcal
Replications	2	12.45	6.23	0.31
Treatments	11			
V	2	529.82	264.91	13.22**
N	3	158.82	52.94	2.64
N_1	1	104.12	104.12	5.20*
N_q	1	6.33	6.33	0.32
N_c	1	48.36	48.36	2.41
VN	6			
VN_1	2	370.29	185.15	9.24**
VN_q	2	201.68	100.84	5.03*
VN_e	2	1.16	0.58	0.03
Error	22	440.97	20.04	
Total	35	1715.19		

* Significant at 5 per cent level.   ** Significant at 1 per cent level.

From Table 16.18 can be observed that N_I, VN_1 and VN_q components were found significant indicating thereby that the yield is expected to increase at the higher doses of nitrogen. This conclusion is not valid since the main. effect of N was not found significant. Further there is significant interaction of varieties with linear and quadratic components of nitrogen.

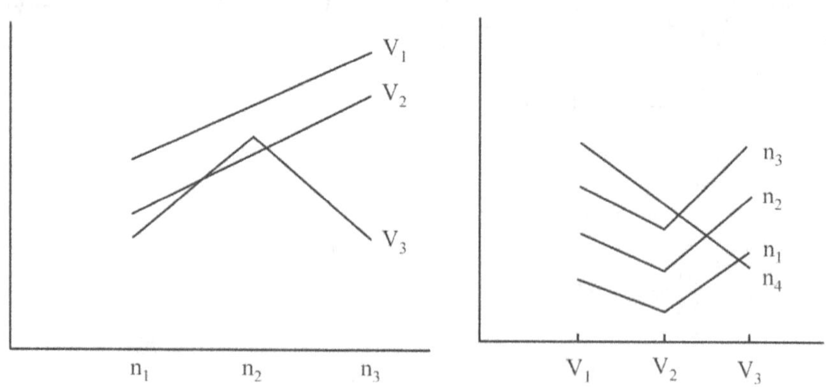

Fig. 16.15(a) Interaction 'V' with 'N'.   **Fig. 16.15(b)** Interaction 'N' with 'V'.

The yields may increase only up to certain stage for the interaction of varieties and nitrogen levels since VN_q is significant. Further examination of the meaning of interaction of V with linear and quadratic components of nitrogen shows that V_1 had a good response with increasing levels of nitrogen whereas V_3 has an increasing trend up to certain level and then decreasing trend for the increasing levels of nitrogen.

MIXED FACTORIAL EXPERIMENT 3 × 3 (N=3,P=3) R(REPLICATIONS) =2

SPSS DATA ANALYSIS PROCEDURE

The data will be entered column wise as follows in the SPSS Excel sheet.

N	P	R	Y
1	0	1	10
1	1	1	13
1	2	1	15
2	0	1	11
2	1	1	9
2	2	1	17
3	0	1	12
3	1	1	19
3	2	1	22
1	0	2	8
1	1	2	14
1	2	2	16
2	0	2	18
2	1	2	24
2	2	2	23
3	0	2	15
3	1	2	18
3	2	2	26

Here the data will be entered column wise as mentioned above. CLICK 'ANALYZE' from the 'TOOL BAR'. Choose 'GENERAL LINEAR MODEL' AND OPT FOR 'UNIVARIATE '. We get a display table .Take 'N',P' 'R' as 'FIXE D FACTORS' AND 'Y' AS DEPENDENT FACTOR' to the Right Hand Side as per indicator. Further Press 'MODEL' AND OPT FOR 'CUSTOM'. & 'INTERACTION' take 'NP' as interaction to RHS by pressing 'CTRL' KEY ALONG WITH 'N' OR 'P' for making 'NP'..Then press 'OPTION 'for getting means of 'N','P' and 'NP'. Finally CLICK 'O.K.' OPTION. We get a table with ANOVA TABLE CONTAINING F-values of 'N','P', NP 'and 'R' and means of them. We can also OPT FOR RESIDUAL PLOTS ETC.

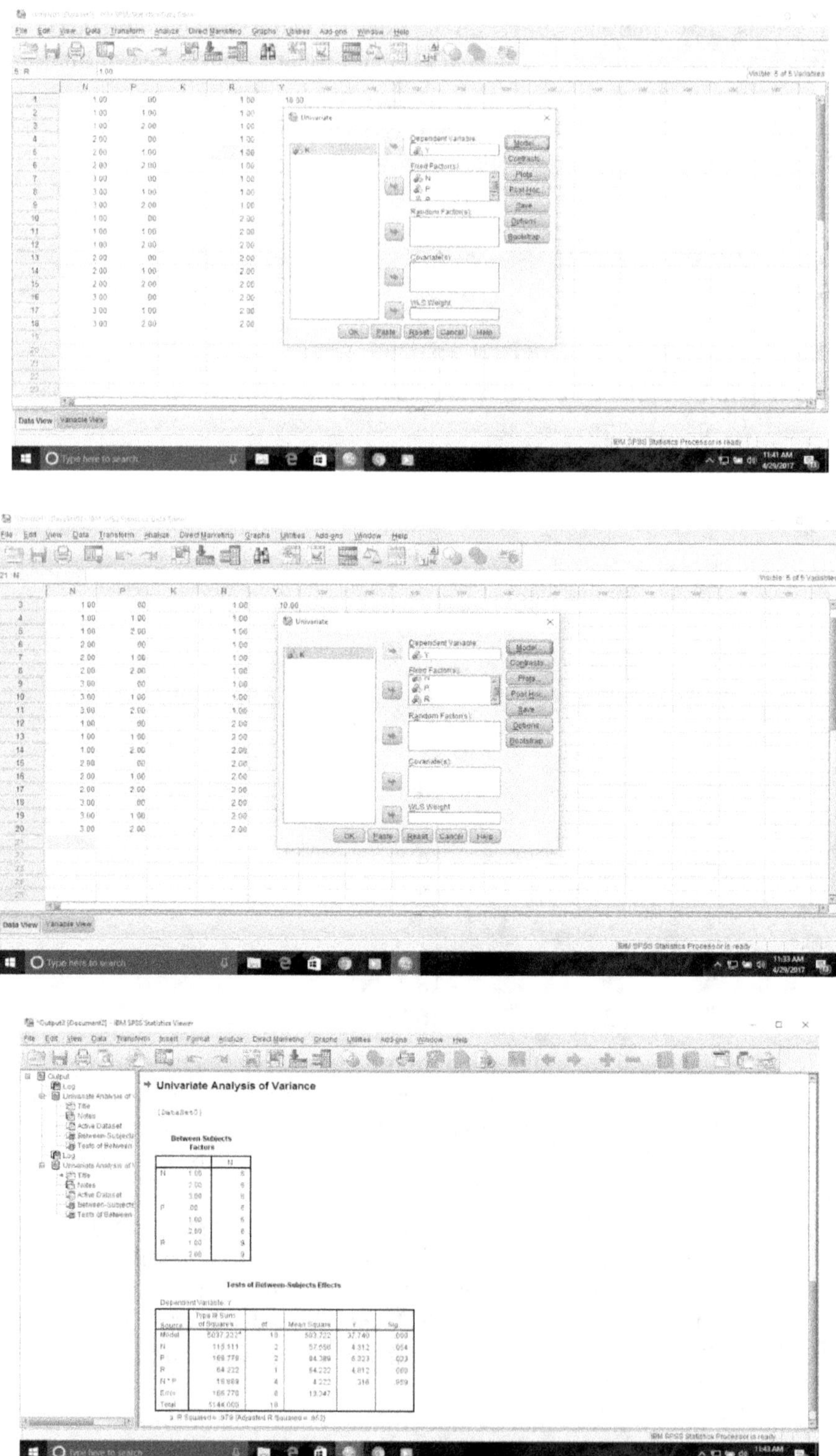

MIXED FACTORIAL EXPERIMENT : 3 × 2 × 2 WITH 2 REPLICATIONS N = 3,P = 2, K = 2, R = 2

SPSS DATA ANALYSIS PROCEDURE

The data will be entered column wise as follows in the SPSS Excel sheet.

N	P	K	R	Y
0	0	0	1	3
1	0	0	1	7
0	1	0	1	5
1	1	0	1	12
0	0	1	1	8
1	0	1	1	14
0	1	1	1	15
1	1	1	1	18
2	0	0	1	13
2	1	0	1	17
2	0	1	1	20
2	1	1	1	24
0	0	0	1	7
1	0	0	1	9
0	1	0	1	8
1	1	0	1	16
0	0	1	1	9
1	0	1	1	18
0	1	1	1	11
1	1	1	1	17
2	0	0	1	11
2	1	0	1	19
2	0	1	1	25
2	1	1	1	24

Here the data will be entered column wise as mentioned above. CLICK 'ANALYZE' from the 'TOOL BAR'. Choose 'GENERAL LINEAR MODEL' AND OPT FOR 'UNIVARIATE '. We get a display table .Take 'N', P,'K'' 'R' as 'FIXE D FACTORS' AND 'Y' AS DEPENDENT FACTOR' to the Right Hand Side as per indicator. Further Press 'MODEL' AND OPT FOR 'CUSTOM'.&'INTERACTION' take 'NP,NK, PK, and 'NPK ' as interaction to RHS by pressing 'CTRL' KEY ALONG WITH 'N' OR 'P',K, for making 'NP, NK, PK, and NPK 'interactions..Then press 'OPTION 'for getting means of 'N','P' and 'NP. NK, PK, and NPK'. Finally CLICK 'O.K.' OPTION. We get a table with ANOVA TABLE CONTAINING F-values of 'N', 'P', NP', NK, PK, and NPK and 'R' and means of them. We can also OPT FOR RESIDUAL PLOTS ETC.

16.10.6 2^3 Confounding

In R.B. design if the number of treatments increases, the size of the block would have to be increased resulting in heterogeneity of the soil within the block which contributes to increase in experimental error. In order to reduce the experimental error in such situations, confounding method was devised by dividing complete replication into incomplete blocks. The division of complete block into incomplete blocks is done in such a way that the certain interaction (preferably higher order Interaction) effect would be made identical to the incomplete block comparisons. In that case the interaction is known to be confounded with block effects. In 2^3 experiment, ABC is the highest order' of interaction and its effect is

$$ABC = \frac{1}{2^2 \times r}(a-1)(b-1)(c-1)$$

$$= \frac{1}{2^2 \times r}(abc - ab - ac + a - bc + c - 1)$$

	R-I		R-II		R-III	
BLOCK	1	2	3	4	5	6
	c	bc	b	ab	bc	c
	abc	(1)	a	ac	ab	b
	b	ab	c	(1)	ac	abc
	a	ac	abc	bc	(1)	a
		ABC		ABC		ABC

Fig. 16.16 2^3 Confounding.

The treatment combinations with positive sign are grouped in one incomplete block and the treatment combinations with negative sign are in another incomplete block as shown in Fig. 16.16.

The difference between totals of two incomplete blocks in each replication in Fig. 16.16 is nothing but an ABC interaction. In other words, ABC interaction is 'totally confounded' with block effects. The information on ABC interaction is lost completely since it was mixed up with the block effects in all the replications. That is why, highest order interactions would generally be confounded since these interactions are generally of less importance as compared to main effects or lower order interactions. 'partial confounding' of certain interactions would be generally adopted instead of total confounding since the interaction would be confounded only in a part of the replications but not in all replications. The information on confounded interaction can still be obtained from the replications where it was not 'confounded'. This method of confounding is known as 'partial confounding'.

16.10.7 2^3 Partial Confounding

Let ABC, AB, AC interactions ,be confounded in 1st, 2nd and 3rd replications respectively and the composition of 2 incomplete blocks in each of the replications is given in Fig. 16.17.

The composition of the blocks will change from replication to replication since the interactions to be confounded are different from replication to replication. In 1st replication ABC is confounded because the difference of totals of 1 and 2 blocks is same as the ABC interaction effect. Similarly are the interaction effects of AB and AC in 2nd and 3rd replications respectively. Since ABC is confounded in 1st replication and not confounded in the remaining replications, it is said to be free from block effects in 2nd and '3rd replications and is partially confounded. The relative information of ABC is of the order of 2/3. The relative informations of AB 'and AC are also equal to 2/3. Unlike in total confounding, the partially confounded interactions would be present in the ANOVA table. The sum of squares due to Replications, Blocks within replications, unconfounded main effects and interactions would be computed in a similar way as was described in Table 16.11 either by Yates' method or through usual method by forming' two-way tables. The sum of squares due to partially confounded interactions are obtained as follows:

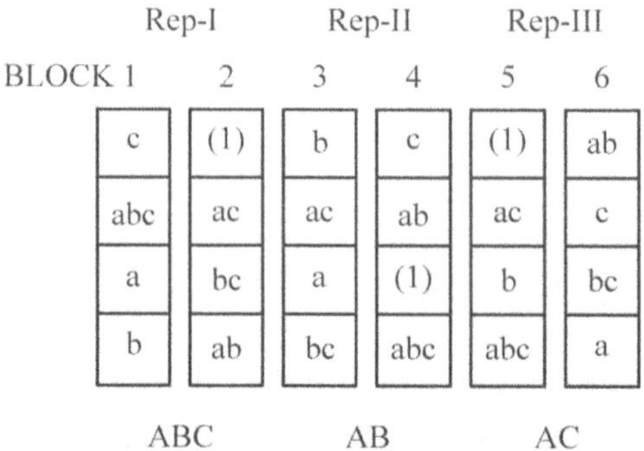

Fig. 16.17 2^3 Partial Confounding.

ABC.S.S. = 1/8 (Sum of the values of the treatments of Block 1 obtained from Replications 2 & 3)^2 + 1/8 (sum of the values of the treatments of Block 2 obtained from Replications 2 & 3)^2 – 1/16 (total of all the values from replications 2 & 3)^2. Similarly AB. S.S = 1/8 (Sum of the values of the treatments of Block 3 obtained from Replications 1 & 3)^2 + 1/8 (sum of the values of the treatments of Block 4 obtained from Replications (1 & 3)^2 – 1/16 (total of all the values from Replications 1 & 3)^2. The sum of squares due to AC· also can be calculated on the lines described above. Alternatively, the sum of squares due to confounded interactions ABC, AB and AC' can be calculated through the Yates method as follows.

ABC. S.S.= 1/16 [The value· in Col. (3) 'of Yates algorithm Table)-(Total of Block 1 - Total of Block 2)^2.

AB. SS = 1/16 [(The value in col (3) of Yates algorithm table) - (Total of Block 4 - Total of Block 3)^2

16.10.8 2^4 Partial Confounding

There would be four factors each at two levels. Since there are in all 16 treatment combinations it would be convenient to form two blocks of 8 treatments for each replication. If there are four blocks in each replication then two interactions can be confounded independently and another interaction (generalized interaction) of the given two would also automatically be confounded. Suppose ACD and BD interactions are confounded in 4 blocks then ABCDD i -.e. ABC is also confounded. The generalized interaction is obtained by writing all the letters together and deleting the letters which are repeated twice,

16.10.9 Das Method

Das (1966) gave method of construction of blocks for confounding any number of interactions, We shall give here the procedure for constructing blocks for $(2^n, 2^k)$ series where n is called the number of factors, k the number of added factors and (n - k) number of basic factors which consists of 2^k blocks in each replication and 2^{n-k} treatment combinations in each block. The levels of each factor are denoted by 0 and 1. The key block would be obtained first and the remaining $(2^k - I)$ blocks would be obtained from the key block. The key block is the one in which the treatment combination consisting of first levels of all the factors occur i.e. (1). The procedure of constructing blocks $(2^4, 2^2)$ experiment is given in Table 16.19 by confounding say, ABC and BD interactions.

16.10.9.1 $(2^4, 2^2)$. Experiment

TABLE 16.19

Independent treat. Combination (n-k)	Basic factor (n-k)		Added factors (k)	
	a	b	c	d
1	1	0	1	0
2	0	1	1	1

In Table 16.19, the independent treatment combinations will be taken as equal to the number of basic factors i.e. n – k = 4 – 2 = 2 and the diagonal matrix will be written underneath the 'basic factors'. The first and second columns under Added factors' will have '0's and' 1's depending upon the interactions to be confounded. In this case the first column refers to ABC and the second column refers to BD. The generalized interaction is ABBCD i.e., ACD. The presence and absence of factors in these columns will be represented by '1' and '0' respectively with the addition of factors 'c' and 'd' under the heading 'Added factors'. The first row in these columns refers to factor A and the second row refers to factor B. The key block confounding ABC, BD and ACD interactions is

[(1), ac, bcd, abd]

where 'ac' refers to the first row of the table, 'bcd' refers to the second row of the table and 'abd' is the multiplication of 'ac' with 'bcd' which is 'abccd' i.e. abd. The other blocks can be constructed from key block as follows.

By multiplying a particular treatment combination, which is not present in the key block to all the treatments in the block new blocks will be obtained. By multiplying 'a' 'b' and 'ab' the new blocks are

(a, c, abcd, bd) ; (b, abc, cd ad) and (ab, bc, acd, d)

Hence, the replication confounding ABC, BD, ACD is given in Fig. 16.18.

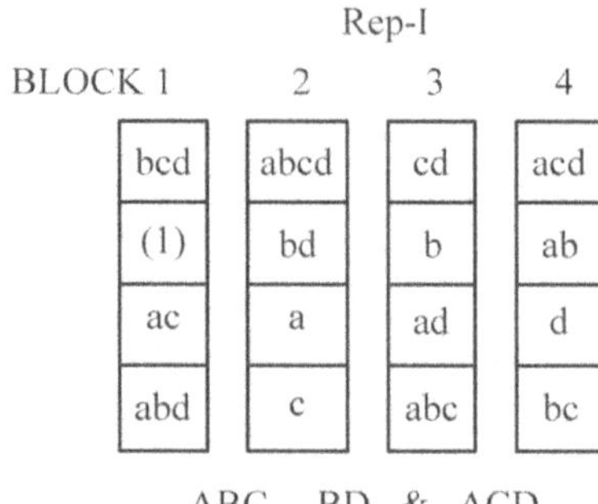

Fig. 16.18 2^4 Confounding.

If AC, AD and CD are to be confounded in the second replication then the composition of key block and other blocks are obtained from the following Table 16.20.

TABLE 16.20

Independent treat. combinations	Basic factors		Added factors	
	a	b	c	d
1	1	0	1	1
2	0	1	0	0

The key block is [(1), acd, b, abcd] and other blocks are obtained by multiplying it with 'a', "c' and d'.

(a, cd, ab, bcd); (c, ad, bc, abd) and (d, ac, bd, abc)

EXAMPLE: To test the effects of nitrogen and phosphorus and their interactions on different varieties and sowing methods, a 2^4 factorial experiment was carried out. There were two replications with four blocks in each. Two different sets of three interactions were confounded in the two replications. The plan and yields are given as

REP.I

Block I	Block II	Block III	Block IV
1100 (36.0)	0000 (23.5)	1000 (27.0)	0100 (30.0)
1010 (30.5)	0110 (29.4)	1110 (42.5)	0010 (20.5)
0001 (26.0)	1101 (31.0)	0101 (18.0)	1001 (30.3)
0111 (24.0)	1011 (32.2)	0011 (13.2)	1111 (27.5)
116.5	116.1	100.7	108.3

REP.II

Block V	Block VI	Block VII	Block VIII
0100 (36.0)	1000 (18.0)	1100 (26.0)	0000 (34.0)
1010 (25.7)	0110 (36.5)	0010 (33.2)	1110 (30.5)
1101 (30.2)	0001 (22.0)	0101 (23.5)	1001 (32.5)
0011 (25.0)	1111 (29.0)	1011 (27.0)	0111 (24.8)
116.9	105.5	109.7	121.8

The figures to the left indicate the levels of the four factors V (varieties), N (nitrogen), P (phosphorus) and S (sowing methods) respectively. (1) Identify the confounded interactions and the corresponding block contrasts. (2) Carry out a complete analysis of the data and interpret the results.

By even Vs odd rule given in Section 16.10.2, the confounded interactions for each replication can be identified. The confounded interactions for Rep. 1 are VNP, VS and NPS and' the corresponding block contrasts are $(- B_1 - B_2 + B_3 + B_4)$, $(- B_1 + B_2 - B_3 + B_4)$ and $(B_1 - B_2 - B_3 + B_4)$ respectively, where B_1, B_2, etc. are the totals of blocks I, II, etc. respectively. The confounded interactions for Rep. 2 are NP, VNS & VPS and the corresponding block contrasts are $(- B_5 + B_6 - B_7 + B_8)$; $(+ B_5 + B_6 - B_7 - B_8)$ and $(- B_5 + B_6 + B_7 - B_8)$ respectively where B_5, B_6, etc. are the totals for Blocks V, VI, etc. respectively.

TABLE 16.21 YATES ALGORITHM

Treats	Rep	Rep	treat	Columns			
	1	2	total	1	2	3	4
1	23.5	34.0	57.5	102.5	230.5	479.3	895.5
v	27.0	18.0	45.0	128.0	248.8	416.2	56.3
n	30.0	36.0	66.0	109.9	213.5	−6.9	54.3
vn	36.0	26.0	62.0	138.9	202.7	63.2	4.7
p	20.5	33.2	53.7	110.8	-16.5	54.5	7.5
vp	30.5	25.7	56.2	102.7	9.6	−0.2	20.3
np	29.4	36.5	65.9	97.4	34.5	13.1	19.5
vnp	42.5	30.5	73.0	105.3	28.7	−8.4	−22.1
s	26.0	22.0	48.0	−12.5	25.5	18.3	−63.1
vs	30.3	32.5	62.8	−4.0	29.0	−10.8	70.1
ns	18.0	23.5	41.5	2.5	−8.1	26.1	−54.7
vns	31.0	30.2	61.2	7.1	7.9	−5.8	−21.5
ps	13.2	25.0	38.2	14.8	8.5	3.5	−29.1
vps	32.2	27.0	59.2	19.7	4.6	16.0	−31.9
nps	24.0	24.8	48.8	21.0	4.9	−3.9	12.5
vnps	27.5	29.0	56.5	7.7	−13.3	−18.2	−14.3

Total S.S. = 1152.58, Blocks S.S. = 86.5

Replication S.S = 4.73

Blocks within replication S.S=Block S.S-Repl. S.S=: 81.77

$$\text{V.S.S} = \frac{(56.3)^2}{2^4 \times 2} = 99.05, \quad \text{S.SS} = \frac{(-63.1)^2}{2^4 \times 2} = 124.43$$

$$\text{N.S.S.} = \frac{(54.3)^2}{2^4 \times 2} = 92.14, \quad \text{NS.SS} \frac{(-54.7)^2}{2^4 \times 2} = 93.50$$

$$\text{VN.S.S} = \frac{(4.7)^2}{2^4 \times 2} = 0.69, \quad \text{PS.SS} = \frac{(-29.1)^2}{2^4 \times 2} = 26.46$$

$$\text{P.S.S.} = \frac{(7.5)^2}{2^4 \times 2} = 1.76, \quad \text{VNPS. S.S} = \frac{(-14.3)^2}{2^4 \times 2} = 6.39$$

$$\text{VP S.S} = \frac{(20.3)^2}{2^4 \times 2} = 12.88$$

$$\text{NP*SS} = \frac{\left[(np) - (-B_5 + B_6 - B_7 + B_8)\right]^2}{2^4 \times 1}$$

$$\text{NP*SS} = \frac{\{19.5 - (-116.9 + 105.5 - 109.7 + 121.8)\}^2}{16} = 22.09$$

$$\text{VNP*.SS} = \frac{\{[vnp] - (-B_1 - B_2 + B_3 + B_4)\}^2}{2^4 \times 1}$$

$$= \frac{\{(-22.1) - (-116.5 - 116.1 + 100.7 + 108.3)\}^2}{16} = 0.14$$

$$\text{VS**.SS} = \frac{\{[vs] - (-B_1 + B_2 - B_3 + B_4)\}^2}{2^4 \times 1}$$

$$= \left\{ \frac{\{(70.1) - (-116.5 + 116.1 - 100.7 + 108.3)\}^2}{16} = 247.28 \right.$$

$$\text{VNS*.SS} = \frac{\{[vns] - (+B_5 + B_6 - B_7 - B_8)\}^2}{2^4 \times 1}$$

$$= \frac{\{(-21.5) - (116.9 + 105.5 - 109.7 - 121.8)\}^2}{16} = 9.61$$

$$\text{VPS*.SS} = \frac{\{[vps] - (-B_5 + B_6 + B_7 - B_8)\}^2}{2^4 \times 1}$$

$$= \frac{\left\{(-31.9) - (-116.9 + 105.5 + 109.7 - 121.8))\right\}^2}{16} = 4.41$$

$$\text{NPS*.SS} = \frac{\left\{[\text{nps}] - (B_1 - B_2 - B_3 + B_1)\right\}^2}{2^4 \times 1}$$

$$= \frac{\left[(12.5) - (116.5 - 116.1 - 100.7 + 108.3)\right]^2}{16} = 1.27$$

Error S.S is obtained by subtraction.

TABLE 16.22 ANOVATABLE

Source	d.f.	S.S	M.S	Fcal
Replications	1	4.73	4.73	0.13
Blocks within Replications	6	81.77	13.63	0.38
Treats	15			
V	1	99.05	99.05	2.75
N	1	92.14	92.14	2.56
VN	1	0.69	0.69	0.02
P	1	1.76	1.76	0.05
VP	1	12.88	12.88	0.36
NP*	1	22.09	22.09	0.61
VNP*	1	0.14	0.14	0.004
S	1	124.43	124.43	3.46
VS*	1	247.28	247.28	6.87
NS	1	93.50	93.50	2.60
VNS*	1	9.61	9.61	0.27
PS	1	26.46	26.46	0.73
VPS*	1	4.41	4.41	0.12
NPS*	1	1.27	1.27	0.04
VNPS	1	6.39	6.39	
Error	9	323.98	35.99	0.18
Total	31	1152.58		

The means for VS* interaction which is found to be significant from Replication 2 is given in Table 16.23

TABLE 16.23

	V_0	V_1
S_0	30.39	29.53
S_1	22.06	29.96

$$C.D = t_{tab..9d.f.} \times \sqrt{\dfrac{2(E.M.S)}{2^2 \times 1}}$$

$$= 2.262 \times \sqrt{\dfrac{2 \times 35.99}{4}} = 9.59$$

16.10.10 3^2 Partial confounding

In this experiment there are two factors each at three levels and in all there will be 9 treatment combinations in each replication. If each replication is sub-divided into three incomplete blocks of three treatments each then an interaction of 2 d.f. can be confounded in each replication. Supposing that nitrogen and phosphorus each at three levels are denoted by 0, 1 and 2. The NP interaction will have 4 d.f. which will be partitioned into NP (I) and NP (J) each with 2 d.f. so that NP (I) can be confounded in one replication and NP (J) in another replication. It may be noted that this experiment does not need confounding 'any interaction but the method illustrated here is a general one.' The I and J-components of NP interaction are obtained from Table 16.24.

TABLE 16.24

	p_0	p_1	p_2
n_0	00	01	02
n_1	10	11	12
n_2	20	21	22
n_0	00	01	02
n_1	10	11	12

The composition of blocks and replications will be as follows. The differences among I-components are confounded with block comparisons in Rep. I and the differences among J-components are confounded with block comparisons in Rep. II.

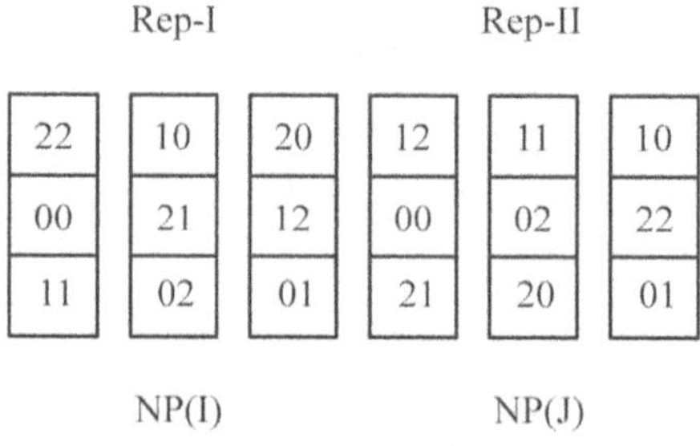

Fig. 16.19 3^2 Confounding.

16.10.11 3^3 Partial Confounding

In this experiment there would be three factors each at three levels. Since there would be in all 27 treatment combinations in a single replication it would be appropriate to confound three factor interaction by dividing each replication into 3 incomplete blocks of 9 treatments each. The three factor interaction having 8 d.f. would be divided into 4 components such as ABC, ABC^2, AB^2C and AB^2C^2 or ABC (I-I), ABC (I-J), ABC (J-I) and ABC (J-J) each with 2 d.f. so that each part of the ABC interaction can be confounded in each replication. 3^3 partial confounding design is widely used in field experiments.

1st Method: The construction of blocks for each replication is on the similar lines of 3^2 partial confounding experiment given in subsection 16.10.10.

2nd Method: The construction of blocks for confounding the four parts of ABC interaction would be done by a general method through group theory. Let x_1, x_2 and x_3 be the levels of three factors A,B and C respectively then the elements of blocks for confounding ABC are obtained from the equation

$$x_1 + x_2 + x_3 = 0 \bmod 3$$

That is, the levels of three factors satisfying the three equations $x_1 + x_2 + x_3 = 0$, $x_1 + x_2 + x_3 = 1$ and $x_1 + x_2 + x_3 = 3$ would form the elements of three blocks confounding ABC part of three factor interaction. Let 0, 1 and 2 be the levels of each factor then the elements of block satisfying $x_1 + x_2 + x_3 = 0$ are (000, 012, 021, 102, 111, 120, 201, 210, 222). The above elements were obtained by substituting the levels of factors in the equation. If the, total value is 3 or multiple of 3 then it is taken as zero otherwise the remainder would be considered by dividing by 3. For example, if $x_1 = 0$, $x_2 = I$, $x_3 = 2$ then $x_1 + x_2 + x_3 = 3$ i.e. o. Similarly if $x_1 = 2$, $x_2 = 2$ and $x_3 = 2$ then $x_1 + x_2 + x_3 = 6$, i.e., o.

The other two blocks confounding ABC part of three factor interaction are obtained by considering equations

$$x_1 + x_2 + x_3 = 1 \bmod 3$$

(001, 010, 022, 100, 112, 121, 202, 211, 220) and $X_1 + X_2 + X_3 = 2 \bmod 3$ (002, 020, 011,101, 110,122, 200, 212, 221)

The composition of blocks for confounding ABC^2 part of ABC interaction are obtained from the equations

$$x_1 + x_2 + 2x_3 = 0 \bmod 3$$

$$1 \text{ ''}$$

$$2 \text{ ''}$$

$$(000, 011, 022, 101, 112, 120, 202, 210, 221)$$

$$(002, 010, 021, 100, 111, 122, 201, 212, 220)$$

$$(001, 012, 020, 102, 110, 121, 200, 211, 222)$$

The composition of blocks for confounding AB^2C part of the ABC interaction are obtained from the equations

$$x_1 + 2x_2 + x_3 = 0 \bmod 3$$

$$1 \text{ ”}$$

$$2 \text{ ”}$$

$$(000, 011, 022, 102, 110, 121, 201, 212, 220)$$

$$(001, 012, 020, 100, 111, 122, 202, 210, 221)$$

$$(002, 010, 021, 101, 112, 120, 200, 211, 222)$$

The composition of blocks for confounding AB^2C^2 are obtained from the equations

$$x_1 + 2x_2 + 2x_3 = 0 \bmod 3$$

$$1 \text{ ”}$$

$$2 \text{ ”}$$

$$(000, 012,021, 101, 110, 122, 202, 211, 220)$$

$$(002, 011, 020, 100, 112, 121, 201, 210, 222)$$

$$(001, 010, 022, 102, 111, 120, 200, 212 , 221)$$

EXAMPLE: An experiment was conducted to test the levels of nitrogen, phosphorus and potash each at three levels in the layout of 3^3 partial confounding by confounding NPK interaction partially. The yields of paddy (kgs) were recorded in parentheses as follows:

Rep. I				Rep. II			
120(16)	022(7)	101(12)		010(9)	022(5)	121(15)	
012(10)	211(12)	110(14)		212(19)	202(16)	001(8)	
210(18)	001(6)	011(8)		111(12)	101(14)	200(20)	
000(2)	202(12)	200(15)		201(11)	000(6)	110(17)	
111(9)	112(14)	221(19)		122(13)	120(19)	222(13)	
021(8)	010(8)	020(11)		100(8)	011(13)	012(8)	
102(12)	121(13)	212(16)		002(6)	221(18)	102(11)	
201(13)	100(9)	002(9)		220(18)	210(15)	211(14)	
222(11)	220(17)	122(11)		021(7)	112(15)	020(13)	
99	98	115	312	103	121	119	343

NPK Rep. III				NPK Rep. IV		
110(13)	012(17)	021(6)		112(13)	022(7)	110(16)
212(19)	202(16)	222(15)		210(16)	010(11)	220(19)
121(15)	100(11)	010(8)		011(12)	221(17)	021(8)
220(21)	122(14)	120(18)		201(17)	111(13)	122(14)
000(5)	020(13)	200(22)		222(20)	200(21)	211(16)
102(15)	210(21)	112(13)		002(7)	102(10)	101(15)
201(16)	001(8)	211(16)		020(12)	212(20)	000(7)
011(10)	221(20)	002(7)		100(9)	001(9)	012(9)
022(6)	111(14)	101(15)		121(14)	120(17)	202(13)
120	134	120	374	120	125	117	362

Fig. 16.20

$$C.F. = \frac{(1391)^2}{108} = 17915.56$$

Block S.S $1/9[(99)^2+(98)^2+ \ldots +(117)^2]- 17915.56 =141.22$.

$$\text{Replication S.S.} = \frac{1}{9 \times 3} [(312)^2+ (343)^2+ \ldots +(362)^2] - 17915.56 = 81.22$$

Blocks within replication S.S. = Block S.S - Replication S.S. =60.00

For computing the main effects and two factor interactions the following two-way tables are formed.

TABLE 16.25

	p_0	p_1	p_2	
n_0	80	123	103	306
n_1	141	163	179	483
n_2	192	202	208	602
	413	488	490	1391

$$N.S.S = \frac{1}{4 \times 9} \left[(306)^2 + (483)^2 + (602)^2 \right] - 17915.56 = 1232.47$$

$$P.S.S. = \frac{1}{4 \times 9} \left[(413)^2 + (488)^2 + (490)^2 \right] - 17915.56 = 107.02$$

$$\text{Table (16.25) S.S.} = \frac{1}{4 \times 3} \left[(80)^2 + (123)^2 + \ldots + (208)^2 \right] - 17915.56$$

$$= 1381.19$$

NP.S.S = Table (16.25) S.S. – (N.S.S. + P.S.S.) = 41.70

TABLE 16.26

	k_0	k_1	k_2	
p_0	135	144	134	413
p_1	166	149	173	488
P_a	194	160	136	490
	495	453	443	1391

$$\text{K.S.S.} = \frac{1}{4\times9}\left[(495)^2 + (453)^2 + (443)^2\right] - 17915.56 = 42.30$$

Table (16.26) S.S

$$= \frac{1}{4\times9}\left[(135)^2 + (149)^2 + ... + (136)^2\right] - 17915.56 = 279.02$$

PK.S.S = Table (16.26) S.S. – (P.S.S + K.S.S) = 129.70

TABLE 16.27

	k_0	k_1	k_2	
n_0	105	103	98	306
n_1	167	161	155	483
n_a	223	189	190	602
	495	453	443	1391

$$\text{Table (16.27) S.S} = \frac{1}{4\times3}\left[(105)^2 + (103)^2 + .1.. + (190)^2\right]$$

$$- 17915.56 = 1303.02$$

NK.S.S = Table (16.27) S.S. –(N.S.S+ K.S.S) = 28.25

Since NPK component of NPK interaction was confounded in 1st replication and free from confounding in the other three replications in Fig. 16.20. For example, the total of the treatment combinations in Block 1 of Rep. I (i.e. 120, 012, 210, 000, 111, 021, 102, 201, 222) would be obtained from Rep. II, III and IV.

$$\text{NPK S.S} = \frac{1}{27}\left[(346)^2 + (333)^2 + (400)^2\right] - \frac{(346+333+400)^2}{27\times3}$$

$$= 93.51$$

Similarly S.S. for NPK^2, NP^2K and NP^2K^2 are obtained from the replications I, III and IV; I, II and IV and I, II and III using the block compositions of replications II, III and IV respectively as

$$\text{NP }K^2\text{ S.S} = \frac{1}{27}\left[(334)^2 + (349)^2 + (365)^2\right] - \frac{(334+349+365)^2}{27\times3}$$

$$= 17.80$$

$$NP^2KS.S = \frac{1}{27}[(339)^2 + (328)^2 + (350)^2$$

$$-(339 + 328 + 350)^2 / (27 \times 3) = 8.96$$

$$NP^2K^2S.S = \frac{1}{27}[(336)^2 + (364)^2 + (329)^2 - \frac{(336 + 364 + 329)^2}{27 \times 3} = 25.41$$

Total S.S = 2077.44

Error S.S is obtained by subtracting all the remaining sum of squares from Total S.S.

In the case of three factor interaction, the treatment means are adjusted with block effects. In order to adjust the treatment mean, the block effects are adjusted as follows. The least squares effect of any block effect can be computed as 1/27 [(4 block total) - (total of treatments appearings in that block)]. For example, the block effect of 1st block in the 1st replication is given as from Fig. 16.20 we have 1/27 [4(99) – (445)] = – 1.81.

TABLE 16.28 ANOVA TABLE

Source	d.f.	S.S	M.S.	Fcal
Replications	3	81.22	27.07	8.96**
Blocks(within) replications	8	60.00	7.50	2.48**
N	2	1232.47	616.24	204.05**
P	2	107.02	53.51	17.72**
NP	4	41.70	10.43	3.45**
K	2	42.30	21.16	7.01**
NK	4	28.25	7.06	2.34
PK	4	129.70	32.43	10.74**
NPK	2'	93.51	46.76	15.48**
NPK^2	2'	17.80	8.90	2.94**
NP^2K	2'	8.96	4.48	1.48
NP^2K^2	2'	23.41	11.71	3.88*
Error	70	211.08	3.02	
Total	107	2077.44		

**Significant at 1 per cent level, *Significant at 5 per cent level.

Similarly the estimates of block effects for other blocks were computed and are presented in the following Table 16.29

TABLE16.29 BLOCKS EFFECTS
REPLICATIONS

		1	2	3	4
	1	−1.81	−0.93	0.78	0.89
Block	2	−1.44	0.52	2.74	0.41
	3	−2.04	−0.30	0.37	0.81

For finding the adjusted treatment totals, the sum of the blocks effects in which a treatment appears is subtracted from the corresponding treatment total.

The adjusted total of $n_2p_1k_0$ is

$$70 - (-1.81 + 0.52 + 2.74 + 0.89) = 67.66$$

where the values in parenthesis are the blocks effects of (Rep. I, block 1), (Rep. II, block 2), (Rep. III, block 2), (Rep. IV, block 1) respectively since $n_2p_1k_0$ appears in these blocks.

The adjusted mean of $n_2p_1k_0$ is $\dfrac{67.66}{4} = 16.92$

Similarly all the adjusted treatment means can be calculated.

16.11 SPLIT-PLOT DESIGN

Treatments requiring large experimental material in combination with treatments requiring less experimental material are to be tested, split plot design is used. In this design the experimental units would be further sub-divided into sub units so that more treatments could be tested within each main treatment. For example if different levels of nitrogen are to be tested at each and different levels of moisture depletion, then the depletion levels will be tested in main plots and levels of nitrogen in subplots within each main plot. The additive model of Analysis of variance for this design is

$$Y_{ijk} = \mu + \alpha_I + \beta_j + E_{ij} + \gamma_\kappa + (\alpha\gamma)_{ik}, + \delta_{ijk} \qquad .....(16.16)$$

where Y_{ijk} be the observational value on the k-th sub-treatment in i-th main-treatment belonging to j-th replication for $i = 1, 2, ..., m$; $j = 1, 2, ... , r$; $k = 1, 2, ...,' n$, $E_{ij} \sim N(0, \sigma_\propto)$; $\delta_{ijk'} \sim N(0, \sigma)$ and E_i's are experimental errors for main plots and δ_{ijk}'s are experimental errors for sub plots. The lay out of split plot design with 4 levels of depletion, 5 levels of nitrogen and 4 replications is given in Fig. 16.21. The levels of depletion are denoted by d_0, d_1 d_2 and d_3 and nitrogen by n_0, n_1, n_2, n_3 and n_4

Rep. I

d_3	d_0	d_2	d_1
n_2	n_0	n_3	n_2
n_4	n_2	n_4	n_0
n_3	n_1	n_0	n_3
n_0	n_3	n_2	n_4
n_1	n_4	n_1	n_1

....

Rep. IV

d_2	d_1	d_3	d_0
n_4	n_0	n_2	n_3
n_0	n_3	n_4	n_1
n_2	n_1	n_0	n_0
n_1	n_4	n_1	n_4
n_3	n_2	n_3	n_2

Fig. 16.21 Split plot design.

The depletion levels are randomized in main plots and the nitrogen levels are randomized in sub plots in each replicate. It is obvious that the depletion levels effects would be compared with less precision and the nitrogen levels would be compared with a more precision since the experimental errors for each sub plot would be added for the main treatments comparison and subtracted for the sub-treatments comparison. The experimental error for main treatments will be usually larger than the experimental error for sub treatments and for interaction of main and sub-treatments. It can accommodate more number of treatments in each replication than randomized block design by reducing the precision of comparison of main treatments and increasing the efficiency of comparison of sub treatments and interaction. The moisture levels, dates of sowing, varieties of a crop, milching machines in dairy experiments, plant protection equipment in Entomology and Pathology experiments, dairy animal in feeding experiment, storage models in post-harvest experiments, etc., would require large experimental material and therefore they would be used as main treatments in this design analysis.

In conducting experiments on grasses, the yield cuttings at different intervals of time for the same variety of grass are taken as sub-treatments and different varieties of grasses as main treatments. Different types of pressure cookers can be compared as main treatments and different varieties of rice cooked by each pressure cooker as sub treatments with respect to taste, appearance, donness, etc. Different edible oils are compared as main treatments and different recipes prepared in each medium of edible oils as sub treatments with respect to loss of nutrients, .etc. Different types of cloth can be compared as main 'treatments and different types of garments prepared from each type of cloth as sub treatments with respect to durability, expenditure involved, etc.

This design can also be viewed as a confounded design where main effects are confounded with main plots. Here the main plots are considered as incomplete blocks and the differences between main treatments are same as the differences between incomplete blocks and the sub treatments and interactions are free from block effects. Let M be the main treatment with m' levels and N be the sub-treatment with n' levels having 'r' replications, the different sum of squares are computed as follows.

$$'C.F = G^2/r.m.n$$

where G is the grand total and r, m, n be the number of replications, number of main treatments and number of sub treaments respectively.

$$T.S.S. = \sum_{i.j.k} Y_{ijk}^2 - C.F$$

Where Y_{ijk} is the observational value for (i, j, k)-th experimental unit and the 'Σ'extends over all the experimental units.

$$R.S.\ S = \frac{1}{m.n}(R_1^2 + R_2^2 + ... + R_r^2) - C.F.$$

Where R_i be the total of i-th replication.. The Error (1) S.S. is obtained from the following two-way table of replications and main treatments.

TABLE 16.30 REPLICATIONS

Main treats	1	2	...r	
1	m_1r_1	m_1r_2	m_1r_r	M_1
2	m_2r_1	m_2r_2	m_2r_r	M_2
:	:	:	:	:
m	m_mr_1	m_mr_2	m_mr_r	M_m
	R_1	R_2	R_r	G

Main treatment S.S. (M.S.S) = $\dfrac{1}{r.n}(M_1^2 + M_2^2 + ... + M_m^2) - C.F.$

Table 16.30 S.S. = $1/n\,[(m_1r_1)^2 + (m_1r_2)^2 + ... + (m_mr_r)^2] - C.F.$

Error (1) S.S, (E$_1$.S.S.) = Table 16.30 S.S –(R.S.S+M.S.S.). The main treatment × sub treatment S.S is computed by considering the following two-way table of sub and main treatments.

TABLE 16.31 MAIN TREATMENTS

Sub treats	1	2	...	m	
1	n_1m_1	n_1m_2	...	n_1m_m	N_1
2	n_2m_1	n_2m_2	...	n_2m_m	N_2
:	:	:	:	:	:
n	n_nm_1	n_nm_2	...	n_nm_m	N_n
	M_1	M_2	...	M_m	G

Sub-treatment S.S, (N.S.S) = $\dfrac{1}{r.m}\left[(N_1)^2 + (N_2)^2 + ... + (N_n)^2\right] - C.F.$

Table 16.31 S.S = $\dfrac{1}{r}\left[(n_1m_1)^2 + (n_1m_2)^2 + ... + (n_nm_m)^2\right] - C.F.$

Sub treat X Main treat S.S, (MN.SS) = Table 16.31 S.S-(N.S.S +M.S.S)
Error (2) S.S. (E$_2$.S.S) = Total S.S-(Table 16.30 S.S+N.S.S +MN.SS).

TABLE 16.32 ANOVA TABLE

Source	d.f	F_{cal}	F_{tab}d.f
Replications	r–1		
M	m–1	M.M.S/E$_1$,M.S	(m–1), (r–1)(m–1)
Error 1	(r–1)(m–1)		
N	n–1	N.M.S/E$_2$,M.S	(n–1), (m (r–1)(n–1)
MN	(m–1)(n–1)	MN.M.S/E$_2$,MS	(m–1) (n–1), m(r–1) (n–1)
Error 2	m (r–1)(n–1)		
Total	rmn-1		

The standard errors for the comparison of the different treatment means are presented in the Table 16.33

TABLE 16.33

Difference between	*Notation*	*S.E.*
Two M means	$\overline{mi} - \overline{mj}$	$\sqrt{\dfrac{2E_1}{r.n}}$
Two N means	$\overline{ni} - \overline{nj}$	$\sqrt{\dfrac{2E_2}{r.m}}$
Two N means at the same level of M	$\overline{minj} - \overline{m_in_k}$	$\sqrt{\dfrac{2E_2}{r}}$
Two M means at same or different levels on N	$\left.\begin{array}{c}\overline{min\,j} - \overline{m_kn_j}\\[4pt]\overline{min\,j} - \overline{m_kn_k}\end{array}\right\}$	$\sqrt{\dfrac{2\left[(n-1)E_2 + E_1\right]}{r\,n}}$

where E_1, E_2 are E_1 M.S and E_2.M.S·respectively

In the case of treatment means of two main treatments at the same level of sub-treatment or different levels of sub-treatment, the ratio of the difference between two treatment means and its S.E. does not follow Student's t-distribution. Let t_1, t_2 be the tabulated values of t corresponding to d.f, for E_1 and E_2 respectively then the tabulated value of t for the above said comparison is

$$t^1 = \frac{(n-1)E_2t_2 + E_1t_1}{(n-1)E_2 + E_1} \qquad \qquad .....(16.17)$$

16.11.1

It may be noted that the split plot design can also be laid out in Latin square design by making the number of whole units equal to the number of replications which are in turn equal to number of rows and columns.

EXAMPLE: An experiment was conducted to test the 4 methods of application of fertilizer in main plots and nitrogen at 3 levels, phosphorus and potash at 2 levels each' in sub plots in split-plot design for sugar cane crop. The yields for 4 replications are presented here. The methods are denoted by m_o, m_1, m_2, and m_3, nitrogen levels by n_o, n_1, n_2 phosphorus by P_o, P_1 and potash by k_o, k_1.

REP.I

m_2	m_0	m_3	m_1
(20)110	010(11)	200(22)	101(19)
(16)100	100(15)	001(15)	201(21)
(18)200	111(19)	101(18)	010(16)
(17)201	211(23)	111(26)	210(28)

REP.I Contd...

(12)010	000(4)	210(29)	211(30)
(6)000	101(13)	100(22)	001(13)
(24)210	201(15)	000(12)	110(24)
(12)011	001(8)	110(25)	011(16)
(15)101	210(20)	211(32)	000(13)
(22)111	011(10)	010(18)	100(20)
(26)211	110(18)	011(14)	200(23)
(10)001	200(16)	201(20)	111(24)
198	172	253	247 870

REP.II

m_1	m_3	m_0	m_2
(22)201	(17)010	(18)200	(13)001
(26)110	(13)000	(14)101	(17)100
(23)100	(19)101	(12)010	(16)200
(31)210	(24)201	(10)001	(10)010
(15)010	(27)111	(23)210	(9)000
(12)001	(34)211	(7)000	(20)210
(29)211	(16)001	(25)211	(17)101
(19)011	(32)210	(20)111	(14)001
(26)200	(24)200	(16)100	(23)211
(11)000	(22)110	(18)201	(19)110
(18)101	(15)011	(22)110	(15)201
(26)111	(25)100	(12)011	(24)111
258	268	197	197 920

REP.III

m_1	m_0	m_3	m_2
(18)011	(8) 000	(26)200	(26)211
(25)110	(13)011	(33)210	(15)200
(27)111	(16)201	(14)011	(20)110
(25)201	(26)211	(13)001	(13)010
(31)211	(14)100	(21)101	(22)210
(24)100	(16)200	(31)211	(14)001
(14)010	(9) 001	(11)000	(18)100
(30)210	(24)110	(23)201	(25)111
(19)101	(20)210	(28)111	(18)201
(24)200	(14)010	(15)010	(17)011
(13)001	(19)111	(27)100	(19)101
(12)000	(13)101	(24)110	(11)000
262	192	266	218 938

REP.IV

m_3	m_0	m_1	m_2
(25)111	(14)011	(20)101	(20)210
(29)100	(20)111	(16)001	(13)200
(14)010	(15)201	(32)210	(16)011
(21)201	(28)211	(27)110	(17)001
(19)101	(10)001	(17)010	(28)111
(13)000	(26)110	(24)201	(24)211
(17)011	(22)210	(32)211	(16)201
(30)211	(11)000	(26)111	(20)100
(25)200	(13)100	(10)000	(12)000
(10)001	(15)200	(19)011	(19)110
(36)210	(13)010	(25)200	(21)101
(25)110	(15)101	(28)100	(13)010
264	202	276	219 961

$$\text{C.F.} = \frac{(3689)^2}{4 \times 4 \times 12} = 70878.76$$

$$\text{Total S.S} = \left[(20)^2 + (16)^2 + \ldots + (13)^2 \right] - \text{C.F} = 7870.24$$

Replication S.S, (R.S.S) =

$$\frac{1}{4 \times 12} \left[(870)^2 + (920)^2 + \ldots + (961)^2 \right] - 70878.76$$

$$= 93.43$$

From two-way table of replications and methods of application, we have

$$\text{M.S.S} = \frac{1}{4 \times 12} \left[(763)^2 + (1043)^2 + \ldots + (1051)^2 \right] - 70878.76 = 1347.14$$

$$\text{Table S.S.} = 1/12 \left[(172)^2 + (197)^2 + \ldots + (264)^2 \right] - 70878.76 = 1474.32$$

Error(1) S.S = Table S.S - (R.S.S + M..S.S) = 33.75

The two-way table of methods of application and 'npk' fertilizer is given in Table 16.34.

$$\text{Fertilizer S.S} = \frac{1}{4 \times 4} \left[(163)^2 + (199)^2 + \ldots + (450)^2 \right] - 70878.76$$

$$= 5533.43$$

The fertilizer S.S can be further split into main effects and inter actions as follows.

TABLE 16.34

Fertilizer	Method of application				Total
	m_0	m_1	m_2	m_3	
$n_0p_0k_0$	30	46	38	49	163
$n_0p_0k_1$	37	54	54	54	199
$n_0p_1k_0$	50	62	48	64	224
$n_0p_1k_1$	49	72	59	60	240
$n_1p_0k_0$	58	95	71	103	327
$n_1p_0k_1$	55	76	72	77	280
$n_1p_1k_0$	90	102	78	96	366
$n_1p_1k_1$	78	103	99	106	386
$n_2p_0k_0$	65	98	62	97	322
$n_2p_0k_1$	64	92	66	88	310
$n_2p_1k_0$	85	121	86	130	422
$n_2p_1k_1$	102	122	99	127	450
	763	1043	832	1051	3689

From two-way table for 'nitrogen' and 'phosphorus', we have

$$\text{N.S.S.} = \frac{1}{4\times4\times2\times2}\left[(826)^2 + (1359)^2 + (1504)^2\right] - 70878.76$$

$$= 3983.32$$

$$\text{P.S.S} = \frac{1}{4\times4\times3\times2}\left[(1601)^2 + (2088)^2\right] - 70878.76$$

$$= 1235.25$$

$$\text{Table S.S.} = \frac{1}{4\times2\times2}\left[(362)^2 + (464)^2 + ... + (872)^2\right] - 70878.76$$

$$= 5374.40$$

NP. S.S.= Table S.S– (N.S.S.+P.S.S) = 155.83

From two-way table of 'nitrogen' .and 'potash', 'we have

$$\text{K.S.S.} = \frac{1}{4\times4\times3\times2}\left[(1824)^2 + (1865)^2\right] - 70878.76$$

$$= 8.75$$

$$\text{Table S.S} = \frac{1}{4\times4\times2}\left[(387)^2 + (439)^2 + ... + (760)^2\right] - 70878.76$$

$$= 4040.96$$

(N.K.S.S+) = Table S.S– (N.S.S + K.S.S) = 48.89

From two-way table of 'phosphorus' and 'potash', we have

$$\text{Table S.S} = \frac{1}{4 \times 4 \times 3}\left[(812)^2 + (789)^2 + ... + (1076)^2\right] - 70878.76$$

$$= 1283.43$$

P K.S.S = Table S.S. – (P.S.S + K.S.S) = 39.43

NPK.S.S = Fertilizer S.S – (N.S.S + P.S.S + K.S.S + NP.S.S + NK.S.S + PK.S.S) = 61.96

From two-way table of 'methods' and 'nitrogen', we have

MN.S.S = Table S.S-(M.S.S + N.S.S) = 214.59

From two-way table of 'methods' and 'phosphorus', we have

MP.S.S = Table S.S-(M.S.S + P.S.S) = 17.39

From two-way table of 'methods' and 'potash', we have

MK.S.S = Table S.S-(M.S.S + K.S.S) = 98.73

The three-way table of methods, nitrogen and phosphorus is

TABLE 16.35

Methods	n_0		n_1		n_2		
	p_0	p_1	p_0	p_1	p_0	p_1	
m_0	67	99	113	168	129	187	763
m_1	100	134	171	205	190	243	1043
m_2	92	107	143	177	128	185	832
m_3	103	124	180	202	185	257	1051
	363	464	607	752	632	872	3689

MNP.S.S = Table 16.35 S.S - (M.S.S + P.S.S + N.S.S + MP.SS. + MN.SS + NP.S.S)· = 46.1

The three way table for methods, nitrogen and potash is

TABLE 16.36

Methods	n_0		n_1		n_2		
	k_0	k_1	k_0	k_1	k_0	k_1	
m_0	80	86	148	133	150	166	763
m_1	108	126	197	179	219	214	1043
m_2	86	113	149	171	148	165	832
m_3	113	114	199	183	227	215	1051
	387	439	693	666	744	760	3689

MNK.S.S = Table 16.36 S.S – (M.S.S + N.S.S + K.SS + MN.SS. + MK.S.S + NK.S.S) = 36.95

The three-way table of methods, phosphorus and potash is

TABLE 16.37

Methods	p_0		p_1		
	k_0	k_1	k_0	k_1	
m_0	153	156	225	229	763
m_1	239	222	285	297	1043
m_2	171	192	212	257	832
m_3	249	219	290	293	1051
	812	789	1012	1076	3689

MPK.S.S = Table 16.37 S.S – (M.S.S + P.S.S + K.S.S + MP.S.S + MK.S.S + PK.S.S) = 12.·80

The MNPK.S.S. is obtained by subtracting the sum of all the 'main effects, two factor and three factor interactions sums of squares from the Table 16.34 S.S (or Fertilizer .S.S) ,

MNPK.S.S = 63.36

Error (2)S.S = 372·.57

The standard errors for the difference between treatment means are given in Table 16.39.

The S.E values multiplied by corresponding tabulated values of t would give values of critical differences.

TABLE 16.38 ANOVA TABLE

Source	d.f.	S.S	M.S.	Fcal
Rplication	3	93.43	31.14	
Methods (M)	3	1347.14	449.05	119.75**
Error (1)	9	33.75	3.75	
N	2	3983.32	1991.66	706.26**
P	1	1235.25	1235.25	438.03**
K	1	8.75	8.75	3.10
NP	2	155.83	77.92	27.63**
NK	2	48.89	24.45	8.67**
PK	1	39.43	39.43	13.98**
NPK	2	61.96	30.98	10.99**
MN	6	214.59	35.77	12.68**
MP	3	17.39	5.80	2.06
MK	3	98.73	32.91	11.67**
MNP	6	46.10	7.68	2.72*
MNK	6	36.95	6.16	2.18*
MPK	3	12.80	4.27	1.51
MNPK	6	63.36	10.56	3.74**
Error (2)	132	372.57	2.82	
Total	191	7870.24		

*Significant at 5 per cent level, **Signt at 1 per cent level

TABLE 16.39 STANDARD ERRORS

Difference between	S.E	S.E. value
$\overline{m_2} - \overline{m_1}$	$\sqrt{\dfrac{2 \times 3.75}{4 \times 3 \times 2 \times 2}}$	0.40
$\overline{n_2} - \overline{n_0}$	$\sqrt{\dfrac{2 \times 2.82}{4 \times 4 \times 2 \times 2}}$	0.30
$\overline{p_1} - \overline{p_0}$ or $\overline{k_1} - \overline{k_0}$	$\sqrt{\dfrac{2 \times 2.82}{4 \times 4 \times 3 \times 2}}$	0.24
$\overline{n_1 p_0} - \overline{n_2 p_1}$ or $\overline{n_2 k_1} - \overline{n_0 k_1}$	$\sqrt{\dfrac{2 \times 2.82}{4 \times 4 \times 2}}$	0.42
$\overline{p_1 k_0} - \overline{p_0 k_1}$	$\sqrt{\dfrac{2 \times 2.82}{4 \times 4 \times 3}}$	0.34
$\overline{m_1 p_0 k_1} - \overline{m_1 p_1 k_0}$	$\sqrt{\dfrac{2 \times 2.82}{4 \times 3}}$	0.69
$\overline{m_1 n_2 p_0} - \overline{m_1 n_1 p_1}$	$\sqrt{\dfrac{2 \times 2.82}{4 \times 2}}$	0.84
$\overline{m_2 n_1 p_0} - \overline{m_2 n_0 p_1}$	$\sqrt{\dfrac{2\left[(6-1)2.82 + 3.75\right]}{4 \times 6 \times 2}}$	0.86
$\overline{m_1 n_2 p_1 k_1} - \overline{m_2 n_2 p_1 k_0}$	$\sqrt{\dfrac{2\left[(12-1)2.82 + 3.75\right]}{4 \times 12}}$	1.20

SPLIT PLOT DESIGN ANALYSIS

SPSS DATA ANALYSIS PROCEDURE

In this design we take I (irrigation levels) =3 as Main treatments) and N (nitrogen   levels) = 2 as sub treatments and R (replications) = 2.Y = observational value.

The data will be entered column wise as follows in the SPSS Excel sheet.

I	N	R	Y
0	0	1	8
1	0	1	11
2	0	1	15
0	1	1	10
1	1	1	14
2	1	1	18
0	0	2	10
1	0	2	14

2	0	2	17
0	1	2	12
1	1	2	18
2	1	2	21

FROM 'TOOL BAR', Choose 'GENERAL LINEAR MODEL' AND OPT FOR 'UNIVARIATE '. We get a display table .Take I, ,' 'R' and 'N' as 'FIXE D FACTORS', AND 'Y' AS DEPENDENT FACTOR'. Further Press 'MODEL' AND OPT FOR 'CUSTOM'.& 'INTERACTION' take 'R','I'' IR' 'in that order then take N,IN, to RHS by pressing 'CTRL' KEY ALONG WITH factors for interaction after clicking 'interaction' option

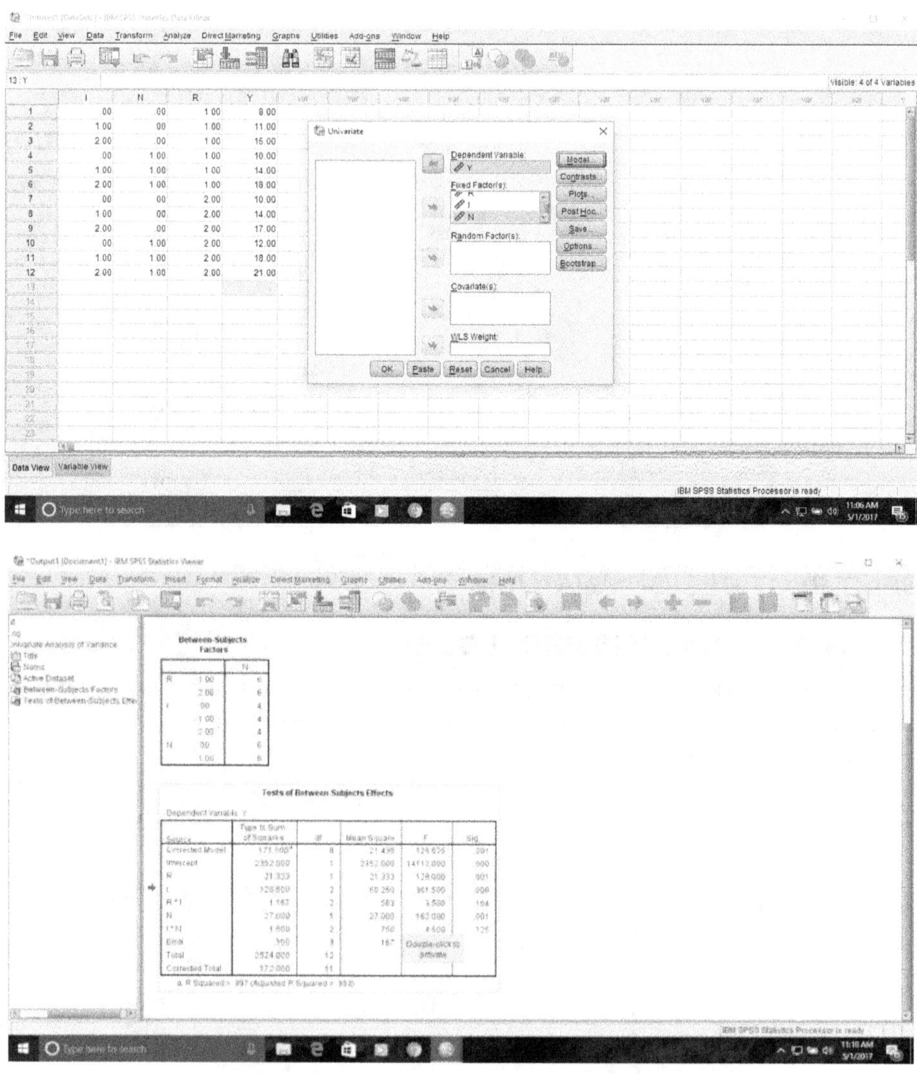

using two-factor option. Finally CLICK 'O.K.' OPTION. We get a table with ANOVA TABLE CONTAINING F-values of 'I' 'R' 'N','IN'. Here we have to use 'IR' M.S. would be used for testing I,R and Error M.S. will be used for testing N,IN factors . In the ANOVA table we will not get IR.M.S directly. Whatever we get in the table we cannot use for testing 'I','R' in the Split Plot analysis. The IR.S.S. we get from the ANOVA table is the Error(a) or Error(1) of the usual Split plot analysis. The Error S.S. from the ANOVA table is Error (b) or Error(2) of the usual Split plot analysis.

16.12 SPLIT-SPLIT PLOT DESIGN

The sub plots in split plot design can further be divided into sub plots to accommodate one more factor. For example, depletion levels will be taken as main treatments, levels of nitrogen as sub-treatments and methods of application of nitrogen as sub-sub-treatments. The method of analysis of this design is an extension of the one given in split-plot design with three error components instead of two. The Analysis of variance Table is given in Table 16.40 with replications (r), main treatments (m), sub treatments (n) and sub-sub treatments (p), In general Error (1) $_{M.S}$ $\geq$ Error (2) $_{M.S}$ $\geq$ Error (3) M.S in this design as a simple extension' of split-plot design. Hence sub-sub treatments are more precisely compared than sub treatments which are in turn more precisely compared than main treatments. The different sum of squares are computed on the basis of sub-sub unit. The divisor at each stage will be the number of sub-sub units involved in the numerator.

TABLE 16.40 ANOVA TABLE

Source	*d.f.*	*Fcal*	*Ftab (d.f)*
Replications	r–1		
M	m–1	M.M.S/E_1MS	(m–1),(r–1)(m–1)
Error(1)	(r–1)(m–1)		
N	n–1	N.M.S/E_2MS	(n–1),m(r–1)(n–1)
MN	(m–1)(n–1)	MN.MS/E_2MS	(m–1)(n–1),m(r–1) (n–1)
Error(2)	m(r–1)(n–1)		
P	p-1	P.M.S/E_3MS	(p–1),mn(p–1)(r–1)
MP	(m–1)(p–1)	MP.M.S/E_3MS	(m–1)(p–1),mn(p–1)(r–1)
NP	(n–1)(p–1)	NP.M.S./E_3MS	(n–1)(p–1),mn(p–1) (r–1)
MNP	(m–1)(n–1)(p–1)	MNP.M.S/E_3MS	(m–1)(n–1) (p–1),mn(p–1),(r–1)
Error(3)	mn(r–1)(p–1)		
Total	mnpr–1		

The standard errors for the difference of two treatment means are same for the whole unit and sub unit treatments as in the case of split plot design except the extra divisor $\sqrt{p}$. The standard errors for the remaining comparisons are presented in Table 16.41.

TABLE 16.41 STANDARD ERRORS

Difference between	*Notation*	*S.E.*
Two P means	$\overline{pi} - \overline{pj}$	$\sqrt{\dfrac{2E_3}{r \cdot m \cdot n}}$
Two P means at the same level of M	$\overline{mipj} - \overline{mip_l}$	$\sqrt{\dfrac{2E_3}{r \cdot n}}$
Two P means at the same level of N	$\overline{nipj} - \overline{nip_l}$	$\sqrt{\dfrac{2E_3}{r \cdot m}}$
Two P means at the same level of MN	$\overline{minjpk} - \overline{minjp_l}$	$\sqrt{\dfrac{2E_3}{r}}$
Two N means at the same or dif. Levels of P	$\left.\begin{array}{c}\overline{nipk} - \overline{njpk} \\ \text{or} \\ \overline{nipk} - \overline{nip_l}\end{array}\right\}$	$\sqrt{\dfrac{2\left[(p-1)E_3 + E_2\right]}{r \cdot m \cdot p}}$
Two NP means at the same level of M	$\overline{minjpk} - \overline{min_lpk}$	$\sqrt{\dfrac{2\left[(p-1)E_3 + E_2\right]}{r \cdot p \cdot}}$
Two M means at the same or different levels of P	$\left.\begin{array}{c}\overline{mipk} - \overline{mjpk} \\ \text{or} \\ \overline{mipk} - \overline{m_jp_l}\end{array}\right\}$	$\sqrt{\dfrac{2\left[\right.}{}}$ Table 16.41 contd…
Two M means at the same level of N & P.	$\overline{m_jn_k p_l} - \overline{m_in_k p_l}$	$\sqrt{\dfrac{2\left[(p-1)E_3 + (n-1)E_2 + E_1\right]}{r \cdot n \cdot p}}$

Where E_1, and E_2 and E_3 are the E_1 M.S., E_2 M.S. and E_3. M.S: respectively.

16.12.1 Missing Data

Let a single sub unit in which the treatment 'm_in_j' occurs is missing in split plot lay out then the estimate of the missing value is

$$Y_{ij} = \frac{rMi + n(min_j) - (mi)}{(r-1)(n-1)} \qquad \ldots\ldots(16.18)$$

where Mi be the total of the whole unit in which missing sub unit occurs, (m_in_j) be the total of all the remaining sub units which receive the sub treatment, m_in_j and (m_i) be the total of all the sub units which receive the whole unit, m_i over all replications.

If one sub unit is missing then one d.f. is subtracted from Error (2) and also from the total. The treatment sum of squares and Error (1)S.S are slightly inflated due to the substitution of missing value.

SPLIT-SPLIT DESIGN ANALYSIS

SPSS DATA ANALYSIS PROCEDURE

In this design M(methods of application),N(nitrogen levels0,P(phosphorus levels), R (replications) are taken as Fixed factors and Y (observational values) as dependent variable. Here M=3,N=2, P=2, R=3 are considered for entering data in SPSS Excel sheet column wise as follows.

M	N	P	R	Y
0	0	0	1	6
1	0	0	1	9
0	1	0	1	11
1	1	0	1	14
0	0	1	1	12
1	0	1	1	15
0	1	1	1	20
1	1	1	1	26
2	0	0	1	17
2	1	0	1	23
2	0	1	1	29
2	1	1	1	31
0	0	0	2	8
1	0	0	2	11
0	1	0	2	13
1	1	0	2	16
0	0	1	2	17
1	0	1	2	19
0	1	1	2	23
1	1	1	2	28
2	0	0	2	16
2	1	0	2	25
2	0	1	2	33
2	1	1	2	37
0	0	0	3	8
1	0	0	3	13
0	1	0	3	14
1	1	0	3	18
0	0	1	3	15
1	0	1	3	16
0	1	1	3	21
1	1	1	3	33
2	0	0	3	37
2	1	0	3	33
2	0	1	3	29
2	1	1	3	34

FROM 'TOOL BAR', Choose 'GENERAL LINEAR MODEL' AND 'OPT' FOR 'UNIVARIATE '. We get a display table .Take M,N,P, 'R' as 'FIXE D FACTORS', AND 'Y' AS DEPENDENT FACTOR' .Further Press 'MODEL' AND OPT FOR 'CUSTOM'& 'INTERACTION'. Take 'R','M'' MR' in that order then take N,MN,MNR,P,NP,MP,MNP to RHS by pressing 'CTRL' KEY ALONG WITH factors for interaction after clicking 'interaction' option   using Main effects, two-factor, three-factor option. Finally CLICK 'O.K.' OPTION. We get a table with ANOVA TABLE CONTAINING F-values of '' 'R',M, MR''N,'MN,MNR, P,NP,MP,MNP,'. Here we have to use 'MR' M.S. would be used for   testing M,R and Error (2) for N,MN, for testing P,NP,MP,MNP we will use the usual Error(3) M.S. Whatever we get in the table we cannot use for testing 'M,''R' ,N,MN, in the Split-Split Plot analysis. The MR.S.S. we get from the ANOVA table is the Error(a) or Error(1) of the usual c Split-Split analysis and MNR.S.S. will not be Error (b) and usual Error S.S. will be Error (c)in Split-Split plot analysis.

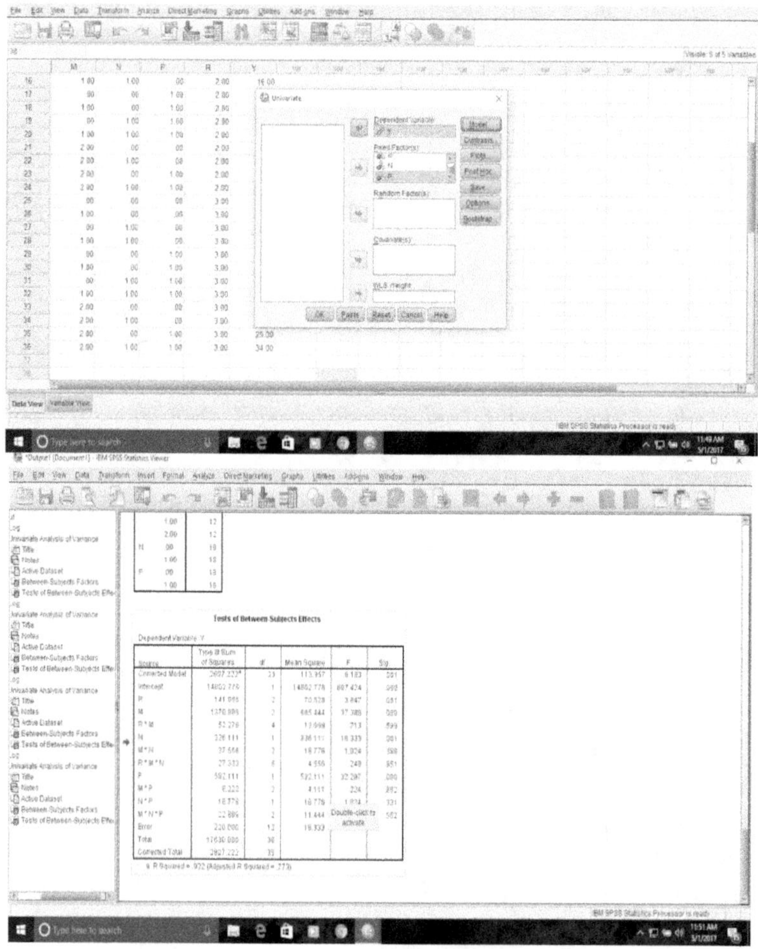

16.13 STRIP PLOT DESIGN

In split plot design if the sub treatments are laid out in strips, then the design is known as strip plot design. This design is used when both main and sub treatments require large experimental material. Here the interaction is more efficiently compared than main and sub treatments due to the fact that main and sub treatments have large experimental material and interaction has less experimental material. If depletion levels and sowing dates are to be tested then the depletion levels may be taken in main plots and dates of sowing in strips as both require large experimental area. The layout of strip plot design with 4 depletion levels and 3 dates of sowing of a crop having 4 replications is given in Fig. 16.22. The depletion levels are denoted by d_0, d_1 d_2 and d_3 and the sowing dates by s_0 s_1 and s_2.

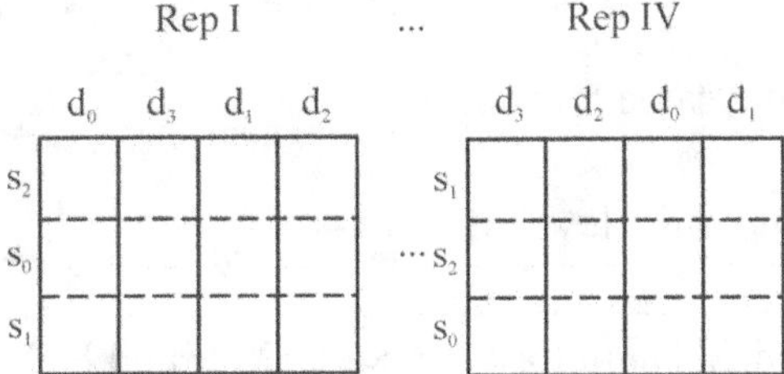

Fig. 16.22 Strip plot design.

In Fig. 16.22 the randomization of dates of sowing is done over the complete strips as in the case of depletion levels for the complete columns. Hence dates of sowing are also considered as main treatments. In split plot design the randomization of sub treatments is done within each main treatment but in this design the randomization is done over complete strip. In general if there are 'm' main treatments; 'n' sub treatments laid out in strips, with 'r' replications, the analysis of variance table is given in Table 16.42

TABLE 16.42 ANOVA TABLE

Source	d.f.	f_{cal}	F_{tab} (d.f)
Replications	r–1		
M	m–1	M.M.S/E_1MS	(m–1),(r–1)(m–1)
Error(1)	(r–1)(m–1)		
N	(n–1)	N.M.S/E_2MS	(n–1),(r–1),(n–1)
Error(2)	(r–1)(n–1)		
MN	(m–1)(n–1)	MN.M.S./E_3M.S.	(m–1)(n–1),(r-1) (m–1)(n–1)
Error(3)	(r–1)(m–1) (n–1)		
Total	rmn–1		

Replication S.S, Main treatment S.S. Sub treatment S.S., Main × Sub treatment S.S are obtained in the usual way as given in split plot design. Error (1) S.S. is obtained from two-way table of Replications and Main treatments, Error (2) S.S. is obtained from two-way table of Replications and Sub treatments and Error (3) S.S. is obtained by subtraction from Total S.S., the remaining all the sums of squares.

TABLE 16.43 STANDARD ERRORS

Difference between	*Notation*	*S.E.*
Two M means	$\overline{mi} - \overline{mj}$	$\sqrt{\dfrac{2(E_1)}{rn}}$
Two N means	$\overline{ni} - \overline{nj}$	$\sqrt{\dfrac{2(E_2)}{r \cdot m}}$
Two M means at the same level of N	$\overline{mini} - \overline{mjni}$	$\sqrt{\dfrac{2\left[(n-1)(E_3) + E_1\right]}{r \cdot n}}$
Two N means at the same level of M	$\overline{mini} - \overline{minj}$	$\sqrt{\dfrac{2\left[(m-1)(E_3) + E_2\right]}{r \cdot m}}$

Where E_1, E_2 and E_3 stand for E_1.M.S., E_2.M.S and E_3.M.S.respectively.

STRIP PLOT DESIGN ANALYSIS

SPSS DATA ANALYSIS PROCEDURE

In this design I (irrigation levels), S (Spacings), R (replications) are taken as Fixed factors and Y (observational values) as dependent variable. Here I=2,S=2, , R=4 are considered for entering data in SPSS Excel sheet column wise as follows.

I	S	R	Y
0	0	1	9
1	0	1	12
0	1	1	10
1	1	1	15
0	0	2	18
1	0	2	22
0	1	2	24
1	1	2	28
0	0	3	11
1	0	3	15
0	1	3	14
1	1	3	18
0	0	4	19
1	0	4	25

0	1	4	29
1	1	4	31

FROM 'TOOL BAR', Choose 'GENERAL LINEAR MODEL' AND OPT FOR 'UNIVARIATE '. We get a display table .Take I,S,''R' as 'FIXE D FACTORS', AND 'Y' AS DEPENDENT FACTOR'. Further Press 'MODEL' AND OPT FOR 'CUSTOM'.&'INTERACTION' take 'R','I'' IR''in that order then take S,IS, to RHS by pressing 'CTRL' KEY ALONG WITH factors for interaction after clicking 'interaction' option using two-factor option. Finally CLICK 'O.K.' OPTION. We get a table with ANOVA TABLE CONTAINING F-values of 'I' 'R' IR''S,'IS'. Here we have to use 'IR' M.S. would be used for testing I,R and RS M.S. will be used for testing S factor .For testing IS we will use the usual Error M.S. given In the last entry of ANOVA table as we should not use usual Error M.S. for testing I,R,S. Whatever we get in the table we cannot use for testing 'I','R' ,S in the Strip Plot analysis. The IR.S.S. we get from the ANOVA table is the Error(a) or Error(1) of the usual Strip plot analysis and RS .S.S. will be Error (b) and usual Error S.S. will be Error (c). of the Strip plot analysis.

16.14 ANALYSIS OF COVARIANCE

Extraneous variables sometimes influence the character under study resulting in misleading interpretation. For example, for testing the different rations in animal feeding experiment, the initial body weights of animals would considerably influence the gain in weights. In Analysis of variance the differences in initial body weights will be added to the error variance which results in decreasing precision for testing of different rations. To remove the influence of initial weights of animals on gain in weights covariance analysis would be adopted. In this method, regression analysis would be used for adjusting the ration means by fixing initial weights at constant level (mean of initial weights). In agronomy experiments, the variation in plant population in plots effects the' yield. For removing, the influence of plant population on yield, the yield is considered as dependent variable and plant population as independent variable in covariance technique. In entomology experiment, for testing the effectiveness of insecticides on (say) cabbage crop, the volume of the bulb will be considered as independent variable and number of-insects as dependent variable since the volume of the bulb is positively correlated with the number of insects on the bulb. The independent variables are also referred here as concomitant variables. Analysis of covariance is not a design but a statistical technique through which the error variance could be reduced for increasing the efficiency of design.

16.14.1 Assumptions

(1) The concomitant variable should be a quantitative variable but not a random variable. (2) The comitant variable should not be influenced by the treatments applied. (3) The dependent variable (or the variable under study) has linear regression on the concomitant variable.

The model of analysis of covariance in the Randomized block lay out with one observation per cell is

$$Y_{ij} = \mu + \alpha_i + \beta_j + \gamma(X_{ij} - X..) + E_{ij} \qquad .....(16.19)$$

for i = 1, 2, ..., t and j = 1,2, ... , r

Where Y_{ij} is the observation on the i-th treatment in the j-th block, μ, is the general mean, α_i the i-th treatment effect, β_j the j-th block effect, γ the regression coefficient, X_{ij} the observation of the concomitant variable on the i-th treatment in the j-th block, X.. the mean of the concomitant variable based on rt units, and Eij s are errors which are randomly and independently distributed with mean zero and constant variance σ^2.

TABLE 16.44 SUM OF SQUARES AND SUM OF PRODUCTS

Source	d.f.	Y	XY	X
Blocks	r–1	B.S.S(Y)	B.S.P(XY)	B.S.S(X)
Treatments	t–1	T_r.S.S(Y)	Tr.S.P(XY)	Tr.S.S(X)
Error	(r–1)(t–1)	E.S.S(Y)	E.S.P(XY)	E.S.S(X)
Treatments +Error	r (t–1)	T_r.S.S(Y) +E.S.S(Y)	Tr.S.S(XY) +E.S.P(XY)	Tr.S.S(X) +E.S.S(X)

The-block sum of squares, treatment sum of squares for characters Y and X are obtained in the usual way as in Analysis of variance for Randomized block design. The sum of products for Blocks and Treatments are computed as

$$\text{B.S.P(XY)} = \Sigma \frac{B_y \cdot B_x}{t} - \frac{G_y \cdot G_x}{rt} \qquad \ldots\ldots(16.20)$$

where B_y, B_x, G_y and G_x, are the block totals and Grand totals for characters Y and X respectively.

$$\text{Tr.S.P(XY)} = \Sigma \frac{T_y \cdot T_x}{r} - \frac{G_y \cdot G_x}{rt} \qquad \ldots\ldots(16.21)$$

$$\text{Tr.S.P(XY)} = \Sigma T, \cdot Tx \_ Gy \cdot GJ < 06.21) \text{ r rt}$$

where T_y, T_x are the totals of each treatment for Y and X respectively. The sum of squares and sum of products in Error line are obtained by subtraction. The (treatments+ Error) line is included in Table 16.44 by adding the corresponding values of treatments and Error.

TABLE 16.45 CORRECTEP ANOVA

Source	Regression		Corrected	
	d.f	S.S	d.f	S.S
Error	1	$\dfrac{[\text{E.SP(XY)}]^2}{\text{E.SS(X)}}$	(r –1)(t –1) –1	S_1
Treatments + Error	1	$\dfrac{[T_r\text{SP(XY)} + \text{ESP(XY)}]^2}{T_r\text{SS(X)} + \text{E.SS(X)}}$	r(t –1) –1	S_2
Treatments			t–1	$(S_2 – S_1)$

$$\text{where } \quad S_1 = \text{E.SS(Y)} - \frac{[\text{E.SP(XY)}]^2}{\text{E.S.S(X)}}$$

$$S_2 = \{T_r.\text{S.S(Y)} + \text{E.SS(Y)}\} - \frac{[T_r\text{S.P(XY)} + \text{E.S.P(XY)}]^2}{\{T_r\text{S.S(X)} + \text{E.S.S(X)}\}}$$

It may be noted that the d.f, for error and (treatment +error) will be reduced by one for each concomitant variable. To test the null hypothesis that the adjusted treatment means are equal, the ratio $\dfrac{(S_2 - S_1)/t - 1}{S_1/(r-1)(t-1) - 1}$ is used as F with (t-1) [(r-l) (t-l)-l]d.f.

CONCLUSION: If $F_{cal} > F_{tab}$ with (t–1), (r–l) (t–l)–l d.f. at chosen level of significance, the null hypothesis is rejected. Otherwise the null hypothesis is accepted.

If the null hypothesis is rejected, each pair of adjusted treatment means $\hat{\alpha}_p$, $\hat{\alpha}_q$ are to be tested with the help of student's t-test using. S.E as'

$$\sqrt{\text{Corrected E.M.S}\left[\frac{2}{r} + \frac{\left(X_{p\cdot} - X_{q\cdot}\right)^2}{\text{E.S.S}(X)}\right]} \qquad \ldots\ldots(16.22)$$

Where $X_{p\cdot}$ and $X_{q\cdot}$ are the means of p-th and q-th treatments on the concomitant variable. The S.E. for adjusted treatment mean $\hat{\alpha}_p$ is given as

$$\sqrt{\text{Corrected E.M.S}\left[\frac{1}{r} + \frac{\left(X_p - X_{\cdot\cdot}\right)^2}{\text{E.S.S}(X)}\right]} \qquad \ldots\ldots(16.23)$$

EXAMPLE: The following is a 3 x 2 factorial experiment laid out in 4 randomized blocks. The first factor is a nitrogenous fertiliser applied at 3 levels and the second factor phosphorus applied at 2 levels to a wheat crop. The yields (kg) (Y) and plant population (X) are presented in Table 16.46.

TABLE 16.46

Treatments	Blocks					
	(I)		(II)		(III)	
	Y	X	Y	X	Y	X
n_0p_0	6.0(–3.0)	40(0)	6.5(–2.5)	37(–3)	6.4(–2.6)	39(–1)
n_0p_1	8.5(–0.5)	39(–1)	8.0(–1.0)	42(2)	8.7(-0.3)	38(–2)
n_1p_0	9.0(0)	38(–2)	8.7(–0.3)	40(0)	9.2(0.2)	45(5)
n_1p_1	10.0(1.0)	41(1)	9.8(0.8)	43(3)	10.4(1.4)	42(2)
n_2p_0	8.2(–0.8)	40(0)	9.0(0)	41(1)	9.8(0.8)	38–(2)
n_2p_1	11.5(2.5)	43(3)	10.9(1.9)	40(0)	11.2(2.2)	40(0)
	–0.8	1	–1.1	3	1.7	2

Table 16.46 can be rewritten by taking deviations from arbitrary means for Y (9.0) .and X (40) respectively and presented in Table 16.46 itself.

$$G_Y = -0.2, \quad G_X = 6, \qquad r = 3, \, t = 6$$

$$C.F(Y) = \frac{(-0.2)^2}{18} = .002, \qquad C.F(X) = \frac{(6)^2}{18} = 2.0$$

$$C.F(XY) = \frac{(-0.2)(6)}{18} = 0.067$$

Total S.S(Y) = $\Sigma\ Y^2$ – C.F(Y) = 43.06– 0.002 = 43.058

Total S.S(X) = $\Sigma\ X^2$ – C.F(X) = 76.0 – 2.0 = 74.0

Total S.P (XY) = $\Sigma\ XY$ – C.F. (XY) = 22.3 – .067 = 22.233

Block S.S(Y) = $1/6[0.8)^2 + (-1.1)^2 + (1.7)^2]$ – .002 = 0.788

Block S.S(X) = $1/6\ [(1)^2 + (3)^2 + (2)^2]$ –2 = .0333

Block S.P(XY) = $1/t\ \Sigma\ B_X.B_y$ – C.F(XY)

$$= 1/6\ [(-0.8)\ (1) + (-1.1)(3) + (1.7)(2)] - 0.067 = 0.767$$

The two-way table for computing the treatment sum of squares is

TABLE 16.47

	n_0		n_1		n_2		*Total*	
	Y	X	Y	X	Y	X	Y	X
p_1	–8.1	–4	– 0.1	3	0	–1	– 8.2	–2
p_0	–1.8	–1	3.2	6	6.6	3	8.0	8
Total	–9.9	–5	3.1	9	6.6	2	–0.2	6

N.S.S(Y) = $1/6\ [(-9.9)^2 + (3.1)^2 + (6.6)^2]$ – 0.002 = 25.195

N.S.S(X) = $1/6\ [(-5)]^2 + (9)^2 + (2)^2]$ – 2.0 = 16.333

N.SP (XY) = $1/r\times2\ \Sigma\ N_X N_Y$ – C.F (XY) = 1/6 [(-9.9) (–5) + (3.1) (9) +

$$(6.6)\ (2)] -0.067 = 15.033$$

P.S.S(Y) = $1/9\ [-8.2)^2 + (8.0)^2]$ –0.002 = 14.580

P.S.S(X) = $1/9\ [(-2)]^2 + (8)^2]$–2.0 = 5.556

P.SP(XY)=$1/r\times3\ \Sigma P_X.P_Y$=C.F(XY)=1/9[(-8.2)(-2) +(8.0) (8)] -0 067 = 8.866

For finding out the sum of squares and products for interaction, the sum of squares and products for Table 16.47 are obtained as follows.

Table S.S (Y) = $1/3\ [(-8.1)^2 + (-0.1)^2 + ...+ (6.6)^2]$–0.002= 40.885

Table S.S(X) = $1/3\ [(-4)^2] + (3)^2 + ...+ (3)^2]$–2.0 = 22.0

Table S.P(XY) = $1/3\ [(-8.1)\ (-4) + (-0.1)\ (3) + ... + (6.6)\ (3)]$ –0.067 = 24.233

NP.S.S.(Y) = Table S.S(Y)–[N.S.S(Y)+P.S.S(Y)] = 1.110 Similarly NP.S.S(X) and NP.SP(XY) are obtained as 0.334 and 0.111 respectively. The error sum of squares and products are obtained by subtraction.

TABLE 16.48 SUM OF SQUARES AND PRODUCTS

Source	d.f.	S.S.(Y)	SP(XY)	S.S(X)
Blocks	2	0.788	-0.767	0.333
N	2	25.195	15.033	16.333
P	1	14.580	8.866	5.556
NP	2	1.110	0.334	0.111
Error	10	1.385	1.233	51.667
N+Error	12	26.580	16.266	68.000
P+Error	11	15.965	10.099	57.223
NP+Error	12	2.495	1.567	51.778

TABLE 16.49 CORRECTED ANOVA

Source	Regression		Corrected			Fcal
	d.f	S.S	d.f	S.S	M.S.	
Error	1	0.029	9	1.356	0.151	
N+ Error	1	3.891	11	22.689		
P + Error	1	1.782	10	14.183		
NP + Error	1	0.047	11	2.448		
N			2	21.333	10.666	70.63**
P			1	12.827	12.827	84.94**
NP			2	1.092	0.546	3.61

**Significant at 1 per cent level.

Here F_{cal} is found to be significant at 1 per cent level in the case of main effects but not significant even at 5 per cent level in the case of interaction. The regression coefficient,

$$\hat{\gamma} = \frac{E.SP(XY)}{E.S.S(X)} = 0.024$$

The adjusted means of N are calculated for comparison within their levels and presented in Table 16.50.

TABLE 16.50 ADJUSTED MEANS OF N

Levels of N	$Y_{i.}$	$X_{i.}$	$(Xi. - X..)$	$\hat{\gamma}\,(Xi. - X..)$	$\dfrac{Yi. - \hat{\gamma}}{(Xi.-X..)}$
n_0	7.35	39.17	−1.16	−0.028	7.378
n_1	9.52	41.50	1.17	0.028	9.492
n_2	10.10	40.33	0	0	10.100

The adjusted means of levels of P are presented in Table 16.51

TABLE 16.51 ADJUSTED MEANS OF P

Levels of p	$Y_{i.}$	$X_{i.}$	$(Xi.-X..)$	$\hat{\gamma}\,(Xi. - X..)$	$\dfrac{Yi.- \hat{\gamma}}{(Xi. - X..)}$
p_0	8.09	39.78	-0.55	-0.013	8.103
p_1	9.89	40.89	0.56	0.013	9.877

$$\text{S.E of the difference between two levels of N} = \sqrt{\frac{2[\text{adj.E.M.S}]}{r \times 2}} = 0.224$$

$$P = \sqrt{\frac{2[\text{adj.E.M.S}]}{r \times 3}} = 0.183$$

16.15 TUKEY'S TEST OF ADDITIVITY

One of the assumptions in all the models for different designs given in this chapter so far is that different factor effects are additive. This assumption may not always be true. If this assumption *is* not true the errors will be inflated. To know whether this assumption holds good or not 'Tukey's test of additivity is used. In other words this test is useful to know whether transformation of given data is necessary for bringing the original observations to additive scale. Let A and B be two factors such that 'A is at's' levels and B *is* at 't' levels and n_{ij} be the number of observations for the (i,j)-th cell where $n_{ij} = 1$. Let γij be the interaction between i-th level of A and j-th level of B. Assuming that γij represents quadratic expression, we have

$$\gamma ij = a + b\alpha_i + c\beta_j + d\alpha^2 + e\beta_j^2 + g\alpha_i\beta_j \qquad .....(16.24)$$

where α_i, β_j, are the effects of i-th level of A and j-th level of B respectively and a, b, c, d, e and g are the constants In the quadratic expression of γ_{ij}.

Using the conditions $\sum_j \gamma_{ij} = 0$ for every i

$$\sum_I \gamma_{ij} = 0 \text{ for every j} \qquad .....(16.25)$$

We finally obtain

$$\gamma_{ij} = g\alpha_i\beta_j.$$

The mathematical model for the two-way classification is

$$Y_{ij} = \mu + \alpha_i + \beta_j + g\alpha_i\beta_j + E_{ij} \qquad \ldots(16.26)$$

The null hypothesis is that there is no non-additivity in the model i.e.

$\gamma_{ij} = 0$ which implies $g = 0$

The sum of squares due to non – additivity is

$$\frac{\left(\sum_i \sum_j Y_{ij}\alpha_i\beta_j\right)^2}{\left(\sum_i \hat{\alpha}_i^2 \sum_j \hat{\beta}_j^2\right)} \qquad \ldots(16.27)$$

where

$$\hat{\alpha}_i = Y_{i.} - Y.. = \text{ and } \hat{\beta}_j = Y_{.j} - Y..; Y_{i.} = \sum_j Y_{ij};$$

$$Y_{.j} = \sum_i Y_{ij}, Y.. = \sum_i \sum_j Y_{ij}; \overline{Y}_{i.} = \frac{1}{ni.}\sum_j Y_{ij};$$

$$\overline{Y}_{.j} = \frac{1}{n.j}\sum_i Y_{ij}; \overline{Y}.. = \frac{1}{n..}\sum_i \sum_j Y_{ij} \text{ and}$$

$$ni = \sum_i nij; \quad n.j = \sum_i nij; n.. = \sum_i \sum_j nij$$

The analysis of variance table for testing g=0 is given in Table 16.52

TABLE 16.52 ANOVA TABLE

Source	*d.f.*	*S.S.*	*M.s.*	*Fcal*
A	s-1	$\sum_i \dfrac{Y_j^2.}{t} - \dfrac{Y^2..}{st}$	AMS	
B	t-1	$\sum_j \dfrac{Y^2._j}{s} - \dfrac{Y^2..}{st}$	BMS	
Non additivity (g)	1	$\dfrac{\left(\sum_i \sum_j Y_{ij}\hat{\alpha}_i\hat{\beta}_j\right)^2}{\left(\sum_i \hat{\alpha}_i^2 \sum_j \hat{\beta}_j^2\right)}$	GMS	GMS/EMS
Error	st-s-t	subtraction	EMS	
Total	st-1	$\sum_i \sum_j Y_{ij}^2{}' - \dfrac{Y^2..}{st}$		

Conclusion: If F (calculated) > F (tabulated) with 1, (st-s-t) d.f, at chosen level of significance, the null hypothesis is rejected, Otherwise, it is accepted. If the null hypothesis is rejected there exists non-additivity in the data and hence one of the transformations like logarithmic, angular or square root may be applicable for bringing the data into additivity.

16.16   RANDOM EFFECTS MODELS

For random effects model all the observations have same expectation, while they have different expectation for a fixed effects model. For a fixed effects model all observations are not all independent. The analysis of variance model for a Randomized block design is

$$Y_{ij} = \mu + \alpha_I + \beta_j + E_{ij}. \qquad \dots\dots(16.28)$$

If α_i's and βj's are unknown constants and are known as fixed effects. If α_i's and β_j's are random variables except the general mean, then it is called as 'random effects' model. If one of them is a random variable and the other fixed effect in addition to general mean then the model is called 'mixed effects' model.

16.16.1 One-way Classification

Consider an experiment in dairy with I breeds which are assumed to have come from a population of breeds. Let there be 'J' cattle within each breed in the experiment. There will be variation between the breeds as well as the variation between the cattle within each breed with respect to milk yield. Let Y_{ij} be the milk yield of j-th cattle in the i-th breed, then the analysis of variance model is

$$Y_{ij} = \mu + a_i + e_{ij} \qquad \dots\dots(16.29)$$

Here ai's and e_{ij}'s are independent from each other, ai's are identically distributed with zero mean and variance. σ_A^2 e_{ij}'s are identically distributed with zero mean and variance σ_e^2.

Hence V(Yij) = V(μ+ai+eij)

$$\sigma_Y^2 = \sigma_A^2 + \sigma_e^2$$

σ_A^2 and σ_e^2 are called the 'variance components of σ_Y^2.

The Analysis of variance Table for testing the different variance components in the model is given in Table 16.53.

TABLE 16.53 ANOYA TABLE

Source	d.f.	S.S.	M.S.	E(M.S)
Breeds	1–1	$\sum\limits_{i}(Y_{i.} - Y..)^2$	A.M.S	$\sigma_e^2 + j\sigma_A^2$
Error	1(j–1)	$\sum\limits_{i}\sum\limits_{j}(Y_{ij} - Y_{i.})^2$	E.M.S	σ_e^2
Total	1j–1	$\sum\limits_{i}\sum\limits_{j}(Y_{ij} - Y..)^2$		

where $\quad Y_{i.} = I\,/\,J\sum_i Y_{ij}, Y.. = I\,/\,IJ\sum_i\sum_j Y_{ij}$

Null Hypothesis: $H_0: \sigma_A^2 = 0$, $H_1: \sigma_A^2 \neq 0$

$$F = \frac{A.M.S}{E.M.S}$$

Conclusion: If F (calculated) > F(tabulated) with (1-1), I(J-1) d.f, at chosen level of significance, the null hypothesis is rejected. Otherwise, it is accepted.

Estimation of variance components: Let $\hat{\sigma}_A^2$ and $\hat{\sigma}_e^2$ be the unbiased estimates of σ_A^2 and σ_e^2 respectively. Equating the expresions of E.(M.S.) with M.S. in Table 16.53 by replacing σ_A^2 and σ_e^2 with $\hat{\sigma}_A^2$, and $\hat{\sigma}_e^2$, we have

$$A.M.S. = J\,\hat{\sigma}_A^2 + \hat{\sigma}_e^2, E.M.S = \hat{\sigma}_e^2$$

$$\hat{\sigma}_A^2 = \frac{A.M.S - E.M.S}{J}$$

16.16.2 Two-way Classification

Let there be I milking machines and J breeds and each machine is assigned to each breed for K cattle. The' milking machines are assumed to be of the same make and model and I machines are the random sample from a large population of machines. Let Y_{ijk} be the yield of milk obtained by using the i-th machine, on the k-th cattle belonging to the j-th breed. J breeds are assumed to be a random sample from a population of breeds. If i-th machine is obtained from a random sample of machines and j-th breed is obtained from the random sample of breeds, i and j are assumed to be statistically independent. In other words, i-th machine was selected irrespective of the breed for which it was to be applied. The mathematical model is

$$Y_{ijk} = \mu + a_i + b_j + c_{ij} + e_{ijk} \qquad\qquad .....(16.30)$$

where a_i, b_j, c_{ij} are the effects of i-th machine j-th breed, and (i, j)-th interaction respectively.

Here $\qquad E(a_i) = E(b_j) = E(c_{ij}) = 0$

where $E(a_i)$ is denoted as the expected value of a_i and so on.

$$V(a_i) = \sigma_A^2,\ V(b_j) = \sigma_B^2,\ V(c_{ij}) = \sigma_{AB}^2$$

a_i, b_j and c_{ij} are statistically independent. Further a_i's, b_j's, c_{ij}s are independently normally distributed with zero means and variances $\hat{\sigma}_A^2$, $\hat{\sigma}_B^2$, $\hat{\sigma}_{AB}^2$, and $\hat{\sigma}_e^2$ respectively.

where
$$Y_{ij.} = 1/K \sum_i \sum_k Y_{ijk}, \quad Y_{i..} = 1/JK \sum_i \sum_j \sum_k Y_{ijk},$$

$$Y_{.j.} = 1/IK \sum_i \sum_k Y_{ijk}, \quad Y... = 1/IJK \sum_i \sum_j \sum_k Y_{ijk},$$

(i) *Null Hypothesis:* $H_0: \sigma^2_{AB} = 0 \quad H_1: \sigma^2_{AB} \neq 0$

F = AB. M.S./E.M.S

TABLE 16.54 ANOVA TABLE

Source	d.f.	S.S.	M.S.	E(M.S)
Machines (A)	I–1	$JK \sum_i (Y_{i..} - Y...)^2$	A.M.S	$\sigma^2_e + K\sigma^2_{AB} + JK\sigma^2_A$
Breeds (B)	J–1	$IK \sum_j (Y_{.j.} - Y...)^2$	B.M.S	$\sigma^2_e + K\sigma^2_{AB} + IK\sigma^2_B$
AB	(I–1)(J–1)	$K \sum_i \sum_j (Y_{ij.} - Y_{i...} - Y_{.j.} + Y...)^2$	AB.M.S	$\sigma^2_e + K\sigma^2_{AB}$
Error	IJ(K–1)	$\sum_i \sum_j \sum_k (Y_{ijk} - Y_{ij.})^2$	E.M.S	σ^2_e
Total	IJK–1	$\sum_i \sum_j \sum_k (Y_{ijk} - Y...)^2$		

Conclusion: If F (calculated) > F(tabulated) with (I – 1) (J – 1), IJ(K – I) d.f, at chosen level of significance, the null hypothesis is rejected. Otherwise, it is accepted.

If F (calculated) is found to be significant, then we test $\sigma^2_A = 0$ and $\sigma^2_B \sim = 0$.

(ii) *Null Hypothesis:* $H_o: \sigma^2_A = 0. \quad H_i: \sigma^2_A \neq 0$

F = A.M.S./AB.M.S.

Conclusion: If F (calculated) >F (tabulated) with (I-1), (I-I) (J-I) d.f. at chosen level of significance, the null hypothesis is rejected. Otherwise, it is accepted.

(iii) *Null Hypothesis:* $H_0: \sigma^2_B = 0$, $H_1: \sigma^2_B \neq 0$

F = B.M.S/AB.M.S.

Conclusion: If F(calculated) > F (tabulated) with (J-1), (I-1) (J-1) d.f. at chosen level of significance, the null hypothesis is rejected. Otherwise, it is accepted,

It may be noted that if there is only one animal for each combination of breed and machine, E.S.S. and its d.f. would be zero. In that case AB.S.S would' substitute for E.S.S.

Estimation of variance components:

$$\hat{\sigma}_A^2 = \frac{(A.M.S - AB.M.S)}{JK}, \hat{\sigma}_B^2 = \frac{(B.M.S - AB.M.S)}{IK}$$

$$\hat{\sigma}_{AB}^2 = \frac{(AB.M.S - E.M.S)}{K}, \hat{\sigma}_e^2 = E.M.S.$$

16.17   MIXED MODELS

In the example given in sub section 16.16.2 if 'the 'milking machines could be considered as fixed and the breeds of cattle as random sample from a population of breeds then it becomes 'mixed model'. Let Y_{ijk} be the milk yield of the k-th cattle belonging to the j-th breed using i-th machine. The mathematical model is

$$Y_{iik} = \mu + \alpha_i + b_j + c_{ij} + e_{iik} \qquad \qquad .....(16.31)$$

where μ be the general mean, α_i be the effect of i-th machine, b_j, be the j-th breed effect, c_{ij} be the effect of j-th breed using i-th machine and e_{ijk}'s are errors which are independently and identically distributed with zero mean and same variance as σ_e^2. The b_j's, c_{ij}'s, e_{ijk}'S are jointly normal, b_i's and c_{ij}'s have zero means and variances and covariances.

TABLE 16.55 ANOVA TABLE

Source	d.f.	S.S.	M.S	E(M.S)
A (fixed)	1–1	$JK\sum_i(Y_{i..} - Y...)^2$	A.M.S	$\sigma_e^2 + K\sigma_{AB}^2 + JK\sigma_A^2$
B (random)	J–1	$IK\sum_j(Y_{.j.} - Y...)^2$	B.M.S	$\sigma_e^2 + IK\sigma_B^2$
AB	(I–1) (J–1)	$\sum_i\sum_j(Y_{ij.} - Y_{i..} - Y_{.j} - Y...)^2$	AB.M.S	$\sigma_e^2 + K\sigma_{AB}^2$
Error	IJ(K–1)	$\sum_i\sum_j\sum_k(Y_{ijk} - Y_{ij.})^2$	E.M.S	σ_e^2
Total	IJK–1	$\sum_i\sum_j\sum_k(Y_{ijk} - Y...)^2$		

Tests of hypothesis: To test the null hypothesis $\sigma_B^2 = 0$ and $\sigma_{AB}^2 = 0$,

F = B.M.S/E.M.S and F = AB.M.S./E.M.S would be used and compared with F (tabulated) with $(J-1)$, $IJ(K-1)$ d.f, and $(I-1)(J-1)$, $IJ(K-I)$ d..f. respectively. However for testing the null hypothesis $\sigma_A^2 = 0$,

F = A.M.S./AB.M.S would be compared with F (tabulated) with $(I-1,)$ $(I-1)(J-1)$ d.f. as an approximate test. For further reading on this topic please refer to Scheffe (1967).

16.18 HENDERSON METHODS

Henderson (1953) gave three methods for estimating variance components for random models, mixed models and random models through fitting of constants method in unbalanced case since estimation procedures for variance components for unbalanced data are not as straight forward as in the balanced case. The Method-I is described in detail and the Method-II and III are dealt with briefly.

16.1.8.1 Henderson Method-I

This method is used for Random models. The estimation procedure is essentially the same as that of analysis of variance method in balanced case except that instead of mean squares their quadratic forms would be equated with their expected values. Mean squares in balanced case are always positive but their equivalent quadratic forms in unbalanced case may be positive or negative.

16.18.1.1 One-way Classification

Here it is assumed that there are unequal numbers in each level of factor A. Let Y_{ij} be the value on the j-th unit of i-th level of A for i = 1, 2, ..., I and j = 1, 2, ..., ni. The mathematical model is

$$Y_{ij} = \mu + a + e_{ij} \qquad \qquad .....(16.32)$$

where μ be the general mean, ai be the effect of i-th level of A and e_{ij}'s are the errors. ai's are assumed to have zero means and variance as σ_A^2 and e_{ij}s are assumed to have zero means' and variance as σ_e^2 ai's and e_{ij}'s are assumed to be uncorrelated with each other.

TABLE 16.56 ANOVA TABLE

Source	d.f.	S.S.	M.S.	E(M.S)
A	1–1	$\sum_i \dfrac{Y_{i.}^2}{n_i} - \dfrac{Y_{..}^2}{N}$	A.M.S	$\sigma_e^2 + k_0 \sigma_A^2$
Error	N–1	$\sum_i \sum_j Y_{ij}^2 - \sum_i \dfrac{Y_{i.}^2}{n_i}$	E.M.S	σ_e^2
Total	N–1	$\sum_i \sum_j Y_{ij}^2 - \dfrac{Y_{..}^2}{N}$		

where $N = \sum_i n_i$, $Y_{i.} = \sum_j Y_{ij}$ $Y_{..} = \sum_i \sum_j Y_{ij}$. The expectation

of $\sum_i \sum_j Y_{ij}^2, \sum_i \dfrac{Y_{..}^2}{n_i}$ and $\dfrac{Y_{..}^2}{N}$ could be worked out and could be shown that

$$E(E.M.S.) = \sigma_e^2 \text{ and}$$

$$E(A.M.S) = \sigma_e^2 + \frac{1}{(a-1)}\left(N - \frac{\sum_i n_i^2}{N}\right)\sigma_A^2 = \sigma_e^2 + K_0\sigma_A^2$$

where $\qquad k_0 = \dfrac{1}{(a-1)}\left(N - \dfrac{\sum_i n_i^2}{N}\right)$

Since k_o is known and equating observed M.S and E(M.S), we have

$$\hat{\sigma}_e^2 = E.M.S \text{ and } \hat{\sigma}_A^2 = \frac{1}{k_0}(A.M.S. - E.M.S)$$

16.18.1.2 Nested Two-way Classification with Unequal Sub Class Numbers

Let Y_{ijk} be the value on the (i, j, kj-th unit for i = 1,2, .. I; j = 1,2, …. Ji and k = 1, 2, …, n_{ij}. The mathematical model is

$$Y_{ijk} = \mu + a_i + b_{ij} + e_{ijk} \qquad\qquad .....(16.33)$$

where μ be the general mean, a_i and b_{ij} be the effects of i-th level 'of A, j-th level of B within i-th level of A respectively and e_{ijk} be the error on the (i, j, k)-th unit. All the effects except μ are random variables with zero means and covariances and finite variances as $\sigma_A^2, \sigma_B^2, \sigma_e^2$.

Let $N_{i.} = \sum_j n_{ij}$, $N.. = \sum_i \sum_j n_{ij}$, $J = \sum_i j_i$

TABLE 16.57 ANOVA TABLE

Source	d.f.	S.S	M.S	E(M.S)
A	1–1	$\sum_i N_i(Y_{i.} - Y...)^2$	A.M.S	$\sigma_e^2 + q_1\sigma_B^2 + q_2\sigma_A^2$
B(With in A)	J–1	$\sum_i \sum_j n_{ij}\left(Y_{ij.} - Y_{i..}\right)^2$	B.M.S	$\sigma_e^2 + q_0^2\sigma_B^2$
Error	N.. –J	$\sum_i \sum_j \sum_k \left(Y_{ijk} - Y_{ij.}\right)^2$	E.M.S	σ_e^2
Total	N..–1	$\sum_i \sum_j \sum_k \left(Y_{ijk}^2 - \dfrac{Y^2...}{N}\right)$		

where $\qquad Y_{ij.} = \dfrac{1}{n_{ij}}\sum_k Y_{ijk}$, $Y_{i..} = \dfrac{1}{N_{i.}}\sum_j \sum_k Y_{ijk}$,

$$Y = 1/N.. \sum_i \sum_j \sum_k Y_{ijk}$$

It could be shown that
$$q_0 = \frac{N.. - \sum_i \left(\sum_j \frac{n_{ij}^2}{N_{i.}} \right)}{(J-I)} = \sum \sum n_{ij}^2 f_{ij}$$

where
$$f_{ij} = \frac{\left(\dfrac{1}{n_{ij}} - \dfrac{1}{N_{i.}} \right)}{J-1}$$

$$q_i = \frac{\sum_i \sum_j \left\{ \dfrac{n_{ij}^2}{N_{i.}} - \dfrac{n_{ij}^2}{N..} \right\}}{(I-1)} = \sum_i \sum_j n_{ij}^2 f_i$$

where
$$f_i = \frac{\left(\dfrac{-1}{N_{i.}} - \dfrac{1}{N..} \right)}{(I-1)} \quad \text{and} \quad q_2 = \frac{\left(N.. - \sum_i \dfrac{N_{i.}^2}{N..} \right)}{(I-1)} = \sum_i N_{i.}^2 f_i$$

Thus q_o, q_1 and q_2 are known and hence equating observed and E(M.S), $\hat{\sigma}_e^2$, $\hat{\sigma}_B^2$ and $\hat{\sigma}_A^2$ can be obtained.

16.18.1.3 Two-way classification without interaction in unbalanced case

Let Y_{ijk} be the value on the (i, j, k)-th unit for i =. 1, 2, ..,. I; j = 1, 2, ... , J and k = 0, 1, 2, ... , n_{ij}. The mathematical model is

$$Y_{ijk} = \mu + a_i + b_j + e_{ijk} \qquad \dots\dots(16.34)$$

where μ be the general mean, a_i, b_j, are the effects on the i-th level of A and j-th level of B respectively and e_{ijk}'s are errors. The effects a_i, b_j, e_{ijk} are random variables with zero means and variances as σ_A^2, σ_B^2 and σ_e^2 respectively.

TABLE 16.58 ANOVA TABLE

Source	d.f.	S.S.	M.S	E(M.S)
A	1–1	$\sum_i \dfrac{Y_{i..}^2}{N_{i.}} - \dfrac{Y^2...}{N..}$	A.M.S	$\sigma_e^2 + r_5 \sigma_B^2 + r_6 \sigma_A^2$
B	J–1	$\sum_j \dfrac{Y_{.j.}^2}{N.j} - \dfrac{Y^2...}{N..}$	B.M.S	$\sigma_e^2 + r_3 \sigma_B^2 + r_4 \sigma_A^2$
Error	(N.. –I–J+1)	Sub traction	E.M.S	$\sigma_e^2 + r_1 \sigma_B^2 + r_2 \sigma_A^2$
Total	N.. – 1	$\sum_i \sum_j \sum_k Y_{ijk}^2 - \dfrac{Y^2...}{N..}$		

It could be obtained that

$$r_1 = \frac{1}{(N.. - I - J + 1)} \left\{ \sum_i \frac{N_{.j}^2}{N..} - \sum_i \sum_j \frac{n_{ij}^2}{N..} \right\}$$

$$r_2 = \frac{1}{(N.. - I - J - 1)} \left\{ \sum \frac{N_{i.}^2}{N..} - \sum_j \frac{n_{ij}^2}{N_{.j}} \right\}$$

$$r_3 = \frac{1}{(J-1)} \left(N.. - \sum_j \frac{(N_{.j})^2}{N..} \right)$$

$$r_4 = \frac{1}{(J-1)} \left\{ \sum_i \sum_j \frac{n_{ij}^2}{N_j} - \sum_i \frac{N_{i.}^2}{N..} \right\}$$

$$r_5 = \frac{1}{(I-1)} \left\{ \sum_i \sum_j \frac{n_{ij}^2}{N_{i.}} - \sum_j \frac{N_{.j}^2}{N..} \right\}$$

$$r_6 = \frac{1}{(I-1)} \left\{ N.. - \sum_i \frac{N_{i.}^2}{N..} \right\}$$

Thus knowing r_1, r_2,..., r_6 and equating observed M.S with E(M.S), $\hat{\sigma}_e^2, \hat{\sigma}_A^2$ and $\hat{\sigma}_B^2$ can be obtained.

16.18.1.4 Two-way classification with interaction and unequal sub class numbers

Let Y_{ijk} be the value on the (i, j, k)-th unit for i = 1,2..., I; j = 1, 2,..., J and k = 0, 1, 2,..., nij, The mathematical model is given as

$$Y_{ijk} = \mu + a_i + b_j + (ab)_{ij} + e_{ijk} \qquad .....(16.35)$$

where μ be the general mean, ai, bj, $(ab)_{ij}$ and e_{ijk} are random variables with zero means and variances as $\sigma_A^2, \sigma_B^2, \sigma_{AB}^2$ and σ_e^2 respectively. The correlations between any two variables (not the same) are assumed to be zero. Let $N_{i.} = \sum_j n_{ij}$, $N_{.j} = \sum_j n_{ij}$, $N.. = \sum_i \sum_j n_{ij}$ and N.. be the number of sub classes filled in a two-way table with A and B factors.

TABLE 16.59 ANOVA TABLE

Source	d.f.	S.S.	M.S.	E(M.S)
A	1–1	$\sum \dfrac{Y_{i..}^2}{N_{i.}} - \dfrac{Y^2...}{N..}$	A.M.S	$\sigma_e^2 + k_7 \sigma_{AB}^2 k_8 \sigma_A^2 + k_9 \sigma_B^2$
B	J–1	$\sum_j \dfrac{Y_{.j.}}{N_{.j}} - \dfrac{Y^2...}{N..}$	B.M.S	$\sigma_e^2 + k_4 \sigma_{AB}^2 + k_5 \sigma_A^2 + k_6 \sigma_B^2$

Table 16.59 Contd...

Source	d.f.	S.S.	M.S.	E(M.S)
AB	$(N!.I-J+1)$	$\sum_i \sum_j \dfrac{Y_{ij.}^2}{n_{ij}} - \sum_i \dfrac{Y_{i..}^2}{N_{i.}}$	AB.MS	$\sigma_e^2 + k_1\sigma_{AB}^2 + k_2\sigma_A^2 + k_3\sigma_B^2$
Error	$(N..- N!.)$	$-\sum_j \dfrac{Y_{.j.}^2}{N_{.j}} + \dfrac{Y^2...}{N..}$	E.M.S	
		$\sum_i \sum_j \sum_k Y_{ijk}^2 - \sum_i \sum_j \dfrac{Y_{ij.}^2}{n_{ij}}$		σ_e^2
Total $N..1$		$\sum_i \sum_j \sum_k Y_{ijk}^2 - \dfrac{Y^2...}{N..}$		

where

$$k_1(I-1)(J-1) = \left(N.. - \sum_i \sum_j \frac{n_{ij}^2}{N_{i.}} - \sum_i \sum_j \frac{n_{ij}^2}{N_{.j}} + \sum_i \sum_j \frac{n_{ij}^2}{N..} \right)$$

$$k_2(I-1)(J-1) = \sum_i \frac{N_{i.}^2}{N..} - \sum_i \sum_j \frac{n_{ij}^2}{N_{.j}}$$

$$k_3(I-1)(J-1) = \sum_i \frac{N_{.j}^2}{N..} - \sum_i \sum_j \frac{n_{ij}^2}{N_{i.}}$$

$$k_4(J-1) = \sum_i \sum_j \frac{n_{ij}^2}{N_{.j}} - \sum_i \sum_j \frac{n_{ij}^2}{N..}$$

$$k_5(J-1) = \sum_i \sum_j \frac{n_{ij}^2}{N_{.j}} - \sum_i \frac{N_{i.}^2}{N..}$$

$$k_6(J-1) = N.. - \sum_i \frac{N_{.j}^2}{N..}$$

$$k_7(I-1) = \sum_i \sum_j \frac{n_{ij}^2}{N_{i.}} - \sum_i \sum_j \frac{n_{ij}^2}{N..}$$

$$k_8(I-1) = N.. - \sum_i \frac{N_{i.}^2}{N..}$$

$$k_9(I-1) = \sum_i \sum_j \frac{n_{ij}^2}{N_{i.}} - \sum_i \frac{N_{.j}^2}{N..}$$

The κ–functions were defined for filled cells only. Since the sum of squares given in Table 16.59 being quadratic forms which may or may not be positive. Since the values of k-functions were known the estimates of variances $\sigma_A^2, \sigma_B^2, \sigma_{AB}^2$ and σ_e^2 can be obtained by equating M.S and E(M.S) and solving for the estimates of variances.

EXAMPLE: An experiment was conducted with two rations on gain in weight of pigs with three sires and the results are given as follows.

Ration No.	Pig No.	Sire No.		
		1	2	3
1	1	5	2	3
	2	6	3	–
	3	–	5	–
	4	–	6	–
	5	–	7	–
2	1	2	8	4
	2	3	8	4
	3	–	9	6
	4	–	–	6
	5	–	–	7

Estimate the variance components with the help of Henderson Method-I.

Using the model of 16.18.1.4 of Henderson Method-I, the sum of squares, mean squares and *k-functions* were computed and presented in analysis of variance table in Table 16.60.

$$C.F = (94)^2/18$$

$$\text{Ration S.S} = \left[\frac{(37)^2}{8} + \frac{(57)^2}{10}\right] - \frac{(94)^2}{18} = 5.14$$

$$\text{Sire S.S} = \left[\frac{(16)^2}{4} + \frac{(48)^2}{8} + \frac{(30)^2}{6}\right] - \frac{(94)^2}{18} = 11.11$$

Ration × Sire S.S = Table S.S – (Ration S.S + Sire S.S)

$$\text{Table S.S} = \left[\frac{(11)^2}{2} + \frac{(23)^2}{5} + ... + \frac{(27)^2}{5}\right] - \frac{(94)^2}{18} = 51.04$$

Ration × Sire S.S = 51.04 – (5.14 + 11.11) = 34.79.

Total S.S. = 77.11

Error S.S. = 77.11 – 51.04 = 26.07

TABLE 16.60 ANOVA TABLE

Source	d.f.	S.S.	M.S.	E(M.S)
Rations	1	5.14	5.14	$\sigma_e^2 + k_7\sigma_{RS}^2 + k_8\sigma_R^2 + k_9\sigma_S^2$
Sires	2	11.11	5.56	$\sigma_e^2 + k_4\sigma_{RS}^2 + k_5\sigma_R^2 + k_6\sigma_S^2$
Rations x sires	2	34.79	17.40	$\sigma_e^2 + k_1\sigma_{RS}^2 + k_2\sigma_R^2 + k_3\sigma_S^2$
Error	12	26.07	2.17	σ_e^2
Total	17	77.11		

where σ_e^2, σ_R^2, σ_S^2 and σ_{RS}^2 are the variances for the error, rations, sires and rations × sires respectively, and

$$k_1 = \frac{1}{(I-1)(J-1)}\left[N.. - \sum_i \sum_j \frac{n_{ij}^2}{N_{i.}} - \sum_i \sum_j \frac{n_{ij}^2}{N_{.j}} + \sum_i \sum_j \frac{n_{ij}^2}{N..}\right]$$

$$= \frac{1}{2}\left[18 - (7.55) - (10.58) + (3.78)\right] = 1.825$$

$$k_2 = \frac{1}{(I-1)(J-1)}\left[\sum_i \frac{N_{i.}^2}{N..} - \sum_i \sum_j \frac{n_{ij}^2}{N_{.j}}\right]$$

$$= \frac{1}{2}(9.11 - 10.58) = -0.735$$

$$k_3 = \frac{1}{(I-1)(J-1)}\left[\sum_j \frac{N_{.j}^2}{N..} - \sum_i \sum_j \frac{n_{ij}^2}{N_{i.}}\right]$$

$$= \frac{1}{2}(6.44 - 7.55) = -0.555$$

$$k_4 = \frac{1}{(J-1)}\left[\sum_i \sum_j \frac{n_{ij}^2}{N_{.j}} - \sum_i \sum_j \frac{n_{ij}^2}{N..}\right]$$

$$= \frac{1}{2}(10.58 - 3.78) = 3.40$$

$$k_5 = \frac{1}{(J-1)}\left[\sum_i \sum_j \frac{n_{ij}^2}{N_{.j}} - \sum_i \frac{N_{.j}^2}{N..}\right]$$

$$= \frac{1}{2}(10.58 - 9.11) = 0.735$$

$$k_6 = \frac{1}{(J-1)}\left[N.. - \sum_j \frac{N_{.j}^2}{N..}\right]$$

$$= \frac{1}{2}(18 - 6.44) = 5.78$$

$$k_7 = \frac{1}{(I-1)}\left[\sum_i \sum_j \frac{n_{ij}^2}{N_{i.}} - \sum_i \sum_j \frac{n_{ij}^2}{N_{..}} \right]$$

$$= \frac{1}{2}(7.55 - 3.78) = 3.77$$

$$k_8 = \frac{1}{(I-1)}\left[N_{..} - \sum_i \frac{N_{i.}^2}{N_{..}} \right]$$

$$= \frac{1}{1}(18 - 9.11) = 8.89$$

$$k_9 = \frac{1}{(I-1)}\left[\sum_i \sum_j \frac{n_{ij}^2}{N_{i.}} - \sum_j \frac{N_{.j}^2}{N_{..}} \right]$$

$$= \frac{1}{1}(7.55 - 6.44) = 1.11$$

The estimates of variances are obtained as

$$\hat{\sigma}_e^2 = 2.17$$

$$k_i \hat{\sigma}_{RS}^2 + k_2 \hat{\sigma}_{R^2} + k_3 \hat{\sigma}_S^2 = 15.23$$

$$1.825\,\hat{\sigma}_{RS}^2 - 0.735\hat{\sigma}_R^2 - 0.555\hat{\sigma}_S^2 = 15.23$$

$$k_4 \hat{\sigma}_{RS^2} + k_5 \hat{\sigma}_{R^2} + k_6 \hat{\sigma}_{S^2} = 3.93$$

$$3.40\hat{\sigma}_{RS}^2 + 0.735\hat{\sigma}_R^2 + 5.78\hat{\sigma}_S^2 = 3.93$$

$$k_7 \hat{\sigma}_{RS}^2 + k_8 \hat{\sigma}_R^2 + k_9 \hat{\sigma}_S^2 = 2.97$$

$$3.77\hat{\sigma}_{RS}^2 + 8.89\hat{\sigma}_R^2 + 1.11\hat{\sigma}_S^2 = 2.97$$

Solving equation (16.36), we have

$$\hat{\sigma}_{RS}^2 = 6.59, \quad \hat{\sigma}_R^2 = -2.08, \quad \hat{\sigma}_S^2 = -3.02 \hspace{2cm} .....(16.36)$$

Here the estimates of variances for rations and sires have come out to be negative and which is possible since quadratic forms are considered for mean squares.

16.18.2 Henderson method- II

This is a method for correcting the bias in E(M.S) when mixed model was used. The bias in E(M.S) occurs since they contain functions of fixed effects which differ from line to line of the analysis of variance table. Therefore,

Henderson's Method −1 cannot be applied successfully to estimate variance components in mixed model.

This method uses the data to first estimate fixed effects of the model and then using these estimators, the data are adjusted. Henderson's Method-I would be used for the adjusted data to estimate the variance components. Now these estimates of variance components are unbiased. However, the method of analysis of data adjusted for fixed effects cannot be uniquely defined. For further reading please refer Henderson (1953).

16.18.3 Henderson Method-III

In this method Random effects model is considered for non-orthogonal case. For example, the model for two-way classification with interaction is

$$Y_{ijk} = \mu + a_i + b_j + c_{ij} + e_{ijk} \qquad .....(16.37)$$

where Y_{ijk} be the value on the (i, j, k)-th unit for i = 1, 2, ... , r, j = 1, 2 ... , t; k = 1, 2, .., nij. In this case E[A (elim. B)M.S], E[B (elim, A)MS] and E[AB.M.S] would be obtained. For further reading please refer to Henderson (1953).

16.19 COMPACT FAMILY BLOCK DESIGN

In this design not only progenies but also families (or crosses or field selections) would be tested. The families are randomized in each replication and the progenies are randomized within each family. The lay out in this case is similar to that of split plot design as the families are laid out in main plots and the progenies are laid out in sub-plots within each main plot. However, the statistical analysis is slightly different from that of split plot design since the families are analysed separately and the progenies within each family are analysed separately in different ANOVA Tables. Let there be 'p' families and 'q' progenies within each family with 'r' replications. The families are randomized within each replication and the progenies are randomized within each family as shown in Fig. 16.23. In Fig. 16.23 the progenies are represented by 1, 2, ..., q and the families are represented by $F_1., F_2, ... , F_p$. The analysis of variance table for testing the equality of families with respect to performance is given in Table 16.61.

The sum of squares due to replications, families and error are obtained as in the case of Randomized block design analysis.

	REP. I		REP. II		REP. III

3, 4   q

<table>
<tr><td>F_3</td><td>F_2</td></tr>
<tr><td>F_5</td><td>F_6</td></tr>
<tr><td></td><td></td></tr>
<tr><td>F_p</td><td>$F_{(p-2)}$</td></tr>
</table>
<table>
<tr><td>F_6</td><td>$F_{(p-2)}$</td></tr>
<tr><td>F_p</td><td>$F_{(p-1)}$</td></tr>
<tr><td></td><td></td></tr>
<tr><td>F_4</td><td>F_2</td></tr>
</table>
<table>
<tr><td>F_3</td><td>F_p</td></tr>
<tr><td>F_1</td><td>F_5</td></tr>
<tr><td></td><td></td></tr>
<tr><td>$F_{(p-4)}$</td><td>$F_{(p-3)}$</td></tr>
</table>

Fig. 16.23 Compact family block design.

TABLE 16.61 ANOVA TABLE

Source	d.f.	S.S	M.S	F_{cal}
Replications	(r-1)	R.S.S	R.M.S	
Families	(p-1)	F.S.S	F.M.S	F.M.S/E.M.S
Error	(p-1)(r-1)	E.S.S	E.M.S	
Total	pr-1			

The analysis of variance table for testing the equality of progenies within each family is presented in Table 16.62.

TABLE 16.62 ANOVA TABLE

Source	d.f.	S.S	M.S	F_{cal}
Replications	(r-1)	R.S.S		
Progenies	(q-1)	P.S.S	P.M.S	P.M.S/E.M.S
Error	(r-1) (q-1)	E.S.S	E.M.S	
Total	rq-1			

Here the each family is repeated in r replications and hence the total number of progenies in all the replications is qr. The Sum of squares due to progenies would be obtained by considering the total of each progeny in all the r replications. The sum of squares due to replications and error are obtained in the usual way as in 'randomized block' analysis. The equality of progenies would be tested with respect to performance for each of the families as given in Table 16.62.

The comparison of error mean squares within families would provide us an idea about the genetic variation within families. If the error mean squares are not significantly different from each other then it may be said the genetic variation between the progenies within each family is of the same order.

The comparison of error mean squares would be done with the help of Bartlett's test of homogeneity of variances given in the following sub-section 16.19.1.

16.19.1 Bartlett's Test of Homogeneity of Variances

Let s_1^2, s_2^2,..., s_k^2 be k mean squares with d.f $r_1, r_2, \ldots, r_k$ respectively.

Null Hypothesis: $\sigma_1^2 = \sigma_2^2 = \ldots = \sigma_k^2$ where $\sigma_1^2, \sigma_2^2 \ldots, \sigma_k^2$ are the variances for 1st, 2nd, ...k-th populations respectively.

$$\chi^2 = \frac{1}{1 + \dfrac{1}{3(k-1)}\left[\displaystyle\sum_1^k \frac{1}{ri} - \frac{1}{\sum_i ri}\right]} \left\{\left(\sum_i^k ri\right)\log s_c^2 - \sum_i ri \log s_i^2\right\} \qquad \ldots\ldots(16.38)$$

where $s_c^2 = \dfrac{1}{\displaystyle\sum_1 ri}\left\{\displaystyle\sum_i^k risi^2\right\}$ is the combined estimate of mean squares.

Conclusion: If X^2(calculated) > X^2(tabulated) with (k-l) d.f, at chosen level of significance, the null hypothesis is rejected. In other words there is significant difference between variances. Otherwise the null hypothesis is accepted.

16.20   SIMPLE LATTICE DESIGN

In balanced lattice design the number of replications required for conducting an experiment would be large since the replications should be (k+1) for k^2 treatments. If the number of varieties are 81 then the number of replications should be 10 which require not only large experimental field but also the large quantity of experimental seed. In this situation two, three or four replications could be maintained without much loss of experimental precision. If there are two replications then it is called 'simple lattice' three replications 'triple lattice'; four replications 'quadruple lattice' and so on.· These designs are known as 'Partially balanced Lattice' designs. The method of analysis for 'simple lattice' is illustrated by taking an example with 49 varieties and 2 replications,

EXAMPLE: An experiment was conducted with 49 varieties of *bajra* and two replications in a simple lattice layout. Seven varieties were randomized within each incomplete block. The lay out along with yields (kg) are presented here.

Rep. I

Blocks								Total(B)
1	(2)	(4)	(7)	(6)	(5)	(1)	(3)	30.2
	3.0	4.1	2.8	5.7	3.4	6.2	5.0	
2	(9)	(11)	(14)	(8)	(10)	(13)	(12)	37.7
	5.9	6.4	3.3	2.7	9.5	4.8	5.1	
3	(18)	(20)	(17)	(15)	(19)	(21)	(16)	
	3.7	2.9	7.9	8.5	7.5	9.2	8.4	48.1
4	(26)	(24)	(25)	(22)	(27)	(28)	(23)	
	6.5	5.9	4.3	5.7	3.1	4.5	3.8	33.8
5	(29)	(32)	(35)	(30)	(34)	(31)	(33)	
	9.8	9.5	5.4	6.8	7.4	5.8	4.6	49.3
6	(38)	(40)	(42)	(37)	(39)	(36)	(41)	
	4.3	4.0	5.8	5.0	9.4	7.2	6.8	42.5
7	(47)	(44)	(48)	(43)	(45)	(49)	(46)	
	5.2	4.0	3.7	4.5	6.8	9.8	10.2	44.2
								285.8

Rep. II

Blocks								Total(B)
1	(15)	(29)	(8)	(43)	(22)	(36)	(1)	40.8
	9.0	10.4	2.2	5.1	6.0	2.1	6.0	
2	(30)	(16)	(44)	(9)	(23)	(2)	(37)	
	7.5	8.0	5.8	5.1	4.6	4.1	4.2	39.3
3	(17)	(31)	(38)	(3)	(10)	(24)	(45)	
	8.9	4.9	4.6	5.2	9.1	6.2	7.5	46.4
4	(46)	(18)	(4)	(32)	(39)	(11)	(25)	
	10.4	4.3	3.7	9.8	10.2	6.1	3.8	48.3
5	(5)	(19)	(33)	(47)	(12)	(26)	(40)	
	4.0	7.8	4.0	5.9	4.9	7.2	4.3	38.1
6	(41)	(13)	(34)	(6)	(48)	(20)	(27)	
	6.3	4.5	7.8	5.2	4.6	3.2	3.2	34.8
7	(49)	(28)	(7)	(35)	(42)	(14)	(21)	
	10.1	4.4	3.4	5.6	5.6	3.7	10.0	42.8
								290.5

The values in parentheses are the varieties.

The total S.S.., Replication S.S and treatment (unadjusted) S.S. are calculated in the usual way except the blocks with in replications (adj.) S.S

$$\text{Blocks within replications (adj.)S.S.} = \sum - \frac{B^{*2}_{i}}{kr(r-1)} \sum - \frac{Ri^2 *}{k^2 r(r-1)}$$

Where B_i^* = total of all the treatments in i-th block over all replications - r. (total of i-th block)

$R^* =$ total of all B_i^*s in a replication. For example, B_i^* for block 3 and replication 1 is

$$B_3^* = 8.0 + 6.1 + 16.8 + 17.5 + 15.3 + 19.2 + 16.4 - 2(48.1) = 3.1.$$

The values of B_i^* and R_i^* are given Table 16.63

TABLE 16.63 REPLICATIONS

Block	$1\ (B_i^*)$	$11(B_i^*)$
1	1.4	3.8
2	-2.1	-2.4
3	3.1	-1.2
4	1.6	-0.7
5	0.7	-1.8
6	-5.2	-0.4
7	5.2	-2.0
R_1^*	4.7	-4.7

Blocks within replications (adj.) S.S

$$= \frac{1}{7 \times 2(2-1)}\left[(1.4)^2 + (-2.1)^2 + ...(..2.0)^2\right] - \left[\frac{(4.7)^2 + -(4.7)^2}{(7)^2 \times 2 \times (2-1)}\right] = 6.8806$$

TABLE 16.64 ANOVA TABLE

Source	d.f.		S.S	M.S
Replications	1.	$(r-1)$	22.54	0.1397
Treatments(unadj.)	48.	(k^2-1)	465.5882	9.6998
Blocks within Replications(adj.)	12.	$[r(k-1)]\,(k-1)$	6.8806	0.5734 (B.M.S)
Error	36.	$(rk-k-1)$	2.839	0.0789 (E.M.S)
Total	97	rk^2-1	475.5332	

The weighting factor for obtaining adjusted treatment totals and means is given as

$$a = \frac{(BMS) - (EMS)}{k(r-1)BMS} = \frac{0.5734 - 0.0789.}{7(2-1)0.5734} = 0.1232$$

The treatment totals are adjusted by subtracting $[a \times B_i^*]$ of the Blocks in which the treatment appears in different replications from the corresponding treatment total. The correction factors $[a\ B_i^*]$ would add to zero over all replications. For example, the adjusted total of treatment (7) would be $(2.8 + 3.4) - a(B_i^*$ of Block 1 in Rep. I$) - a(B_i^*$ of Block 7 in Rep. II) i.e., $6.2 - .1232\ (1.4) - 0.1232(-2.0) = 6.2739.$

Similarly adjusted totals were obtained for all the treatments and presented in Table 16.65.

TABLE 16.65 ADJUSTED TREATMENT TOTALS

(1)	(2)	(3)	(4)	(5)	(6)	(7)
11.56	7.37	10.18	7.71	7.45	10.78	6.27
(8)	(9)	(10)	(11)	(12)	(13)	(14)
4.69	11.55	19.01	12.84	10.48	9.61	7.51
(15)	(16)	(17)	(18)	(19)	(20)	(21)
16.65	16.31	16.57	7.70	15.14	5.77	19.06
(22)	(23)	(24)	(25)	(26)	(27)	(28)
11.03	8.50	12.05	7.99	13.72	6.15	8.95
(29)	(30)	(31)	(32)	(33)	(34)	(35)
19.65	14.51	10.76	19.30	8.74	15.16	11.16
(36)	(37)	(38)	(39)	(40)	(41)	(42)
9.47	10.14	9.69	20.33	9.16	13.79	12.29
(43)	(44)	(45)	(46)	(47)	(48)	(49)
8.49	9.46	13.81	20.05	10.68	7.71	19.51

S.E. for the difference of two treatment means in the same block is

$$= \sqrt{\frac{2(\text{E.M.S})}{r}[1+(r-1)a]}$$

$$= \sqrt{\frac{2(0.0789)}{2}[1+(2-1)0.1232]}$$

$$= 0.2977$$

S.E for the difference of two treatment means not in the same

$$\text{block} = \sqrt{\frac{2(\text{E.M.S})}{r}(1+ra)} = \sqrt{\frac{2(0.0789)}{2}[1+2(0.1232)} = 0.3136$$

$$\text{S.E of average} = \sqrt{\frac{2(\text{E.M.S})}{r}\left[\frac{1+rka}{(k+1)}\right]}$$

$$= \sqrt{\frac{2(0.0789)}{2}\left[\frac{1+2\times7\times0.1232}{(7+1)}\right]} = 0.1634$$

The ANOVA Table given in Table 16.64 is not useful for testing the adjusted treatment totals by F-test. If the Blocks within replications (adj.) is not found to be significant then the unadjusted treatment M.S. can be tested against Pooled error M.S. which comprises of Blocks within replication M.S. and error M.S.

The F-test for adjusted treatment totals is performed as follows.

The treatment (adj.) S.S. = treatment (unadj.)

$$S.S - k(r-1)a \left[\frac{r}{(r-1)(1+ka)} \right] B.S.S(unadj.) - B.S.S\ (adj.)$$

where B.S.S (unadj.), B.S.S.(adj.) are the sum of squares for unadjusted and adjusted block totals respectively. Since Blocks S.S. (adj.) is already available from ANOVA Table given in Table 16.64, only Blocks S.S (unadj.) has to be calculated.

Blocks S.S (unadj.) = 1/7(24091.63) – 3377.2458 = 64.4156

The treatment (adj.) S.S = 465.5882 – 7(2 – 1)0.1232

$$\left\{ \frac{2}{(2-1)(1+7\times 0.1232)} \right\} 64.4156 - 6.8806$$

$$= 399.0523$$

The ANOVA Table for testing adjusted treatment totals is given in Table 16.66.

TABLE 16.66 ANOVA TABLE

Source	d.f.	S.S.	M.S	Fcal
Treatments(adj.)	48	435.1209	9.0650	22.6060**
Error	36	14.4361	0.4010	

**Significant at 1 per cent level.

The gain in efficiency of simple lattice design over randomized block design is obtained by computing error M.S for R.B. design as the pooled mean square for blocks and error in Table and comparing with the corrected error variance of the simple lattice design.

$$\text{The error M.S of R.B design} = \frac{(5.9292 + 14.4361)}{48} = 0.4243$$

The corrected Variance of Lattice design

$$= E.M.S \left[1 + \frac{rka}{(k+1)} \right]$$

$$= .4010 \left[1 + \frac{2 \times 7 \times 0.1232}{(7+1)} \right] = 0.4875$$

The gain in accuracy over randomized blocks

$$= \frac{.4243}{0.4875} \times 100 = 87\%$$

For analysis of Balanced Lattice design the reader is advised to refer Cochran and Cox (1957).

16.21   COMBINED ANALYSIS OF EXPERIMENTS

It is necessary sometimes to repeat the same experiment at different research stations (places) to arrive at a decision which can be recommended to many places rather than confine to one locality. For example, a variety (varieties) is to be stabilized in a particular region, a field experiment will be conducted at one place and repeated at several places in that region to know whether there is any interaction between places and varieties. If a particular variety consistently shows good response at all the places then that variety can be recommended in that region. Similarly for fixing of fertilizer doses in an agroclimatic region, the experiment has to be repeated in several places in that region. To arrive at a decision the data obtained from different places (or centres) will be pooled for carrying out the statistical analysis.

EXAMPLE: The hypothetical data (total yields) in kg for randomized block layout with 5 replications at 5 different research stations for comparing the 6 varieties of paddy are presented in Table 16.67. The error M.S. per plot for each experiment was obtained and presented in the same table.

TABLE 16.67 RESEARCH STATION

Variety	1	2	3	4	5	Total
Tella Hamsa	75	96	84	95	102	452
Rajendra	82	80	95	97	75	429
I.R.8	70	72	70	89	84	385
Jaya	85	94	69	82	95	425
Padma	60	65	74	75	84	358
Local	50	45	66	58	62	281
Total	422	452	458	496	502	2330
Error m.s	15.2	12.4	10.1	17.5	19.4	
Error d.f.	20	20	20	20	20	

It was found that F-test for treatments showed significant difference among varieties at each research station.

A combined analysis for the data given in Table 16.67 can be worked out since the same Randomized block design was repeated at all the five research stations. Before proceeding for combined analysis it is necessary to test the error M.S for their homogeneity for all the experiments with the help of Bartlett's test given in sub-section 16.19.1. If the Bartlett's test shows homogeneity of error M.S. for all the experiments, we proceed for computing combined analysis by assuming the analysis of variance model as

$$Y_{ij} = \mu + \alpha_i + \beta_j + \gamma_{ij} + \overline{E}_{ij} \qquad \qquad .....(16.39)$$

where $\alpha_i \, \beta_j$, represent the effects of place and variety respectively, γ_{ij}, that of variety × place interaction and $\overline{E}_{ij}$ are the average of the experimental errors on the r plots that receive the variety at that place.

Applying Bartlett's test as given in Section 16.19.1, χ^2 was found to be not significant at 5 per cent level with 4 d.f. Therefore, the error mean squares can be considered as homogeneous, Hence the error mean squares of all experiments can be pooled i.e., $s_c^2 = 14.92$, Places S.S

$$= 1/30\left[(422)^2 + (452)^2 + ... + (502)^2\right] - \frac{(2330)^2}{150} = 146.40$$

Varieties S.S $= 1/25[(452)^2 + (429)^2 + .. + (281)^2]$

$$-\frac{(2330)^2}{150} = 780.13$$

Table S.S $= 1/5[(75)^2 + (82)^2 + ... + (62)^2]$

$$-\frac{(2330)^2}{150} = 1229.73$$

Variety × place S.S = Table S.S – (Place S.S + Varieties S.S)
$$= 303.20$$

TABLE 16.68 COMBINED ANOVA TABLE

Source	d.f.	S.S.	M.S.
Varieties	5	780.13	156.03
Places	4	146.40	36.60
Varieties x place	20	303.20	15.16
Pooled error	100		14.92

For testing the varieties and places we shall ascertain first whether a variety x places interaction is significant or not. If interaction is found to be significant, varieties and places M.S. would be tested against interaction M.S. If interaction is not found significant the interaction M.S. can be pooled with pooled error M.S. by obtaining weighted average. This can be seen by observing the expected mean squares of varieties, places, varieties × places and pooled error from the Table 16.69

TABLE 16.69

Source	d.f.	S.S.	M.S.	E(M.S.)
Varieties	$v - 1 = 5$	780.13	156.03	$\sigma_e^2 + r\,\sigma_I^2 + rp\,\sigma_v^2$
Places	$p - 1 = 4$	146.40	36.60	$\sigma_e^2 + r\,\sigma_I^2 + rv\,\sigma_p^2$
Varieties x places	$(v - 1)(p - 1) = 20$	303.20	15.16	$\sigma_e^2 + r\,\sigma_I^2$
Pooled error	$p(v - 1) \times (r - 1) = 100$		14.92	σ_e^2

where v, p, r be the number of varieties, places and replications respectively and σ_v^2, σ_p^2, σ_I^2 and σ_e^2 are the variances due to varieties, places, varieties × places interaction and experimental error respectively.

From the E(M.S) column of Table 16.69 it can be noted that for testing $\sigma_I^2 = 0$, varieties × places M.S would be tested against pooled error M.S. In the present case F = 15.16/14.92 = 1.02. The F-value was found to be not-significant at 5 percent level and hence the interaction of varieties × places sum of squares can be pooled with pooled error sum of squares as

$$\frac{20 \times 15.16 + 100 \times 14.92}{120} = 14.96$$

Since the interaction was not found to be significant, the varieties M.S. would be tested against the pooled M.S. of interaction and pooled error, Therefore, F = 36.60/14.96 = 2.45.

The F-value was found to be not significant with (4, 120) d.f. at 5 per cent level of significance.

In case the Interaction is found to be significant, the varieties and places M.S. would be tested against interaction M.S.

16.21.1

If error mean squares found to be significantly different from place to place as evidenced by Bartlett's test, then the procedure of testing treatments would be different. If the interaction of treatments × places was found to be significant then the treatments means squares will be tested against interaction mean square. If the interaction is not found to be significant then the weighted analysis of variance has to be followed. In this method each treatment total is divided by the corresponding error mean square of that place. For further reading please refer to Cochran and Cox (1957).

16.22 RESPONSE SURFACE

If in an agricultural experiment, yield is influenced by several factors like height of the plant, length of ear head, temperature, relative humidity, etc., which are all quantitative variables then the yield (or response) is a function of the levels of these variables and is denoted by

$$Y_j = \Phi(X_{1j}, X_{2j}, \ldots, X_{kj}) + E_j \qquad \ldots\ldots(16.40)$$

where j = 1, 2, …, n represent the j-th observation in the factorial experiment and X_{ij}, denotes the level of i-th factor of the j-th observation and E_j's are experimental errors which are assumed to be independent and follow normal distribution with mean zero and variance σ_k^2. The function 'ϕ' is called response surface. If ϕ is known then it is easy to predict (or forecast) the value for knowing the different levels of factors. Further the combination of levels of factors can be arrived at to attain the optimum and maximum response once the function is known.

In the absence of knowledge of the function it can be assumed that the experimental region can be represented by a polynomial of first or second degree. The designs used for fitting the first degree and second degree polynomials are called first order and second order designs respectively. The fitting of second order polynomial is illustrated here with an example.

EXAMPLE: An experiment was conducted with nitrogen at four levels (40, 60, 80, 100 kg/acre) along with phosphorus at three levels (15,30,45 kg/acre) in a layout of randomized block design having three replications for paddy. The hypothetical yields are presented in the following Table 16.70.

TABLE 16.70 REPLICATIONS

Treatment	1	2	3	Total
n_0p_0	8	10	9	27
n_0p_1	10	14	12	36
n_0p_2	11	9	10	30
n_1p_0	13	12	15	40
n_1p_1	15	14	13	42
n_1p_2	18	16	19	53
n_2p_0	14	12	10	36
n_2p_1	20	22	22	64
n_2p_2	24	26	25	75
n_3p_0	16	15	18	49
n_3p_1	22	20	23	65
n_3p_2	28	27	29	84
	199	197	205	601

Fit the response surface for the above data.

The two-way table of nitrogen and phosphorus with plot yield totals of three replications is given in Table 16.71.

TABLE 16.71 PHOSPHORUS

Nitrogen	15	30	45	Total
40	27	36	30	93
60	40	42	53	135
80	36	64	75	175
100	49	65	84	198

The factorial analysis is presented in the ANOVA Table 16.72.

TABLE 16.72 ANOVA TABLE

Source	d.f	S.S.	M.S.	Fcal
Replications	2	2.89	1.45	
Treatments	11			
N	3	711.42	237.14	117.40**
P	2	343.06	171.53	84.92**
NP	6	177.83	29.64	14.67**
Error	22	44.44	2.02	
Total	35	1279.64		

** Significant at 1 per cent level.

To examine the trend in yield for different levels of nitrogen and phosphorus, the linear, quadratic components for phosphorus; linear, quadratic and cubic components for nitrogen as well as for NP interaction were computed as follows. The coefficients of orthogonal polynomials for linear and quadratic components for phosphorus levels are $(-1,0,1)$ and $(1, -2, 1)$ respectively, the coefficients for nitrogen levels for linear, quadratic and cubic components are $(-3, -1, +1, +3)$, $(+1, -1, -1, +1)$. $(-1, +3, -3, +1)$ respectively.

TABLE 16.73 NITROGEN LEVELS

	40	60	80	100	
$P_{L\ (linear)}$	3	13	39	35	90
$P_{Q\ (quadratic)}$	$--15$	9	-17	3	-20

$$P_L.S.\ S. = \frac{(90)^2}{3 \times 4 \times 2} = 337.50, \qquad P_Q.S.\ S. = \frac{(-20)^2}{3 \times 4 \times 6} = 5.56$$

It can be verified that $P_L.S.S + P_Q.S.\ S = P.S.S$

Similarly the linear, quadratic and cubic components for nitrogen are computed as follows.

TABLE 16.74 PHOSPHORUS LEVELS

	15	30	45	
$N_{L\ (Linear)}$	62	109	184	355
$N_{Q\ (quadratic)}$	0	-5	-14	-19
$N_{C\ (cubic)}$	34	-37	-12	-15

$$N_L.\ S.\ S = \frac{(355)^2}{3 \times 3 \times 20} = 700.14$$

$$N_Q.\ S.S. = \frac{(-19)^2}{3 \times 3 \times 4} = 10.03$$

$$N_C.\ S.S. = \frac{(-15)^2}{3 \times 3 \times 20} = 1.25$$

The interactions of P_L with N_L, N_Q, N_c and P_Q with N_L, N_Q, N_c would be computed as follows.

$$P_L\ N_L = -3\,(3) - 1\,(13) + 1\,(39) + 3\,(35) = 122$$

$$P_L N_Q = +1\,(13) - 1\,(13) - 1\,(39) + 1\,(35) = -14$$

$$P_L N_C = -1\,(3) + 3\,(13) - 3\,(39) + 1\,(35) = -46$$

$$P_L N_L.\ S.\ S. = \frac{(122)^2}{3 \times 2 \times 20} = 124.03,$$

$$P_L N_Q.\ S.\ S. = \frac{(-14)^2}{3 \times 2 \times 4} = 8.17$$

$$P_L N_C.\ S.S. = \frac{(-46)^2}{3 \times 2 \times 20} = 17.63$$

$$P_Q L_N = -3\,(-15) - 1\,(9) + 1(-17) + 3\,(3) = 28$$

$$P_Q N_Q = +1(-15) - 1(9) - 1\,(-17) + 1\,(13) = -4$$

$$P_Q N_C = -1\,(-15) + 3\,(9) - 3\,(-17) + 1\,(3) = 96$$

$$P_Q N_L.\ S.\ S = \frac{(28)^2}{3 \times 6 \times 20} = 2.18,$$

$$P_Q N_Q.\ S.\ S. = \frac{(-4)^2}{3 \times 6 \times 4} = 0.22$$

$$P_Q N_C.\ S.\ S. = \frac{(96)^2}{3 \times 6 \times 20} = 25.60$$

The Analysis of variance table given in Table 16.72 is now rewritten as presented in Table 16.75.

TABLE 16.75 ANOVA TABLE

Source	d.f.	S.S	M.S.	Fcal
Replications	2	2.89	1.45	
N	3	711.42		
N_L	1	700.14	700.14	346.60**
N_Q	1	10.03	10.03	4.97
N_C	1	1.25	1.25	
P	2	343.06		
P_L	1	337.50	337.50	167.08**
P_Q	1	5.56	5.56	
NP	6			
$N_L P_L$	1	124.03	124.03	61.40**
$N_L P_Q$	1	2.18	2.18	
$N_Q P_L$	1	8.17	8.17	
$N_Q P_Q$	1	0.22	0.22	
$N_C P_L$	1	17.63	17.63	8.73**
$N_C P_Q$	1	25.60	25.60	12.67**
Error	22	44.44		
Total	35	1279.64		

* Significant at I per cent level.

From Table 16.75 it can be observed that the yield is significantly effected by linear and quadratic trend of nitrogen, linear trend of phosphorus, linear trend of nitrogen with linear trend of phosphorus, cubic trend of nitrogen with linear trend of phosphorus cubic trend of nitrogen with quadratic trend of phosphorus.

The response surface is the mathematical relation taking yield as the dependent variable and the above mentioned factors as independent variables. Let the relation between yield and N_L, N_Q, P_L, $N_L P_L$, $N_C P_L$ and $N_C P_Q$ is given by

$$\hat{Y} = \overline{Y} + b_1 X_1 + b_2 X_2 + b_3 X_3 + b_4 X_4 + b_5 X_5 + b_6 X_6$$

where Y is the estimated value of yield and

$X_1 = N_L$, X_2, N_Q, $X_a = P_L$, $X_4 = N_L P_L$ $X_5 = N_C P_L$, $X_6 = N_C P_Q$ and b_i,'s are regression coefficients.

In order to find out the regression coefficients (b_i.'s), the coefficients of orthogonal polynomials will be used as given in the Table 16.76 since $b_i = \Sigma X_I Y / \Sigma X_i^2$

TABLE 16.76

Nitrogen levels	Phosphors levels	Yield total (Y)	$N_L=0.1$ $(N-70)$	N_Q	PL	N_LP_L	N_CP_L	N_CP_Q
40	15	27	−3	+1	−1	+3	+1	-1
	30	36	−3	+1	0	0	0	+2
	45	30	−3	+1	1	−3	-1	-1
60	15	40	−1	−1	−1	+1	-3	+3
	30	42	−1	−1	0	0	0	-6
	45	53	−1	−1	1	-1	+3	+3
80	15	36	+1	−1	−1	-1	+3	-3
	30	64	+1	−1	0	0	0	+6
	45	75	+1	−1	1	+1	-3	-3
100	15	49	+3	+1	−1	-3	-1	+1
	30	65	+3	+1	0	0	0	-2
	45	84	+3	+1	1	+3	+1	+1
$\Sigma X_i Y$		601	355	−19	90	122	−46	96
ΣX_i^2			60	12	8	40	40	120
b_i			5.9167 -1.5833		11.2500 3.0500		−1.1500	0.800

The response surface between yield and X_i s' is given by

$$\hat{Y} = 50.0833 + 5.91767\ X_1 - 1.5833\ X_2 + 11.2500\ X_3 + 3.0500\ X_4 - 1.1500\ X_5 + 0.8000 X_6$$

The same relation between yield and N_L, N_Q P_L, N_LP_L , N_CP_L and N_CP_Q is rewritten as

$$\hat{Y} = 50.0833 + 5.9167\ N_L - 1.5833\ N_Q + 11.2500\ P_L + 3\ 0500\ N_LP_L - 1.1500\ N_CP_L + 0\ 800\ N_CP_Q$$

The estimated yields can be obtained for given levels of nitrogen and phosphorus,

For example, the estimated yield, when the level of nitrogen is 80 and the level of phosphorus is 45, is obtained by substituting in the fitted equation for $N_L = +\ 1$, $N_Q = -1$, $P_L = +1$, $N_LP_L = 1$, $N_CP_L = -3$ and $N_CP_L = -\ 3$ and $N_CP_Q = -\ 3$ i.e

$$\hat{Y} = 50.0833 + 5.9167\ (+1) - 1.5833\ (-1) + 11.2500\ (+1) + 3.0500\ (+1) - 1.1500\ (-3) + 0.8000\ (-3) = 74.5333.$$

FITTING OF RESPONSE SURFACE

SPSS DATA ANALYSIS PROCEDURE

EXAMPLE: An experiment was conducted with nitrogen at four levels (40,60,80,100kg/h) along with phosphorus at three levels(15,30,45kg/h) in a lay out   RBD   having 3 replications of paddy. The coefficients of orthogonal polynomials for 4 levels of nitrogen are Linear (-3,-1,+1,+3) ,Quadratic (-1,-1,+1,+1) and Cubic (-1,+3,-3,+1) and for 3 levels of P are Linear (-1,0,+1),Quadratic (1,-2,1). The table containing data to be entered into SPSS Excel sheet is givXen as follows. Here $X_1 = N_1, X_2 = N_q, X_3 = P_1$,$X_4 = N_1 P_1, X_5 = NcPL, X_6 = N_C P_Q$

N	P	Y	X_1	X_2	X_3	X_4	X_5	X_6
40	15	27	-3	$+1$	-1	$+3$	$+1$	-1
	30	36	-3	$+1$	0	0	0	$+2$
	45	30	-3	$+1$	1	-3	-1	-1
60	15	40	-1	-1	-1	1	-3	$+3$
	30	42	-1	-1	0	0	0	-6
	45	53	-1	-1	1	-1	$+3$	$+3$
80	15	36	$+1$	-1	-1	-1	$+3$	-3
	30	64	$+1$	-1	0	0	0	$+6$
	45	75	$+1$	-1	1	$+1$	-3	-3
100	15	49	$+3$	$+1$	-1	-3	-1	$+1$
	30	65	$+3$	$+1$	0	0	0	-2
	45	84	$+3$	$+1$	1	$+3$	$+1$	$+1$

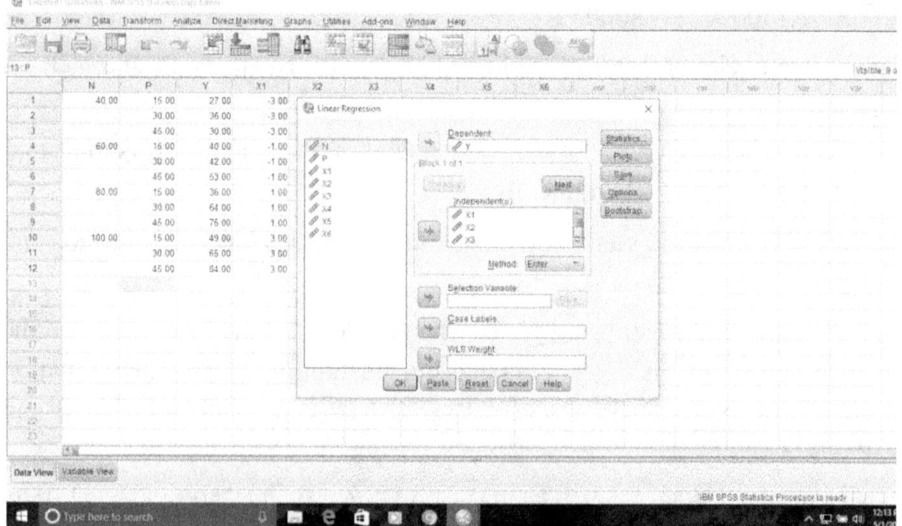

After entering the data column wise in the manner given above in SPSS Excel sheet ,CLICK ANALYSE FROM TOOL BAR and then OPT for 'LINEAR REGRESSION' AND 'LINEAR'. EnterX_1, X_2 ... X_6 as independent variables and Y as dependent variable. Go for 'OPTIONS' IF NECESSARY. Finally CLICK 'O.K.' We get display table of regression analysis containing partial regression coefficients and their test of significance including 't-values' and 'probabilities'. The 'X' symbols can be converted back to N's and Ps in the fitted multiple regression equation. The expected values can be obtained from the fitted equation by substituting the polynomial coefficients accordingly.

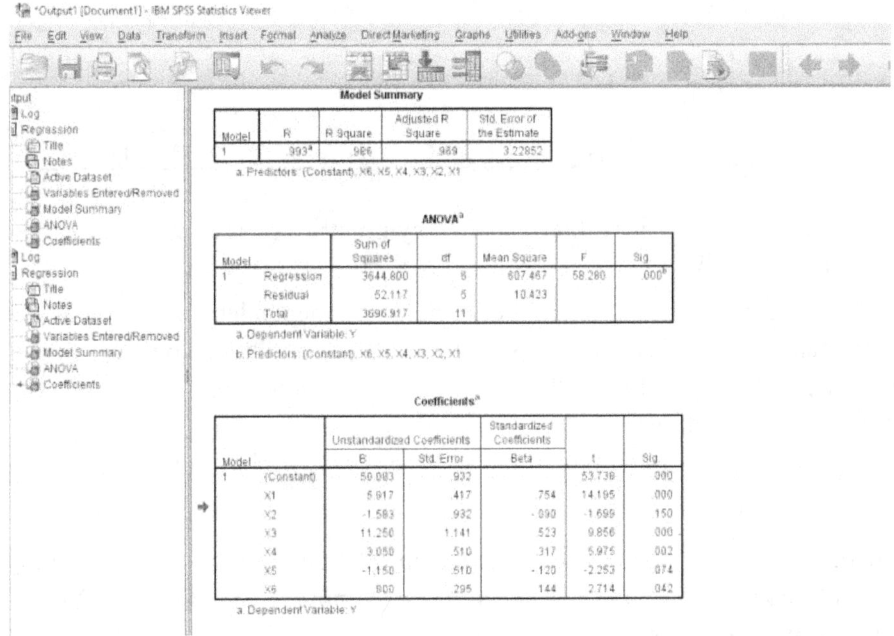

Model Summary

Model	R	R Square	Adjusted R Square	Std. Error of the Estimate
1	.993[a]	.986	.989	3.22852

a. Predictors: (Constant), X6, X5, X4, X3, X2, X1

ANOVA[a]

Model		Sum of Squares	df	Mean Square	F	Sig.
1	Regression	3644.800	6	607.467	58.280	.000[b]
	Residual	52.117	5	10.423		
	Total	3696.917	11			

a. Dependent Variable: Y

b. Predictors: (Constant), X6, X5, X4, X3, X2, X1

Coefficients[a]

Model		Unstandardized Coefficients B	Std. Error	Standardized Coefficients Beta	t	Sig.
1	(Constant)	50.083	.932		53.738	.000
	X1	5.917	.417	.754	14.195	.000
	X2	-1.583	.932	-.090	-1.699	.150
	X3	11.250	1.141	.523	9.856	.000
	X4	3.050	.510	.317	5.975	.002
	X5	-1.150	.510	-.120	-2.253	.074
	X6	.800	.295	.144	2.714	.042

a. Dependent Variable: Y

16.23 PATH COEFFICIENT ANALYSIS

It is observed that there will be not only direct influence' (effect) of independent character on the dependent character but also indirect influence on it through independent characters. For example, in plant breeding experiment, yield is influenced by characters like length of the ear, number of panickles, number of tillers, test weight, height of the plant, etc. Number of panickles, for instance not only influences yield directly but also indirectly through length of the ear-shoot, number of tillers, etc. Therefore, in breeding experiment it is essential to identify the character (characters) which influence the yield directly as well as indirectly by inheritance from generation to generation.

Path coefficient analysis helps us to identify the different independent characters which affect the dependent character directly as well as indirectly.

It gives us the path in which an independent variable is affecting the dependent variable in a given set of independent variables.

Let k independent variables be significantly correlated with dependent variable Y then the correlation matrix representing correlation coefficients (phenotypic or genotypic) is given in Table 16.77.

TABLE 16.77 CORRELATION MATRIX

	Y	1	2	3	....	k
Y	1					
1	r_{1Y}	1				
2	r_{2Y}	r_{21}	1			
3	r_{3Y}	r_{31}	r_{32}	1		
;	:	: :	:	:		
k	r_{ky}	r_{k1}	r_{k2}	r_{k3}	...	1

The matrix given in Table16.77 is a symmetric matrix i.e. $r_{1Y} = r_{Yl}$, $r_{21} = r_{12}$,... r_{k-l}, $k = r_k$, k_{-1}

The correlation coefficient between i-th independent variable X and dependent variable Y is linearly related with the correlation coefficients of i-th independent variable with the remaining independent variables. This relation is denoted by

$$r_{1Y} = P_{1y}r_{i1} + P_{2y}r_{i2} + \ldots + P_{i-1y} r_{i,i-1} + \ldots + P_{iy} + P_{i+1y}r_{1,i+1} + \ldots + P_{ky}r_{ik} \text{ for}$$

$$i = 1,2, \ldots k \qquad \ldots(16.41)$$

where $P_{1Y} P_{2Y}\ldots, P_{ky}$ are the coefficients in the linear relation and are known as path coefficients; r_{il}, $r_{i2},\ldots r_{ik}$ are the simple correlation coefficients (phenotypic or genotypic) among the independent variables and r_{ly} is the simple correlation coefficient between i-th independent variable X_l and the dependent variable Y. P_{iY} is called the direct effect of X_i on Y and P_{1Y}. r_{il}, $P_{2y}r_{i2}\ldots P_{i-1}$, $yr_{i,i-l'}$ $p_{i+1,y} + 1,yr_{i,i+1} \ldots P_{ky}r_{ik}$ are called the indirect effects of X_i, on Y through $X_1,X_2\ldots X_{i-l}$, $X_{1+1} \ldots,X_k$ respectively. Therefore the simple correlation coefficient (Total effect) between X_i and Y is the sum of direct and indirect effects of X_i, on Y. The linear relations are represented by matrix notation as

$$\underset{k \times k}{\begin{bmatrix} 1 & r_{12} & r_{13} & \cdots & r_{1k} \\ r_{21} & 1 & r_{23} & \cdots & r_{2k} \\ \vdots & \vdots & \vdots & \vdots & \vdots \\ r_{k1} & r_{k2} & r_{k3} & \cdots & 1 \end{bmatrix}} \underset{k_{x1}}{\begin{bmatrix} P_{1y} \\ P_{2y} \\ \vdots \\ P_{ky} \end{bmatrix}} = \underset{k_{x1}}{\begin{bmatrix} r_{1y} \\ r_{2y} \\ \vdots \\ r_{ky} \end{bmatrix}}$$

Hence

$$\begin{bmatrix} P_{1y} \\ P_{2y} \\ \vdots \\ P_{ky} \end{bmatrix} = \begin{bmatrix} 1 & r_{12} & r_{13} & \cdots & r_{1k} \\ r_{21} & 1 & r_{23} & \cdots & r_{2k} \\ \vdots & \vdots & & \vdots & \vdots \\ r_{k1} & r_{k2} & r_{k3} & \cdots & 1 \end{bmatrix} \begin{bmatrix} r_{1y} \\ r_{2y} \\ \vdots \\ r_{ky} \end{bmatrix}$$

Therefore, the path coefficients are obtained and hence the direct and indirect effects can be 'obtained.

Further, the residual effect is obtained by the relation

$$P_{Ry} = \sqrt{1 - (P_{1y}r_{1y} + P_{2y}r_{2y} + \ldots + P_{ky}r_{ky})}$$

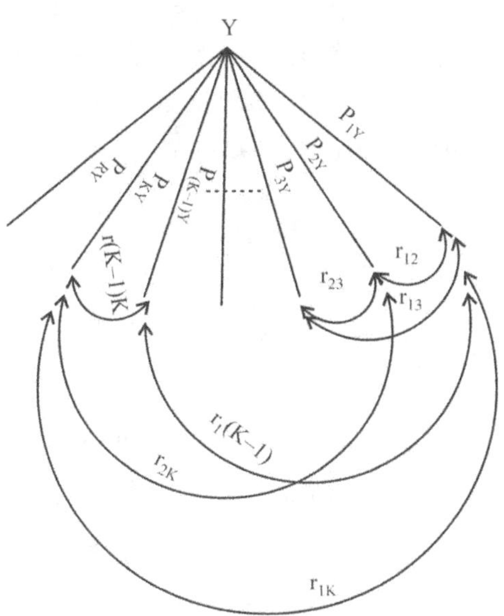

Fig. 16.24 Path coefficient diagram.

EXAMPLE: The following are the significant (hypothetical) phenotypic correlation coefficients between 5 independent variables (height of the plant, length of ear head, test weight, number of panickles, volume of the grain) and yield in bajra based on 21 observations. Obtain the path analysis.

TABLE 16.78 CORRELATION MATRIX

	Y	*1*	*2*	*3*	*4*	*5*
Y	1					
1	0.68	1				
2	0.85	0.59	1			
3	0.92	0.67	0.95	1		
4	0.88	0.76	0.90	0.94	1	
5	0.74	0.68	0.89	0.96	0.87	1

The coefficients are obtained from the following relation in matrix notation.

$$\begin{bmatrix} P_{1y} \\ P_{2y} \\ P_{3y} \\ P_{4y} \\ P_{5y} \end{bmatrix} = \begin{bmatrix} 1 & 0.59 & 0.67 & 0.76 & 0.68 \\ 0.59 & 1 & 0.95 & 0.90 & 0.89 \\ 0.67 & 0.95 & 1 & 0.94 & 0.96 \\ 0.76 & 0.90 & 0.94 & 1 & 0.87 \\ 0.68 & 0.86 & 0.96 & 0.87 & 1 \end{bmatrix}^{-1} \begin{bmatrix} 0.68 \\ 0.85 \\ 0.92 \\ 0.88 \\ 0.74 \end{bmatrix}$$

After inverting the matrix, we have

$$\begin{bmatrix} P_{1y} \\ P_{2y} \\ P_{3y} \\ P_{4y} \\ P_{5y} \end{bmatrix} = \begin{bmatrix} 3.1598 & 1.1499 & 3.8356 & -4.4369 & -2.9941 \\ 1.1499 & 11.3733 & -12.2597 & -1.3923 & 2.0764 \\ 3.8356 & -12.2597 & 51.7796 & -18.6400 & -25.1887 \\ -4.4369 & -1.3923 & -18.6400 & 15.9426 & 8.2805 \\ -2.9941 & 2.0764 & -25.1887 & 8.2805 & 18.1652 \end{bmatrix}$$

$$\times \begin{bmatrix} 0.68 \\ 0.85 \\ 0.92 \\ 0.88 \\ 0.74 \end{bmatrix}$$

Therefore, $P_{1y} = 3.1598 \times 0.68 + 1.1499 \times 0.85 + \ldots -2.9941 \times 0.74 = 0.5347$, $P_{2y} = -0.5183$. Similarly, $P_{3y} = 4.7819$, $P_{4y} = -1.1922$, $P_{5y} = -2.7156$. The direct and indirect effects for height of plant (X_1) are computed as directed effect, $p_{1y} = 0.53477$.

Indirect effect though $\quad X_2 = P_{2Y}r_{12} = 0.5183 \times 0.59 = -.3058$

"   " $X_3 = P_{3Y}r_{13} = 4.7819 \times 0.67 = 3.2039$

"   " $X_4 = P_{4Y}r_{14} = -1.1922 \times 0.76 = -0.9061$

"   " $X_5 = P_{5Y}r_{15} = -2.7156 \times 0.68 = -1.8466$

TABLE 16.79 DIRECT AND INDIRECT EFFECTS

X_1	X_2	X_3	X_4	X_5	r_{iy}
X_1 <u>0.5347</u>	−.3058	3.2039	−0.9061	−1.8466	0.68
X_2 0.3155	<u>−0.5183</u>	4.5428	−1.0730	−2.4169	0.85
X_3 0,3582	−0.4924	<u>4.7819</u>	−1.1207	−2.6070	0.92
X_4 0.4064	−0.4665	4.4950	<u>−1.1922</u>	−2.3626	0.88
X_5 0.3636	−0.4613	4.5906	−1.0372	<u>−2.7156</u>	0.74

It can be verified that the total of direct and indirect effects should be equal to r_{1y} i.e., $0.5347 - 0.3058 + 3.2039 - 0.9061 - 1.8466 = 0.68$ (r_{1y}). Similarly the indirect effects are calculated for other variables and presented in Table 16.79.

The values underlined in Table 16.79 are the direct effects. The residual effect is

$$P_{Ry} = \sqrt{1 - (P_{1y}r_{1y} + P_{2y}r_{2y} + \ldots + P_{5y}r_{5y})} = 0.5135$$

The direct, indirect and residual effects are shown by diagram given in Fig. 16.25.

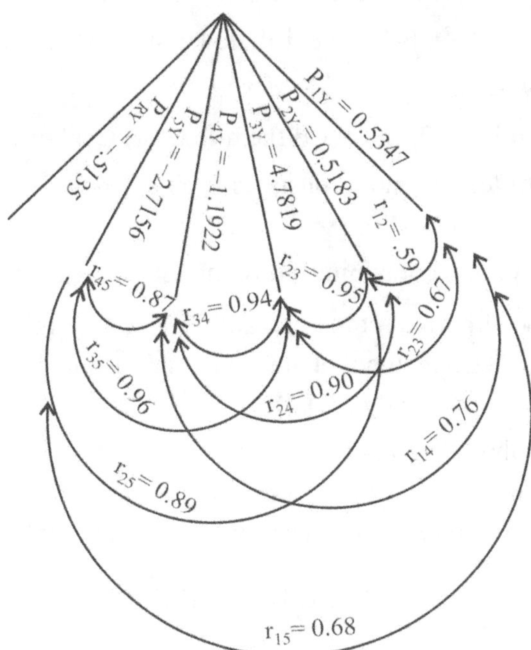

Fig. 16.25 Path coefficient diagram.

16.24 LEAST SQUARES PROCEDURE-TWO-WAY ANALYSIS OF VARIANCE

Kemp (1972) gave the procedure for least squares Analysis of data with unequal sub class numbers. The mathematical model is

$$Y_{IJk} = \mu + \alpha_i + \beta_J + \alpha\beta_{iJ} + \delta_{iJk}$$

for $i = 1, 2, \ldots, a$; $j = 1, 2, \ldots, b$; $k = 1, 2, \ldots, n_{ij}$ and 'a' be the number of levels of factor A, 'b' be the number of levels of factor B and nij be the number of observations for the AB_{ij} combination of factors. Y_{iJk} be the k-th observation on the ij-th treatment combination, μ the overall population mean when equal frequencies exist among the AB sub class, α_i the effect of i-th level of factor A, β_J the effect of j-th level of factor B, $\alpha\beta_{iJ}$ the effect of ij-th AB sub class after average effects of A and B have been removed and δ_{iJk}'S are random errors which are assumed to be normally independently distributed with mean zero and variance ae^2 • The design matrix 'X' is formed as follows:

$X_{i\cdot k} = 1$ for i = 1,2…, a –1 if the level of the A effect is present.

$X_{i\cdot k} = 0$ otherwise, except

$X_{i\cdot k} = -1$ for all i = 1, 2, .., a-1 if a-th (last) level of A is present.

$X_{j\cdot k} = 1$ for j = 1, 2,…, b-1 if the j-th level of the B effect is present

$X_{\cdot jk} = 0$ otherwise, except

$X_{\cdot jk} = -1$ for all j = 1, 2, …, b-1 if the b-th (last) level of B is present.

The estimate of least squares constants is obtained by

$$\hat{\mathscr{F}} = (X^1 X)^{-1} X1Y$$

where $\hat{\mathscr{F}}$ is the vector containing constant estimates $\hat{\alpha}_i, \hat{\beta}_j$ and $\hat{\alpha\beta}_{ij}$ $X^1 X$ is the set of reduced normal equations and $X^1 Y$ is the right side of reduced normal equations. The method of analysis is illustrated with an example hereby taking data from Reddy (1980).

EXAMPLE: In this experiment there are two factors each at two levels. The first factor consists of Brown and Jodipi strains of cows and the other factor as sex and their body weights are presented in the following table.

TABLE 16.80 BREED

Sex	Brown	Jodipi	
Male	23	8	31
Female	27	28	55
	50	36	86

The mathematical model is

$$Y_{ijk} = \mu + \alpha_i + \beta_j + \gamma_{ij} + \delta_{ijk}$$

for $i = 1, 2; j = 1, 2; k = n_{ij}$

For obtaining reduced normal equations the design matrix (X) would be constructed as follows.

TABLE 16.81 DESIGN MATRIX

All obs. X_0	Sex totals X_1	X_1	X_{11}

The X^1X matrix is obtained as the sum of the products of the columns in X matrix.

$$X^1X = \begin{bmatrix} 86 & -24 & 14 & 16 \\ -24 & 86 & 16 & 14 \\ 14 & 16 & 86 & -24 \\ 16 & 14 & -24 & 86 \end{bmatrix}$$

The inverse of X^1X matrix is

$$(X^1X)^{-1} = \begin{bmatrix} 0.015077 & 0.005983 & -0.005012 & -0.005178 \\ 0.005983 & 0.015077 & -0.005178 & -0.005012 \\ -0.005012 & -0.005178 & 0.015077 & 0.005983 \\ -0.005178 & -0.005012 & 0.005983 & 0.015077 \end{bmatrix}$$

The X^1Y column vector is obtained as

$$X^1Y = \begin{bmatrix} 1964.3 \\ 700.3 \\ 209.7 \\ 328.3 \end{bmatrix} \begin{matrix} \rightarrow \text{ Grand total} \\ \rightarrow \text{ Total of male } - \text{ total of female} \\ \rightarrow \text{ Total of Brown } - \text{ total of Jodipi} \\ \rightarrow \text{ (Male Brown } + \text{ Female Jodipi} - \end{matrix}$$

(Female Brown + Male Jodipi)

$$(X^1X)^{-1}.X^1Y = \begin{bmatrix} 22.674900 \\ -1.537282 \\ -1.093052 \\ -0.456828 \end{bmatrix}$$

Estimates of parameters

$\mu = 22.6759$, Breed $= -1.093$

Sex $= -1.5373$, Interaction $= -0.457$

Different effects:

$\hat{\mu} = 22.675$, $\hat{\alpha}_1$ (Male) $= -1.5373$, $\hat{\alpha}_2$ (Female) $= 1.5373$

$\hat{\beta}_1$ (Brown) $= -1.093$, $\hat{\beta}_2$ (Jodipi) $= 1.093$

$\hat{\gamma}_{11} = -0.457$, $\hat{\gamma}_{12} = 0.457$, $\hat{\gamma}_{21} = 0.457$

$\hat{\gamma}_{22} = -0.457$

Least squares Analysis of variance:

$$\text{S.S. due to sex} = (-1.537)\ \frac{1}{0.015077}\ (-1.537) = 156.687$$

$$\text{S.S. due to Breed} = (-1.0.93)\ \frac{1}{0.015077}\ (-1.0.93) = 79.237$$

$$\text{Sex} \times \text{Breed S.S} = (-0.457)\ \frac{1}{0.015077}\ (-0.457) = 13.852$$

Total Crude S. S $= 46220.6$ (without deducting correction factor)

TABLE 16.82 ANOVA TABLE

Source	d.f.	M.S.	F
Sex	1	156.687	16.370*
Breed	1	79.237	8.278*
Sex × Breed	1	13.825	1.447NS
Error	82	9.752	

*Significant at 5 per cent level

Regression S.S $= (X^1X)^{-1}X^1Y.X^1Y$

$$= (22.675 \times 1964.3) + \ldots + (3283 \times -0.457) = 45237.675$$

Eror S.S. $=$ Total S.S. $-$ Regression S.S.

$$= (46220.6 - 45237.675) = 784.924$$

Least square means:

Male $= \mu + \hat{\alpha}_1 = (22.675 - 1.5373) = 21.1377$

Female $= \mu + \hat{\alpha}_2 = (22.675 + 1.5373) = 24.2123$

Brown $= \mu + \hat{\beta}_1 = (22.675 - 1.093) = 21.582$

$$\text{Jodipi} = \mu + \hat{\beta}_2 = (22.675 + 1.093) = 23.768$$

$$\bar{y}_{11} = \mu + \hat{\alpha}_1 + \hat{\beta}_1 + \hat{\alpha\beta}_{11} = 19.58765$$

$$\bar{y}_{12} = \mu + \hat{\alpha}_1 + \hat{\beta}_2 + \hat{\alpha\beta}_{12} = = 22.68775$$

$$\bar{y}_{21} = \mu + \hat{\alpha}_2 + \hat{\beta}_1 + \hat{\alpha\beta}_{21} = 22.66225$$

$$\bar{y}_{22} = \mu + \hat{\alpha}_2 + \hat{\beta}_2 + \hat{\alpha\beta}_{22} = 25.76235$$

Standard errors of means

S.E. $(\mu) = \sqrt{C_{\mu\,\mu}(\text{EMS})}$ where $C\mu\mu$ is the $(1,1)$-th element in inverse matrix $(X'X^{-1})$ and EMS is the Error mean square from ANOVA Table

$$\text{S.E. } (\mu) = \sqrt{0.015077 \times 9.752} = 0.383446$$

$$\text{S.E. (male)} = \sqrt{(C\mu\,\mu + C_{A1A1} + 2C_{\mu A1})\text{EMS}}$$

where C_{A1A1} is the $(2, 2)$-th element, $C_{\mu A1}$ is the $(1, 2)$-th element of $(X'X)^{-1}$ – matrix

$$= \sqrt{(0.015077 + 0.015077 + 2 \times 0.005983)9.572}$$

$$= 0.6350$$

$$\text{SE (Female)} = \sqrt{(C_{\mu\mu} + C_{A2A2} + 2C_{\mu A2})\text{EMS}} = 0.4172$$

$$\text{SE (Brown)} = \sqrt{(C_{\mu\mu} + C_{B1B1} + 2C_{\mu B2})\text{EMS}} = 0.4390$$

$$\text{SE (Jodipi)} = \sqrt{(C_{\mu\mu} + C_{B2B2} + 2C_{\mu B2})\text{EMS}} = 0.6201$$

$$\text{S.E.}(\bar{y}_{11}) = \sqrt{\frac{\text{EMS}}{n_{11}}} = \sqrt{\frac{9.572}{23}} = 0.645$$

$$\text{S.E.}(\bar{y}_{12}) = \sqrt{\frac{\text{EMS}}{n_{12}}} = \sqrt{9.572/8} = 1.093$$

$$\text{S.E.}(\bar{y}_{21}) = \sqrt{\frac{\text{EMS}}{n_{21}}} = \sqrt{9.572/27} = 0.5954$$

$$\text{S.E.}(\bar{y}_{22}) = \sqrt{\frac{\text{EMS}}{n_{22}}} = \sqrt{9.572/28} = 0.5847$$

EXERCISES

1. The numbers of wireworms counted in the plots of a Latin square following soil fumigations (L, M, N, 0, P) in the previous year were:

P (4)	O (2)	N (5)	L (1)	M (3)
M (5)	L (0)	O (6)	N (5)	P (3)
O (4)	M (8)	L (1)	P (5)	N (4)
N (12)	P (7)	M (7)	O (10)	L (0)
L (5)	N (4)	P (3)	M (6)	O (9)

Analyse the data and draw conclusions.

2. An experiment was conducted in R.B. design layout with 4 insecticides and 2 methods of spraying comprising in all 8 treatments (1_1s_1 , 1_1s_2, . ., 1_4s_2) in 3 blocks,

1	2	3
(15) I_1S_1	(10) I_2S_2	(15) I_3S_2
(13) I_1S_2	(12) I_4S_1	(7) I_1S_1
(10) I_2S_2	(17) I_3S_1	(18) I_4S_2
(4) I_2S_2	(10)I_1S_2	(8) I_4S_1
(16) I_4S_2	(19) I_4S_2	(15) I_2S_1
(11) I_3S_1	(16) I_2S_1	( ) I_2S_2
(8) I_1S_2	(6) I_1S_1	(11) I_2S_2
(9) I_4S_1	(13) I_3S_2	(6) I_1S_2

Estimate the missing value and 'complete the analysis.

3. An experiment was conducted to learn about losses of ascorbic acid in soya beans stored at 3 temperatures for 4 periods. 4 packages were assigned at random to each of the 12 treatments.

Temperature	*Weeks of storage*			
	2	4	6	8
0	42	46	44	45
15	40	39	41	35
30	30	25	20	17

Analyse the data factorially and also into linear, quadratic, cubic components and draw the conclusions.

4. An experiment was conducted in the green house to determine the effects of two types of soil (b_o = soil mixed with sand and b_1, = soil with compost added) and two levels of soil moisture (co = dry soil and c_1=wet soil) on the yields (kg) of two varieties a_0 and a_1 of green fodder crop. The scheme of confounding is balanced partial confounding.

(1) Identify the confounded interactions (2) Carry out the complete analysis of data

Block 1

Rep.	I	Rep.	II	Rep.	III	Rep.	IV
000	3	110	14	100	8	111	17
111	12	000	6	001	6	100	8
100	8	111	15	110	16	001	9
011	7	001	9	011	14	010	5

Block 2

010	6	011	7	101	7	000	3
101	10	101	12	010	8	011	7
001	5	010	9	000	6	101	14
110	12	100	11	111	18	110	18

5. The following are the yields of three varieties of fodder (Tons per hectare) with the sub treatments being number of cuttings. A:2 cuttings, B:3 cuttings C:4 cuttings D: 5 cuttings.

Analyse the data and draw conclusions.

Variety	No. of cuttings	Replications 1	2	3	4
1	A	5.2	4.6	4.2	4.8
	B	5.6	5.0	5.4	4.9
	C	4.9	4.5	4.6	4.3
	D	4.6	4.8	4.0	4.4
II	A	4.8	4.7	4.1	4.5
	B	4.9	5.2	4.7	4.8
	C	5.3	5.4	5.6	5.0
	D	4.2	4.0	4.3	3.8
III	A	3.8	3.7	3.4	3.9
	B	4.6	4.8	4.5	4.7
	C	4.9	5.0	5.1	5.3
	D	5.4	5.6	5.2	5.5

Analysis the data draw conclusion.

6. The results of chillee varietal trial at a particular research station with 4 varieties and 5 replications laid out in R.B. design. The data give the yield of green chillees per plot (kgs) and number of plants per unit length.

		Varieties		
Blocks	1	2	3	4
1	10.2(3)	12.7(6)	14.2(9)	13.7(8)
2	10.6(4)	12.4(4)	14.4(8)	12.9(5)
3	10.5(5)	13.1(7)	13.9(6)	12.4(6)
4	10.0(2)	12.9(8)	13.6(5)	12.7(4)
5	10.7(8)	13.2(6)	13.2(7)	13.2(7)

Carry out the analysis of covariance and draw the conclusions.

7. Four rations' (A, B, C and D) are tested on 4 animals at 4 lactations having 4 weight groups. The milk yields were recorded as follows.

		Weight groups		
	B	D	A	C
	(4)	(6)	(10)	(3)
	A	C	B	D
Lactations	(8)	( )	(5)	(7)
	D	A	C	B
	(8)	(11)	(4)	(6)
	C	B	D	A
	(5)	(6)	(7)	(10)

Estimate the missing value and analyse the data and draw conclusions.

8. An experiment was conducted with 4 different strains at 5 different temperatures on sheep for wool production.

Temperature	1	2	3	4
I	5	7	6	8
II	12	10	11	13
III	4	3	2	5
IV	15	18	11	9

Analyse the data and draw the conclusions.

9. The following data swere based on experiment conducted by randomly selecting 4 poultry farms from the population of poultry' farms and 6 birds were randomly selected from each of the selected farm and 3 egg counts were made in a month at random for each selected bird and the part of ANOVA Table is given.

Source	d.f.	S.S	M.S.	E M.S
Farms		125.7		
Birds with in farms		78.6		
Counts within Birds with in farms		14.8		
Total		219.1		

Complete the ANOVA Table and estimate the variance between farm effects and Bird effects and test their significance using Random effects model.

10. An experiment was conducted with 4 depletion levels, 3 sowing dates and 4 replications for a paddy crop with sowing dates taken in strips. The depletion levels are denoted by d_0, d_1, d_2 and d_3, sowing dates by s_0, s_1, and s_2 and the recorded yields in strip plot design are

Rep . I

	d_2	d_0	d_3	d_1
S_2	12	6	16	10
S_0	7	4	13	8
S_1	18	10	16	17

REP. II

	d3	d2	d0	d1
S_1	19	16	11	14
S_2	18	13	9	17
S_0	17	13	6	10

Rep .III

	d0	d2	d3	d1
S_1	13	20	15	16
S_0	8	10	16	11
S_2	7	14	18	12

Rep . IV

	d1	d2	d0	d3
S_0	6	13	5	14
S_1	16	20	7	10
S	11	14	8	19

11. An experiment was conducted by estimating the available protein in four food products such as Idli, Dosa, Upma and Kichidi made from Raagi millet.

Product				
Sample	*Idli*	*Dosa*	*Upma*	*Kichidi*
1	6	8	7	10
2	9	11	8	12
3	12	13	6	11
4	7	9	10	13
5	14	10	9	14

Test the significant difference between food products using one-way analysis of variance at 5% level.

12. An experiment was conducted to estimate the shelf life (Bacteria unit %) of 4 food products made from milk at 5 different durations (days)

Days					
Food product	*2*	*4*	*6*	*8*	*10*
A	4	8	15	20	22
B	2	9	11	16	17
C	3	6	10	11	15
D	5	7	8	13	14

Test the significant difference between days and also between food products using two-way Analysis of variance at 5% level.

13. The following are the scores given by 5 sensory evaluators (E_1, E_2, E_3, E_4, E_5) on 5 food products (P_1, P_2, P_3, P_4, P_5) supplied by 5 caterers (C_1, C_2, C_3, C_4, C_5).

Caterers					
Product	*C_1*	*C_2*	*C_3*	*C_4*	*C_5*
P_1	E_2 (52)	E_5 (60)	E_1 (58)	E_4 (64)	E_3 (49)
P_2	E_5 (65)	E_3 (70)	E_4 (74)	E_2 (61)	E_1 (60)
P_3	E_3 (48)	E_1 (56)	E_2 (59)	E_5 (65)	E_4 (70)
P_4	E_1 (59)	E_4 (69)	E_5 (74)	E_3 (70)	E_2 (54)
P_5	E_4 (65)	E_2 (71)	E_3 (69)	E_1 (75)	E_5 (52)

Test the significant difference between food products, caterers and sensory evaluators using Latin square design analysis at 5% level.

PART III

SAMPLE SURVEYS, ECONOMIC AND NON-PARAMETRIC STATISTICS

CHAPTER 17

SAMPLING METHODS

17.1 INTRODUCTION

Sampling is a part and parcel of our daily life. The housewife uses the technique of sampling in taking a decision whether the rice is cooked properly or not by inspecting a sample of grains from a cooking vessel A businessman inspects a sample of goods for ordering a large consignment. In industry, a sample would be observed to assess the quality of a product (or products). A farmer would estimate his crop prospects by observing a sample of earheads (or the plants). In the above situations, sampling is being followed to save money and time to arrive at an idea of the characteristic in the population. If there would be a considerable variation in the population, sampling adopted in the usual way might not give correct picture about the population.   For example, the consumer wants to purchase rice by inspecting a handful of it from the upper portion of bag. If the quality of rice is not uniform throughout the bag the decision he takes on the basis of inspecting an upper portion of the material may bring him a monetary loss. Similarly the decision taken on only few bags out of large consignment of bags which are not having uniform quality would be of serious consequence. Hence different sampling procedures were evolved for different situations to estimate the population characteristics with minimum risk. These sampling methods were developed based on probability theory,, There is also a sampling method called 'purposive sampling' which do not use probability theory. The main drawback of 'purposive sampling' is that it is not possible to provide the error involved in arriving at an estimate of the population, and also the confidence intervals for the population characteristic.

17.2 SIMPLE RANDOM SAMPLING

In this method every unit in the population will have equal probability of being selected in the sample. Alternatively, the simple random sampling is the method of selecting 'n' sampling units out of total N units such that all the possible $\binom{N}{n}$ samples would have equal chance of being selected.

17.2.1   Sample Random Sampling with Replacement (SRSWR)

A sample is drawn such that every sampling unit drawn would be replaced back in the population. In this way the sample may contain repeated elements and any number of samples could be drawn.

17.2.2   Simple Random Sampling without Replacement (SRSWOR)

A sample is drawn such that every sampling unit drawn would not be replaced back. The sample would contain all distinct elements. If there are N units in the population and n units in the sample, there would be $\binom{N}{n}$ distinct samples by this method.

17.2.3   Selection of a Random Sample

List of units would be prepared by serially numbering all the sampling units from 1 to N and n random numbers would be selected from the column (or row) of a table of random numbers either by SRSWR or SRSWOR. For example, if N=40 and n=5, two columns would be selected from the table of random numbers. The maximum figure in two column table would be 99. The numbers 81 to 99 would be rejected since they have more probability than the numbers from 1 to 80. Supposing that 75 would be selected in the first draw, the actual random number would be the remainder after dividing 75 by 40 i. e., 35. If 80 would be selected in a particular draw the random number selected would be 40 since the remainder would be zero. In this way all the five numbers would be selected either by with or without replacement. The method of providing estimates of population mean, standard error of mean and the confidence intervals for population mean are given as follows. Let. Y_i, be the i-th observational value for the character under study.

Sample	Population
n = size of the sample	N = size of the population
$\overline{Y}_n = \dfrac{1}{n}\sum_{i=1}^{n} Y_i$ = mean of the sample and is an un biased estimate of the population mean $\overline{Y}_N$	$\overline{Y}_N = 1/N \sum_{i=1}^{n} Y_i$ = mean of the population.
$\widehat{Y}_N = N.\overline{Y}_n$ = estimate of the population total, Y and is an unbiased estimate.	$Y = \sum_{i=1}^{N} Y_i$ = Population total
$s^2 = \dfrac{1}{n-1}\sum_{i=1}^{n}(Y_i - \overline{Y})^2$	$S^2 = \dfrac{1}{N-1}\sum_{l=1}^{N}(Y_i - \overline{Y})^2$
$= \dfrac{1}{n-1}\left[\sum_{i=1}^{n} Y_i^2 - \dfrac{\left(\sum_{i=1}^{n} Y_i\right)^2}{n}\right]$ = mean square in the sample and is an unbiased estimate of S^2.	$\dfrac{1}{N-1}\left[\sum_{i=1}^{N} Y_i^2 - \dfrac{(\Sigma Y_i)^2}{N}\right]$ = mean square in the population .
Est. $V(\overline{Y}_n) = \dfrac{N-n}{Nn}s^2$ = estimate of the variance of the sample mean and is an unbiased estimate of $V(\overline{Y}_n)$.	$V(\overline{Y}_n) = \dfrac{N-N}{Nn}S^2$ = variance of sample mean in the population, and $\dfrac{N-n}{N}$ is called the finite
Est. $S.E(\overline{Y}_n) = \sqrt{\text{Est } V(\overline{Y}_n)}$ = estimate of the standard error of sample mean and is an unbiased estimate of $S.E(\overline{Y}_n)$.	population correction. $S.E(\overline{Y}_n) = \sqrt{V(\overline{Y}_n)}$ = standard error of sample mean in the population.

Contd...

Sample	Population
$\text{Est } V(\widehat{Y}) = \dfrac{N^2(N-n)}{Nn}S^2 =$ estimate of variance of the estimate of total and is an unbiased estimate of $V(\widehat{Y})$.	$V(\widehat{Y}) = N^2\dfrac{(N-n)}{Nn}S^2 =$ variance of the estimate of the total in the population.

Confidence limits: If S^2 is not known and the size of sample is small, the confidence limits for population mean, $\overline{Y}_N$ are given as

$$\text{lower limit}: \overline{Y}_n - t_{(n-1)} \times \text{Est S.E.}(\overline{Y}_n)$$

$$\text{and upper limit}: \overline{Y}_n + t_{(n-1)} \times \text{Est S.E.}(\overline{Y}_n)$$

where $t_{(n-1)}$ is tabulated value of student's t-distribution with $(n-1)$ d.f.

Example: A sample of 50 progressive farmers were selected from a district containing 800 progressive farmers by simple random sampling method so as to estimate the total area under high yielding variety of paddy. The list of selected farmers along with corresponding areas under high yielding variety (HYV) is given in Table 17.1. Estimate the mean area under HYV, standard error and confidence limits for the mean area in the district.

TABLE 17.1

Holding	Area (Hectares)	Holding	Area (Hectares)	Holding	Areas (Hectares)
1	3.5	18	4.2	35	2.1
2	3.2	19	6.1	36	2.4
3	2.5	20	1.1	37	1.5
4	4.0	21	1.0	38	1.1
5	3.2	22	1.7	39	0.7
6	2.0	23	2.3	40	3.1
7	2.2	24	5.2	41	3.3
8	1.5	25	4.6	42	2.8
9	2.6	26	0.8	43	2.2
10	2.8	27	1.9	44	4.3
11	3.5	28	2.5	45	3.8
12	3.0	29	2.6	46	6.2
13	1.4	30	3.1	47	5.0
14	1.2	31	6.2	48	0.7
15	1.3	32	5.4	49	0.9
16	3.6	33	3.6	50	1.2
17	3.2	34	4.5		

$$\Sigma Y_i = 142.8, \quad \Sigma Y_i^2 = 517.74$$

$$\overline{Y}_n = \frac{142.8}{50} = 2.86 \quad s^2 = \frac{1}{50-1}\left[517.74 - \frac{(142.8)^2}{50}\right] = 2.24$$

$$\text{Est. } V(\overline{Y}_n) = \frac{800-50}{800 \times 50} \times 2.24 = 0.042$$

$$\text{Est. } SE(\overline{Y}_n) = \sqrt{0.042} = 0.2049$$

Confidence limits of $\overline{Y}_N$;

Lower limit: $2.86 - 1.96 \times 0.2049 = 2.46$

Upper Limit: $2.86 + 1.96 \times 0.2049 = 3.26$

17.3 STRATIFIED RANDOM SAMPLING

In this method the population is divided into different homogenous groups known as strata and a simple random sampling is selected from each of the strata to estimate the population mean (or total), standard error of the estimate and confidence limits for the population mean. It could be seen from the expression of Est. $V(\bar{Y}_n)$ from section 17.2 of simple random sampling, the precision of the estimate depends on size of sample as well as the value of s^2. If the size of sample increase and the value of s^2 decreases, the precision of the estimate ($\bar{Y}n$) increases. Assuming that due to limitation of time and money it would not be possible to increase the size of sample, the only alternative would be to decrease the value of s^2 for increasing the precision of the estimate, ($\bar{Y}n$). Stratified random sampling method provides the scope for decreasing the value of s^2 by dividing the population into homogenous groups such that there would be more heterogeneity between the groups (or strata) and more homogeneity within the groups (or strata) For example, considering the selection of sample of rice from a bag by a consumer, if the bag does not contain uniform quality of the consumer would be risking a loss. Similarly if the adoption of rice yielding varieties is not same throughout the district and differs from one panchayat samiti to another then the estimate of the extent of adoption by farmers in the district would be less efficient assuming that the sample size is not small. Let Y be the character under study. The estimate of population, standard error of the estimate and the confidence limits for the population mean are give as follows.

17.3.1 Proportional Allocation of Sample

If the proportion of the sample size in the i-th stratum is assumed to be same as proportion of population size in the same stratum, we have $n_i/n = N_i/N$ = constant. In other words, the sample size to be allocated for i-th stratum is $n_i = n. N_i/N$ for i = 1, 2,.., k. This type of allocation of sample is known as proportional allocation. By substituting $n_i = n\, N_i/N$ in the mean of the sample, variance of the estimate, estimate of the variance of the estimate in section 17.3 we have

Sample	Population
n = size of the sample	N = size of the population
n_i = size of the sample in the i-th stratum for i = 1, 2, …, k, where k is the number of strata.	N_i = size of the i-th stratum in the population for i = 1, 2,…, k, where k is the number of strata.
$$n = \sum_{i=1}^{k} n_i$$	$$N = \sum_{i-1}^{k} N_i$$
$$\overline{Y}_{ni} = \frac{1}{n_i}\sum_{i=1}^{ni} Y_i = \text{ mean of i-th stratum in the sample and is an unbiased estimate of } \overline{Y}_{Ni}$$	$$\overline{Y}_{Ni} = \frac{1}{N_i}\sum_{i=1}^{k} Y_i = \text{ mean of i-th stratum in the population.}$$
$$\overline{Y}_n = 1/n\sum_{i=1}^{k} n_i \overline{Y}_{ni} = \text{ mean of the sample.}$$	$$\overline{Y}_N = 1/N\sum_{i=1}^{k} N_i \overline{Y}_{Ni} = \text{ mean of the population.}$$
$$\overline{Y}_{st} = 1/N\sum_{i=1}^{k} N_i \overline{Y}_{ni} = \text{ estimate of the population mean in the stratified random sampling method and is an unbiased estimate of } \overline{Y}_N,$$	$$S_i^2 = 1/N_i - 1\sum_{j=1}^{Ni}(Y_{ij} - \overline{Y}_{Ni})^2 = \text{ mean square of i-th stratum in the population for I = i, 2,..k}$$
$$s_i^2 = \frac{1}{n_i - 1}\sum_{j=1}^{ni}(Y_{ij} - \overline{Y}_{ni})^2 = \text{ mean square of i-th stratum in the sample for I = 1, 2, …,k}$$	$$V(\overline{Y}_{st}) = \sum_{i=1}^{k}\frac{N_i^2}{N^2}\frac{(N_i - n_i)}{N_i n_i}S_i^2 = \text{ variance of the estimate}$$
$$\text{Est. } V(\overline{Y}_{si}) = \sum_{i=1}^{k}\frac{N_i^2}{N^2}\frac{(N_i - n_i)}{N_i n_i}s_i^2 = \text{ estimate of variance of the estimate.}$$	$$\text{S..E.}(\overline{Y}_{st}) = \sqrt{V(\overline{Y}_{st})} = \text{ standard error of the estimate.}$$

Contd...

Sample	Population
Est. S.E.$(\overline{Y}_{st}) = \sqrt{Est.V(\overline{Y}_{st})}$ = estimate of the standard error of the estimate *Confidence limits for* $\overline{Y}_N$: If S_i^2 is not known and the size of total sample is small, the limits are given as $\overline{Y}_{st} \pm t_{(n-1)} \times Est.S.E(\overline{Y}_{st})$ as upper and lower limits. $\overline{Y}_{st} = \overline{Y}_n$ = estimate of the population mean in the proportional allocation. Est. $V(\overline{Y}_{st})$ Prop $= \dfrac{N-n}{Nn}\sum_{i=1}^{k}\dfrac{N_1}{N}s_1^2$ estimate of the variance of the estimate in the proportional allocation. Est. S.E. $(\overline{Y}_{st})$ prop. $= \sqrt{Est. V(\overline{Y}_{st})}$ = estimate of the standard error of the estimate in the proportional allocation.	$V(\overline{Y}_{st})$ prop $= \dfrac{N-n}{Nn}\sum_{i=1}^{k}\dfrac{N_i}{N}S_i^2$ = variance of the estimate in the proportional allocation. S.E. $(\overline{Y}_{st})$ prop. $= \sqrt{V(\overline{Y}_{st})}$ prop = standard error of the estimate in the proportional allocation.

Confidence limits for $\overline{Y}_N$ can also be obtained on the similar lines as in Section 17.3.1 for the proportional allocation.

17.3.2 Neyman's /Optimum Allocation of Sample

In this case the total sample would be allocated to different strata in such a way that the variance of the estimate would be minimized with the condition that the total of all the strata samples is equal to the total sample size i.e., $\sum_{i=1}^{k} n_i = n$. By minimizing the variance of the estimate in Section 17.3.1 with the restriction that $\sum_{i=1}^{k} n_i = n$, we have $n_i = n \times \dfrac{N_i S_i}{\sum_{i=1}^{k} N_i S_i}$

On substituting this expression of n_i in the variance of the estimate and estimate of the variance of the estimate in Section 17.3.1, the corresponding expression in optimum allocation, are obtained as follows.

Sample	*Population*

Est. $V(\overline{Y}_{st})$ opt $=$

$$\frac{1}{N^2 n}(\sum_{i=1}^{k} N_i s_i)^2 - \frac{1}{N^2}(\sum_{i=1}^{k} N_i s_i^2) =$$

estimate of
the variance of the estimate in the (optimum) Neyman's allocation.

$V(\overline{Y}_{st})$ opt. $=$

$$\frac{1}{N^2}(\sum_{i=1}^{k} \frac{N_i S_i^2}{n}) - \frac{1}{N^2}(\sum_{i=1}^{k} N_i S_I^2) =$$

variance of the estimate in the (optimum) Neyman's allocation.

17.3.3 Comparison of Stratified Random Sampling with Simple Random Sampling: It was proved that

$$V(\overline{Y}_{st})_{opt} \le V(\overline{Y}_{st})_{prop} \le V(\overline{Y}_n)_{ran}$$

which indicates optimum allocation is more precise in comparison to proportional allocation which in turn more precise to simple random sampling. For further reading on this topic please refer to Cochran (1953).

17.3.4 Selection of a Sample with Probability Proportional to size

Often a situation arises to draw a sample from the population with probability proportional to size. For example, a sample of farmers are to be selected with probability proportional to area under their holding. The procedure of drawing n farmers out of N with probability proportional to area under holding is as follows. Let i-th farmer holds A_i hectares and $\sum_{i=1}^{N} A_i = A$. All the farmers would be arranged serially according to the size of their holding. The third column in Table 17.2 gives cumulative totals of holdings.

TABLE 17.2

S. No. of Farmer *(1)*	*Holding size* *(2)*	*Cumulative total* *(3)*
1	A_1	A_1
2	A_2	$A_1 + A_2$
3	A_3	$A_1 + A_2 + A_3$
$\vdots$	$\vdots$	$\vdots$
N	A_N	$\sum_{i=1}^{N} A_1 = A$

Select a random number (say) r from the table of random numbers out of A. If this number lies between $A_1 + A_2 + \ldots A_{i-1}$ and $A_1 + A_2 \ldots + A_{i-1} + A_i$ then i-th farmer is selected. Select another random number and if it lies between 1 and A_1 then first farmer is selected. In this way all the n farmers can be selected.

17.3.5 Lahiri's Method of Selecting a Sample with Probability Proportional to Size

This method avoids the need of writing cumulative totals for selecting a sample when, the number of units in the population is considerably large. Let N be the number of units in the population and M be the maximum of the sizes of the N units or some number greater than that maximum size (M) and let M_i be the size of the i-th unit. The procedure is as follows :

(i)   Select a number at random from 1 to N, say, i

(ii) Select another number at random from 1 to M, say, R

(iii) Select i-th unit in the sample .if $R \leq M_i$

(iv) Reject i-th unit and repeat the above process if $R > M_i$

For selecting a random sample of n units with probability proportional to size with replacement, the above procedure has to be repeated n times. If the selection is without replacement the above procedure has to be repeated till n distinct units are obtained.

EXAMPLE: A sample survey was conducted to estimate the credit needs of cultivators in a Taluka by considering panchayat samitis in that Taluka as strata. A random sample of 10 villages were selected from each of the three Panchayat Samitis of sizes 150, 100, 120 respectively, The credit needs (in thousand rupees) of the thirty villages are given in Table 17 . 3. Estimate the average credit needs of cultivator in the Taluka, standard error of the estimate and confidence interval for the average credit need.

TABLE 17.3 CREDIT NEED (IN RS THOUSAND) SAMITHI

Village	I	II	III
1	15	16	7
2	12	10	8
3	8	11	5
4	6	9	10
5	10	5	12
6	4	3	9
7	3	7	11
8	16	5	10
9	7	13	6
10	5	4	2

$$n_1 = n_2 = n_3 = 10, \ N_1 = 150, \ N_2 = 100, \ N_3 = 120$$

$$n = \sum_{i=1}^{3} n_i = 30, \ N = \sum_{i=1}^{3} N_i = 370$$

TABLE 17.4

Stratum No.	N_i	N_i	$\bar{Y}_{ni}$	s_i	$\dfrac{N_i^2(N_i - n_i)}{N^2 N_i n_i} s_i^2$	$\dfrac{N_i}{N}\bar{Y}_{ni}$
1	150	10	8.6	20.49	.3135	3.53
2	100	10	8.3	18.01	.1189	2.24
3	120	10	8.0	9.33	.0896	2.99
	370				.5220	8.76

$$s_i^2 = \frac{1}{n_i - 1}\left[\sum_{j=1}^{n_i} Y_{ij}^2 - \frac{\left(\sum_{j=1}^{n_i} Y_{ij}\right)^2}{n_i} \right]$$

$$\bar{Y}_{st} = 1/N \, \Sigma \, N_i \bar{Y}_{ni} = 8.76$$

$$\text{Est. } V(\bar{Y}_{st}) = \sum_{i=1}^{k} \frac{N_i^2}{N^2}\frac{(N_i - n_i)}{N_i n_i} s_i^2 = 0.5220$$

$$\text{Est. S.E } (\overline{Y}_{st}) = 0.7225$$

Confidence limits of $\overline{Y}_N$:

The lower limit: $\overline{Y}_{st} - t_{(n-1)} \times \text{Est. S.E}(\overline{Y}_{st}) = 8.76 - 2.045 \times 0.7225 = 7.28$

The upper limit $= \overline{Y}_{st} + t_{(n-1)} \times \text{S.E}(\overline{Y}_{st}) = 8.76 + 2.045 \times .7225 = 10.24.$

17.4 CLUSTER SAMPLING

In this method the statistical population would be divided into groups of ultimate sampling units called clusters for the process of sampling, For example, the ultimate sampling unit might be farm holding or village or group of villages for estimating the area under high yielding varieties (HYV) in a district. The choosing of ultimate sampling unit as farm holding, etc., depends on the precision required for the estimate and the cost involved in conducting the survey. In the selection of soil samples, the ultimate sampling unit depends on the type and size of tool used for obtaining the soil profile. It was expected that the smaller the size of cluster the better the precision of the estimate but the cost involved might be more due to increase in sample size for covering the entire cross section of the population. On the other hand the larger the cluster the cost involved might be less due to less expensiveness in collecting information on the neighbouring units at the expense of some reduction in precision of the estimate. Further, sometimes the information might be available for clusters but not on the ultimate sampling units, as for example, village-wise data might be available but not on the holdingwise in the Panchayat Samithi record for the area under different crops. In these situations cluster sampling method would be useful for estimating the population characteristics. In this method a sample of clusters were randomly selected out of a population of clusters. The estimate of the population mean, standard error of the estimate and confidence limits of the population mean by cluster sampling method are given as follows assuming that the size of each cluster is same. Let Y be the variable under study and Y_{iJ} be the observational value for the j-th element in the i-th cluster for j = 1, 2 ... M; i = 1, 2,. ., N. Let n clusters be randomly' selected out of N clusters and M be the size of each cluster.

Sample	Population
$\overline{Y}_{iM} = 1/M \sum_{j=1}^{M} Y_{ij}$ = mean of the i-th cluster.	$\overline{Y}_{NM} = 1/N \sum_{i=1}^{N} \overline{Y}_{iM}$ = mean of cluster means in the population.
$\overline{Y}_{cl} = 1/n \sum_{i=1}^{n} \overline{Y}_{iM}$ = estimate of the population mean in the cluster sampling method and is an unbiased estimate of $\overline{Y}_{NM}$.	$\overline{Y}_{NM} = 1/N \sum_{i=1}^{N} Y_{iM} = 1/NM \sum_{i=1}^{N} \sum_{j=1}^{M} Y_{ij}$ = mean in the population. This is true because the size of each cluster is same.
$s_b^2 = 1/n-1 \sum_{i=1}^{n} (\overline{Y}_{iM} - \overline{Y}_{cl})^2$ = mean square between the cluster means in the sample.	$S_b^2 = 1/N-1 \sum_{i=1}^{N} (\overline{Y}_{iM} - \overline{Y}_{NM})^2$ = mean square between the cluster means in the population.
Est. $V(\overline{Y}_{cl}) = \dfrac{N-n}{Nn} s_b^2$ = estimate of the variance and is an unbiased estimate of $V(\overline{Y}_{cl})$.	$V(\overline{Y}_{cl}) = \dfrac{N-n}{Nn} S_b^2$ = variance of the cluster sampling estimate.
Est. $S.E.(\overline{Y}_{cl}) = \sqrt{Est.\ V(\overline{Y}_{cl})}$ = estimate of the standard error of the cluster sampling estimate.	$S.E.(\overline{Y}_{cl}) = \sqrt{V(\overline{Y}_{cl})}$ = standard error of the cluster sampling estimate.
Confidence limits for $\overline{Y}_{NM}$ **:** If S_b^2 is not known and the size of sample is small, the confidence limits for $\overline{Y}_{NM}$ are given by $\overline{Y}_{cl} \pm t_{(n-1)} \times Est\ SE.$ ($\overline{Y}cl$) as the upper and lower limits. If nM ultimate sampling units are randomly selected from NM sampling units in the population, we have	

Contd...

Sample	Population
$\overline{Y}_{nM} = 1/nM \sum_{i=1}^{n} \sum_{j=1}^{M} Y_{ij}$ = mean of nM sampling units in the sample.	$S^2 = \dfrac{1}{NM-1} \sum_{i=1}^{N} \sum_{j=1}^{M} (Y_{ij} - Y_{NM})^2$ = mean square between ultimate sampling units in the population.
Est. $V(\overline{Y}_{nM}) = \dfrac{NM-nM}{NM} 1/nM \left\{ \dfrac{(N-1)Ms_b^2 + N(M-1)s_w^{\,2}}{NM-1} \right\}$ = estimate of the variance of the estimate, where $S_{w^2} = \dfrac{1}{n(M-1)} \sum_{i=1}^{n} \sum_{j=1}^{M} (Y_{ij} - \overline{Y}_{iM})^2$ = mean square within clusters in the sample.	$V(\overline{Y}_{nM}) = \dfrac{NM-nM}{NM} \dfrac{S^2}{nM}$ = variance of the estimate.
	Efficiency: Efficiency of cluster sampling method with cluster as the sampling unit to that of the method with an ultimate sampling unit is given by E where $$E = \dfrac{V(\overline{Y}_{nM})}{V(\overline{Y}_{cl})} = \dfrac{S^2}{MS_b^2}$$

Contd...

Sample	Population
Estimate of efficiency: The estimate of efficiency of clusters sampling method to that of the method with an ultimate sampling unit is given by Est. (E) $$= \frac{(N-1)Ms_b^2 + N(M-1)s_w^2}{(NM-1)Ms_b^2}$$ $$\simeq 1/M + \frac{(M-1)}{M}\frac{s_w^2}{Ms_b^2}$$ *Anova table*: The analysis of variance (ANOVA) table is useful for obtaining the values of S_b^2 and S_w^2 which are in turn required for computing $\text{Est.V}(\overline{Y}_{cl})$ and $\text{Est. }V(\overline{Y}_{nM})$. The ANOVA table is given as	*Intra class correlation coefficient*: The correlation coefficient between the elements of a cluster is denoted by 'ρ' and is known as the intra class correlation coefficient, where $$\rho = \frac{\dfrac{N-1}{N}S_b^2 - \dfrac{S_w^2}{M}}{\dfrac{NM-1}{NM}S^2}$$ *Anova table*: The analysis of variance table (ANOVA) is useful for computing the values of S^2 and S_b^2 which are in turn required in computing $V(\overline{Y}_{nM})$ and $V(\overline{Y}_{cl})$ respectively. The ANOVA table is given as

TABLE 17.5 ANOVA TABLE (SAMPLE)

Source	d.f.	S.S.	M.S.
Between clusters	n-1	$M\sum_{i=1}^{n}(Y_{iM}-Y_{cl})^2$ $=$ C.S.S (say)	$\dfrac{C.S.S}{n-1}=Ms_b^2$
Within clusters	n(M-1)	By subtraction (W.S.S.)	$\dfrac{W.S.S}{n(M-1)}=s_w^2$
Total	nM-1	$\sum_{i=1}^{n}\sum_{j=1}^{M}(Y_{ij}-\overline{Y}_{cl})^2=$ T.S.S(say)	$\dfrac{TSS}{nM-1}=s^2$

TABLE 17.6 ANOVA TABLE (POPULATION)

Source	d.f.	S.S.	M.S.
Between clusters	N-1	$M\sum_{i=1}^{n}(\overline{Y}_{iM}-\overline{Y}_{NM})^2=$ C.S.S$_{(say)}$	$\dfrac{C.S.S}{N-1}=S_b^2$
Within clusters	N(M-1)	By subtraction (W.S.S.)	$\dfrac{W.S.S}{N(M-1)}S_w^2$
Total	NM-1	$\sum_{i=1}^{N}\sum_{j=1}^{M}(Y_{ij}-\overline{Y}NM)^2=$ T.S.S$_{(say)}$ $=$ T.S.S (Say)	$\dfrac{TSS}{NM-1}=S^2$

Formation of clusters: If clusters are formed randomly the variance of the clusters sampling estimate is given by using intra class correlation Coefficient as

$$V(Y_{cl})=\frac{N-n}{n}\frac{(NM-1)S^2)}{M(N-1)nM}\{1+(M-1)\rho\}$$

$$E=\frac{M(N-1)}{NM-1}\frac{1}{\{1+(M-1)\rho\}}$$

EXAMPLE: An investigation was undertaken to estimate the milk consumption per individual in a town by selecting a random sample of 50 households which are considered as clusters of individual members of family from a total of 2000 households. On an average each household was assumed to have a family size of 5 members. Estimate the milk consumption per individual in the town, standard error of the estimate, confidence limits for the population mean given the following data in Table 17.7

TABLE 17.7

S.No.	Milk Consumption of 5 Family members (lits.)					Total Consumption
1	0.5,	0.8,	0.5,	0.5,	0.6	2.9
2	0.4,	0.8,	0.9,	0.5,	0.5	3.1
3	0.6,	0.5,	0.5,	0.6,	0.4	2.6
4	0.5,	0.4,	0.7,	0.3,	0.5	2.4
5	0.2,	0.5,	0.1,	0.5,	0.7	2.0
6	0.3,	0.4,	0.3,	0.2,	0.4	1.6
7	0.5,	0.6,	0.5,	0.7,	0.1	2.4
8	0.4,	0.5,	0.8,	1.0,	0.7	3.4
9	0.3,	0.2,	0.3,	0.4,	0.5	1.7
10	0.2,	0.7,	0.5,	0.6,	0.4	2.4
11	0.3,	0.2,	0.6,	0.7,	0.2	2.0
12	0.4,	0.3,	0.2,	0.4,	0.7	2.0
13	0.6,	0.4,	0.5,	0.7,	0.8	3.0
14	1.0,	0.7	0.5,	0.4,	0.2	2.8
15	0.2,	0.4,	0.8,	0.3,	0.4	2.1
16	0.3,	0.4,	0.3,	0.2,	0.4	1.6
17	0.3,	0.1,	0.5,	0.4,	0.7	2.0
18	0.5,	0.4,	0.3,	0.2,	0.1	1.5
19	0.3,	0.2,	0.8,	0.2,	0.6	2.1
20	0.7,	0.5,	0.6,	0.3,	0.4	2.5
21	0.5,	0.6,	0.9,	0.3,	0.5	2.8
22	0.4,	0.7,	0.5,	0.4,	0.2	2.2
23	0.6,	0.7,	0.8,	0.3,	0.2	2.6
24	0.4,	0.5,	0.5,	0.1,	0.3	1.8
25	0.2,	0.4,	0.3,	0.6,	0.1	1.6
26	0.4,	0.2,	0.5,	0.1,	0.6,	1.8
27	0.3,	0.1,	0.5,	0.7,	0.5	2.1
28	0.4,	0.5,	0.6,	0.3,	0.2	2.0
29	0.6,	0.5,	0.5,	0.2,	0.1	1.9
30	0.6,	0.5,	0.3,	0.4,	0.2	1.9
31	0.2,	0.4,	0.8,	0.3,	0.2	1.9
32	0.6,	0.5,	0.8,	0.7,	0.6	3.2
33	0.4,	0.3,	0.5,	0.2,	0.3	1.7
34	0.6,	0.7,	0.3,	0.2,	0.4	2.2
35	0.7,	0.1,	0.5,	1.0,	1.2	3.5
36	1.0,	0.5,	0.8,	0.7,	0.2	3.2
37	0.7,	0.8,	0.5,	1.0,	0.7	3.7
38	0.1,	0.2,	0.8,	0.3,	0.1	1.5

Table 17.7 Contd...

S.No.	Milk Consumption of 5 Family members (lits.)					Total Consumption
39	0.6,	0.1,	0.8,	0.7,	0.8	3.0
40	0.8,	0.9,	0.5,	0.3,	1.0	2.5
41	0.6,	0.5,	0.2,	0.6,	0.2	2.1
42	0.4,	0.1,	0.3,	0.9,	1.0	2.7
43	1.0,	0.5,	0.5,	0.4,	0.2,	2.6
44	0.6,	0.8,	1.0,	0.2,	0.1	2.7
45	0.3,	1.0,	0.9,	0.4,	0.5	3.1
46	1.0,	0.8,	0.5,	1.0,	0.2	3.5
47	1.0,	0.4,	0.3,	0.0,	1.2	2.9
48	0.5,	0.1,	0.8,	1.0,	0.7	3.1
49	0.4,	0.6,	0.5,	1.0,	0.6	3.1
50	0.7,	0.3,	0.8,	1.0,	0.4	3.2

TABLE 17.8 ANOVA TABLE

Source	d.f.	S.S.	M.S.
Between clusters	$(n-1) = 50 - 1 = 49$	4.50	$0.0918 = Ms_b^2$
Within clusters	$249 - 49 = 200$	8.86	$0.0443 = s_w^2$
Total	$(nM-1) = 250 - 1 = 249$	13.36	$0.0537 = s^2$

$$Y_{cl} = \frac{125.6}{250} = 0.5024, \quad s_b^2 = \frac{.0918}{5} = 0.18$$

$$\text{Est.} V(\overline{Y}_{cl}) = \frac{N-n}{Nn}.s_b^2 = \frac{2000-50}{2000 \times 50} \times .0184 = .0004$$

$$\text{Est. S.E.}(\overline{Y}_{cl}) = \sqrt{.0004} = 0.02$$

Confidence limits for $\overline{Y}_{NM}$:

Lower limit: $0.5024 - 1.96 \times .02 = 0.4632$

Upper limit: $0.5024 + 1.96 \times .02 = 0.5416$

If nM ultimate sampling units are randomly selected from NM units, we have

$$\overline{Y}_{nM} = 1/nM \sum_{i=1}^{n} \sum_{j=1}^{M} Y_{ij} = \frac{125.6}{250} = 0.5024$$

$$\text{Est.}(\overline{Y}_{nM}) = \frac{NM-nM}{NM} 1/nM \left\{ \frac{(N-1)Ms_b^2 + N(M-1)s_w^2}{NM-1} \right\}$$

$$= .00039 \frac{(2000-1) \times .0918 + 2000(5-1) \times .0443}{2000 \times 5 - 1}$$

$$= .0002$$

$$\text{Est. S.E.}(\overline{Y}_{nM}) = \sqrt{.0002} = 0.0141$$

Estimate of efficiency:

$$\text{ESt. (E)} = \frac{(N-1)Ms_b^2 + N(M-1)s_w^2}{(NM-1)Ms_b^2} = 0.5860$$

The estimate of the efficiency of cluster sampling method to that of the method with ultimate sampling unit is 58.60%. In the given study the cluster sampling method is less efficient than the method with ultimate sampling unit.

17.5 TWO-STAGE SAMPLING

Sometimes it might be uneconomical as well as less efficient to enumerate complete clusters in the case of cluster sampling. The elements in each cluster might be homogeneous due to geographical continguity or due to any other reason. In such case enumerating all the elements in the cluster might prove to be uneconomical. Further the larger the size of the cluster the lesser the precision of the estimate as was observed in cluster sampling method. For example, to estimate the area under maize crop in a Panchayat Samiti it would be more economical and precise to select randomly more number of villages at first stage and select randomly fewer households from each of the selected village at the second stage rather than selecting randomly few villages and enumerating completely all the households from each of the selected villages as was done in the case of cluster sampling method. The former method is known as two stage sampling method (or sub-sampling), because the selection was done at two-stages with villages as first stage sampling units and households as second stage sampling units. The two-stage sampling method and sometimes stratified two stage sampling method are quite applicable in most of the socio economic surveys and this method is useful when complete frame is not available.

Let Y be the variable under study and Y_{iJ} be the observational value of the j-th second stage unit of the i-th first stage unit for j=1,. 2,.,., M; i=1, 2, ... , N in the population assuming that each first stage unit contains M second stage units. In other words there would be in all NM second stage units in the population.

Let n first stage units be selected at random out of N and m second stage units be selected at random from each of the selected first stage units in the sample. The estimate of the population mean, standard error of the estimate, confidence limits for the population mean are given as:

Sample	Population

Sample

$\overline{Y}_{im} = 1/m \sum_{j=1}^{m} Y_{ij}$ = mean of i-th first stage unit in the sample.

$\overline{Y}_{ts} = 1/n \sum_{i=1}^{n} \overline{Y}_{im} = 1/nm \sum_{i=1}^{n} \sum_{j=1}^{m} Y_{ij}$ = mean per second stage unit in the sample and also the estimate of the population mean by two-stage sampling method. This is an unbiased estimate of $\overline{Y}_{NM}$

Est. $V(\overline{Y}_{ts}) = (1/n - 1/N)s_{b^2} + 1/N(1/m - 1/M)\overline{s}_w{}^2$ estimate of the variance of the estimate in two-stage sampling method and is an unbiased estimate of $V(\overline{Y}_{ts})$, where

$s_b{}^2 = 1/n - 1 \sum_{i=1}^{m} (\overline{Y}_{im} - \overline{Y}_{ts})^2$ = mean square between the first stage units means in the sample.

Population

$\overline{Y}_{iM} = 1/M \sum_{j=1}^{M} Y_{ij}$ = mean of i-th first stage unit in the population.

$\overline{Y}_{NM} = 1/N \sum_{i=1}^{N} \overline{Y}_{iM}$ = mean of first stage unit means,

$\overline{Y}_{NM} = 1/NM \sum_{i=1}^{N} \sum_{j=1}^{M} Y_{ij}$ = mean per second stage unit in the population.

$V(\overline{Y}_{ts}) = (1/n - 1/N)s_{b^2} + (1/m - 1/M)\dfrac{\overline{S}_w{}^2}{n}$ = variance of the estimate in two-stage sampling method, where

$S_{b^2} = 1/N - 1 \sum_{i=1}^{N} (\overline{Y}_{iM} - \overline{Y}_{NM})^2$ = mean square between first stage unit means $S_i^2 = 1/M - 1 \sum_{i=1}^{M} (Y_{ij} - \overline{Y}_{iM})^2$ = mean square between second stage units within the i-th first stage unit in the population.

Contd...

$$s_i^2 = \frac{1}{m-1}\sum_{j=1}^{m}(Y_{ij} - \overline{Y}_{im})^2 = \text{mean square between the second}$$

stage units within the i-th first stage unit in the sample.

$$\overline{s}_w^2 = 1/n(m-1)\sum_{i=1}^{n}(m-1)S_{i^2}$$

Est. S.E. $(\overline{Y}_{ts}) = \sqrt{\text{Est.}V(\overline{Y}_{ts})}$

$$\overline{S}_w^2 = 1/N\sum_{i=1}^{N}S_{i^2}$$

S.E. $(\overline{Y}_{ts}) = \sqrt{V(\overline{Y}_{ts})}$

Confidence limits for $\overline{Y}_{NM}$

If S_b^2 and S_w^{-2} are not known and total size of sample is small

the confidence limits for $\overline{Y}_{NM}$ are given by $\overline{Y}_{ts} \pm t_{(nm-1)} \times$ Est.

S.E. $(\overline{Y}_{ts})$ where $t_{(nm-1)}$ is the table value of student's t with

(nm–1) d.f

EXAMPLE: A sample survey was conducted to estimate the total production of milk by selecting a random sample of 10 dairy farms out of 40 dairy farms in a district at the first stage. Further a random sample of 5 animals were selected at random from each of the selected dairy farm at the second stage assuming that the each dairy farm contains on an average of 25 animals (buffalo). The milk yields were recorded for all the 50 sampled animals for a week and the average yield per day is recorded in Table 17.9. Estimate the total milk production (buffalo) in a district, standard error of the estimate and confidence limits for the total milk production in a district.

TABLE 17.9 (AVERAGE MILK YIELD PER DAY IN LITRES)

Dairy farm / Animal	1	2	3	4	5	6	7	8	9	10
1	9.2	12.2	1.5	10.0	16.2	9.8	8.7	7.0	14.2	10.5
2	12.5	8.6	9.0	14.7	12.0	16.7	12.6	11.8	10.7	8.6
3	10.0	10.5	8.4	15.2	10.6	15.1	7.5	9.7	11.3	7.9
4	15.1	9.4	6.8	10.0	14.3	12.6	6.9	8.9	8.7	11.5
5	13.5	15.0	7.5	8.8	13.2	10.5	11.2	13.1	7.6	9.6
	60.3	55.7	43.2	58.7	66.3	64.7	46.9	50.5	52.5	48.1

TABLE 17.10 ANOVA TABLE

Source	d.f	S.S.	M.S	
Between First stage units	$n-1=(10-1)$ $=9$	108.41	12.05 $= ms^2_b$	$\overline{Y}_{ts} = 1/nm \sum\limits_{i=1}^{n} \sum\limits_{j=1}^{m} Y_{ij}$ $= \dfrac{1}{10 \times 5}(546.9) = 10.94$
Within first stage units	$49-9$ $=40$	232.71	$5.82 = s^{-2}_w$	$\text{Est.v}\,(\overline{Y}_{ts}) = (1/n - 1/N)$ $s_b{}^2 + 1/N(1/m - 1/M)$ $(1/10 - 1/4$ $1/40(1/5 -$ $= 1.0202$
Total	$nm-1$ $=50-1$ $=49$	341.12		

Est. $S.E.(\overline{Y}_{ts}) = \sqrt{1.0202} = 1.01$

Confidence limits for $\overline{Y}_{NM}$:

Lower limit: $10.94 - 1.96 \times 1.01 = 8.96$

Upper limit: $10.94 + 1.96 \times 1.01 = 12.92$

Estimate for total milk production in a district

$$\hat{Y}_{ts} = NM \, \overline{Y}_{ts} = 40 \times 25 \times 10.94 = 10,940 \text{ litres}$$

$$\text{Est. } V\overline{Y}_{ts} = N^2 M^2 (1/n - 1/N)s_b^2 + NM^2(1/m - 1/M)s_w^{-2}$$

$$= 1306.9671 + 116400.00 = 117706.97$$

$$\text{Est. } S.E.(\hat{Y}_{ts}) = \sqrt{\text{Est.V}(\hat{Y}_{ts})} = 343.08$$

Confidence limits for total production:

Lower limit $= 10,940 - 1.96 \times 343.08 = 10,267.56$

Upper limit: $10, 940 + 1.96 \times 343.08 = 11,612.44$

17.6 SYSTEMATIC SAMPLING

In this method, lf one unit is selected at random the other units would be selected automatically. For example to estimate the area under a particular crop, one household would be selected randomly and the remaining households in the sample would be selected in a systematic manner by listing all the households in a serial order. Let N be the size of population and be the multiple of size of sample i.e, N = nk where n is a size of sample and k is a positive integer. Let one unit be selected at random out of k units and the remaining (n - 1) units be selected at equally spaced intervals of k units. Let 4-th unit was selected out of k units at random then the remaining units in the sample are k +4, 2k + 4, ... , (n - 1) k +4. This could be well understood from the following Table 17.11

TABLE 17.11

1	2	3	4	.. .. k
k+1	k+2	k+3	k+4	.. .. 2k
2k+1	2k+2	2k+3	2k+4	.. .. 3k
:	:	:	:	
(n-1)k+1	(n-1)k+2	(n-1)k+3	(n-1)k+4	.. ..nk

This method resembles stratified random sampling method though there is a subtle difference that one unit is selected at random from each of n strata. But the randomness was observed only for the first stratum but not for the other strata whereas this was not so in stratified random sampling. However, this method is equivalent to cluster sampling wherein one cluster is selected at random out of k clusters. This method is useful in forest research for estimating the volume of timber, total production of tamarind, lack, honey, etc. by serially numbering the trees, estimation of fish in a sea coast, etc. For further reading please refer to Sukhatme (1953).

17.7 NON-SAMPLING ERRORS

The standard errors of the estimates for the different sampling methods given in previous sections are known as sampling errors. The errors other than due to sampling are called non-sampling errors. The non-sampling errors might occur through (i) observational errors and (ii) incomplete samples (or non-response).

17.7.1 Observational Errors

In socio-economic surveys, the observational value of the character under study may differ from investigator to investigator and from time to time even for the same investigator. In the sampling methods dealt in the previous sections, the assumption was 'that the value of the observation is unique but that is not so in practice. For example, estimation of a crop would differ from person to person, field to field and time to time. Further the recording of information by the investigator supplied by third person might be far away from true value. These errors form part of observational errors. For further reading please refer to Sukhatme (1953).

17.7.2 Incomplete samples

The non-sampling errors occur also through incomplete schedules furnished by field investigators or false type of information provided by respondents or investigators. If the information is not available for a complete sample the estimates based on incomplete sample are biased. Further, the cost of the survey would be increased in order to obtain the information again on the incomplete sample. For further reading please refer to Sukhatme (1953).

17.8 TOLERANCES IN THE TESTING OF SEEDS

Rao and Apte (1972) gave a review on tolerances in the testing of seeds. By testing a sample of seeds in a laboratory the purity percentage would vary from scientist to scientist, sample to sample, core to core in a seed bag and finally from a lot to lot. The only way out in order to arrive at a confirmed and accurate result would be to give the confidence limits in which the true value of the purity percentage lies. The difference between upper and lower limits is known as 'Tolerance'. The tolerance value would be obtained by taking into all types of variations which might creep in at the time of drawing a sample of seeds at the time of testing the sample in the laboratory.

The tolerance is expressed in terms of probability. The tolerances change according to the probability assumed. If the tolerance is calculated at 0.05

probability then five samples out of hundred may give results outside the expected variation.

17.8.1   Procedure of Selecting a Sample

The bags containing seeds, the cores within bags, the samples within cores would be selected randomly at each stage so that the drawn samples of seeds would become representative of the whole lot. If the random sampling procedure was not adopted at each stage it would be difficult to calculate the standard error of the estimate and thereafter the values of tolerances could not be estimated precisely by statistical methods. After selecting the sample by random sampling procedure it would be sent to laboratory for testing purity percentage (or germination percentage), other crop seeds, noxious, weed seeds, etc. The samples would be labelled along with tolerance values on each sample after they were tested and would be released to market.

17.8.2   Methods of Computing Tolerances

1 st method G.N. Collins proposed formulae to compute tolerance for non-chaffy and chaffy seeds based on Binomial distribution. For calculating tolerances for non-chaffy seeds (Purity and germination percentage) we have

$$T_1 = 0.6 + \frac{(0.2 \cdot X_1 / X_2)}{100}$$

where 'X_1' be the percentage of the component under consideration and $X_2 = 100 - X_1$. The values of X_1 and X_2 could be obtained by testing the samples either for purity or for germination. These percentages could vary up to an extent of T_2.

Similarly, the tolerances for other crop seeds, weed seeds and Inert matter were calculated by the same author using the formula

$$T_2 = 0.2 + \frac{(0.2 \cdot X_1 X_2)}{100}$$

In T_2, the first component of variation was found to be small, while the other component of variation remained constant.

Since chaffy seeds would not mix well, the range of variation for purity, etc. was expected to be more compared to non chaffy seeds. The tolerance value for chaffy seeds was given by the formula

$$T_3 = T_1 \frac{(X_1 \text{ or } X_2)}{100}$$

where T_1 be the tolerance value for non-chaffy seeds, X_1 or X_2 be used in the formula whichever is less.

2nd method: S.R. Miles, A.S carter and L.C. Shenberger gave the following formula for tolerance of seeds.

$$T_4 = 1.414 \times t \times \sqrt{\frac{N-n}{n}\left[\frac{s_B^2}{n} + \frac{s_C^2}{n} + \frac{s_W^2}{n} + \frac{s_A^2}{n} + \frac{s_T^2}{n}\right]}$$

where t be the tabulated value of student's 't' for one 'tailed test at 5 per cent (or 1 per cent level), $s_B^2, s_C^2, s_W^2, s_A^2, and s_T^2$ are the means square among bags of seeds, cores within bags, working samples taken from the same submitted sample, Analysts, timings of testing the sample by the Analyst, respectively and n be the size of sample in each case. This formula could be used for both, the chaffy and non-chaffy seeds and this was applicable to a given' percentage of pure seed, other crop seeds, weed seeds or inert matter. The tolerances obtained for pure seed by this formula are somewhat small compared to tolerances obtained by previous formulae either in the case of chaffy and non-chaffy seeds. But in the case of other crop seeds, weed seeds and inert matter, the tolerances obtained by this formula are somewhat greater than previous tolerances due to small percentage of the component under consideration.

3rd method: The method for calculating tolerance for weed seeds is given here. Tests for seeds of noxious, weeds are quite different from the usual tests since the number of weed seeds per Kg (or Lb) in a lot should be determined. Usually the number of weed seeds would be quite small so the Poisson distribution is used in obtaining tolerances. These tolerances are dependent on the number of weed seeds found in each sample. They would give the maximum interval in which the number of weeds lie in each sample taking into consideration of variation due to sampling. Let T_5 be the tolerance value, we have

$$T_5 = (m+1) + 1.96\sqrt{m}$$

where m' be the number of weed seeds found from a random sample of seeds and $\sqrt{m}$ be the standard deviation of the variable 'm' in a Poisson distribution. Normally, the sample selected for finding the weed seed tolerance would be 10 times of the sample taken for purity and germination percentage due to small percentage of weed seeds found in an ordinary sample.

EXERCISES

1. A sample of 20 adult women were selected from a locality containing 200 households by simple random sampling to estimate the average protein intake in a diet in that locality. The hypothetical data of average intakes of protein in a diet in a week by 20 adult women are presented here. Give the estimate of average intake in that locality and estimate of standard error and confidence limits for the population mean.

S. No.	Average protein intake (gm)	S.No.	Average protein intake (gm)
1	57	4	45
2	33	5	52
3	47	6	51
7	56	14	35
8	40	15	58
9	48	16	41
10	38	17	46
11	37	18	52
12	49	19	47
13	42	20	36

2. A sample survey was conducted in a locality by dividing the households into four income groups, for estimating the average height of adult males along with standard error. The sample was selected based on proportional allocation. The hypothetical data are presented. Estimate the average height of adult male along with standard error and confidence limits.

Income group	N_i	n_i	Yn_i	si^2	$N_i/N.si^2$
I	100	10	150	20.4	6.80
II	80	8	157	24.8	6.61
III	70	7	164	36.2	8.45
IV	50	5	169	45.3	7.55

3. A sample survey was conducted to estimate the total egg production in a district by cluster sampling method by randomly selecting 6 clusters of villages from the 40 clusters in a district. There are *5* poultry farms in each cluster and the egg count (in 10's) (hypothetical) in each Poultry farm is given.

Poultry farm	Clusters					
	1	*2*	*3*	*4*	*5*	*6*
1	42	18	18	14	13	3
2	38	37	39	17	9	13
3	26	43	42	28	31	18
4	20	25	50	16	16	24
5	19	21	35	25	19	16

Estimate the total egg production, estimate the standard error and the confidence limits for the total in a district.

4. A sample survey was conducted to estimate the total area under the high yielding varieties of paddy in a district with the stratified random

sampling method with proportional allocation and the fallowing results were obtained.

Stratum	Stratum size (Ni)	Sample size (n_i)	Si^2	$\overline{Y}_{ni}$ ni (in 100 hectares)
1	5	15	3.5	10
2	8	25	4.2	12
3	7	20	5.4	14

Estimate the total area, standard error and confidence limits for the total area under high yielding varieties.

CHAPTER 18

ECONOMIC STATISTICS

18.1 INTRODUCTION

Series of observations generated through time on a character under study is known as time series. Population of India for different years, coal output in India for different months, rainfall in a region for different weeks, temperature for different days, sale of umbrellas in a super bazaar for different months, etc. are some of the examples of a time series data. Let Y_1, Y_2, ..., Yn be the time dependent observed values for the years 1, 2, 3, ... , n. and time, t may be years, months, weeks, days, hours, etc. which may not be necessarily in equal intervals. Time-series is said to be continuous if time is continuous variable otherwise it is discontinuous series. In this chapter only discontinuous series is discussed. The observations in time series should be comparable among themselves though time variable may not have equal duration. For example, some months have 31 days, others have 30 days and February have 28 or 29 days, it would be desirable to have monthly wise data as such rather than day wise.

18.2 ANALYSIS OF TIME-SERIES DATA

The objective of analysis of time series is to know the different sources of variation effecting the behaviour of the observations. Further, the analysis is useful for deeper understanding of the inner behaviour of the character under study as well as for forecasting the future time dependent observed value. Also the analysis would be useful to compare one series with another. The observed value of the time series is effected by four components: (i) longtime (or secular) trend, (ii) seasonal variation, (iii) cyclical variation, and (iv) random fluctuation. The relation between observed value and components of time series is

$$Y = T * S * C * R$$

Where '*' denotes convolution which represent '+' sign whenever the series follows additive law and '×' sign whenever the series follows exponential law and Y, T, S, C and R denote observed value, secular trend, seasonal variation, cyclical variation and random fluctuation respectively. Secular trend would indicate the general behaviour of the observed value over a long time as whether the trend is increasing or decreasing or remaining stationary.

Seasonal variation would be the effect of season extending to a period less than a year, cyclical variation would be the effect of cycles extending to a period more than a year and random fluctuations are the fluctuations due to unforeseen causes due to wars, floods, droughts etc. The analysis would attempt to isolate these factors separately and study them. This isolation of different causes of variation would help the planners and policy makers for taking proper decisions. These four components of time series are depicted in Fig. 18.1 (a), (b) and (c).

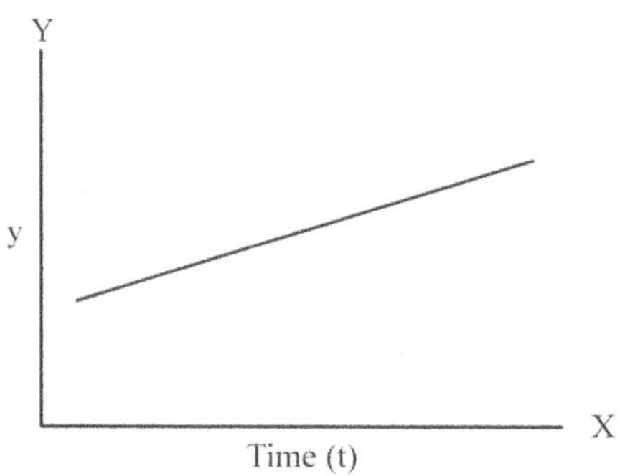

Fig. 18.1(a) Long time trend.

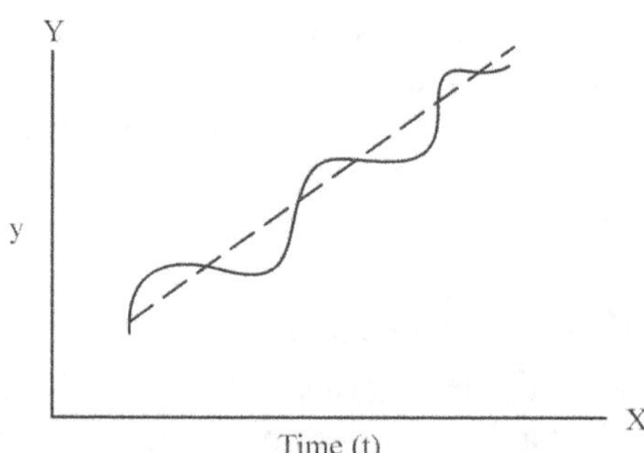

Fig. 18.1(b) Long time trend with cyclical variation.

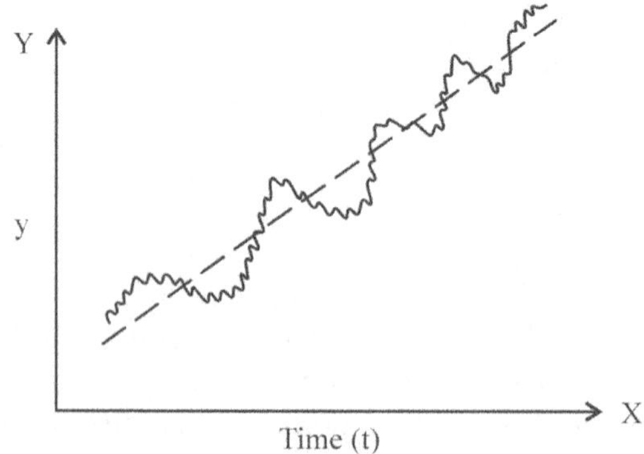

Fig. 18.1(c) Long time trend, cyclical variation and seasonal variation.

18.2.1   Secular Trend

It would be desirable to know the movement of the observation over a long period given a time dependent series. For example, it is desirable to know the production trend of paddy over a long period whether it is in increasing or decreasing trend or remains stationery. The term 'long' is subjective and the range of time depends on objective under study. First a graph would be drawn by plotting the points with time as independent variable on the X-axis and observed value on the Y-axis. The figure obtained by joining these points with a scale is known as 'historigram' and is shown in Fig. 18.2

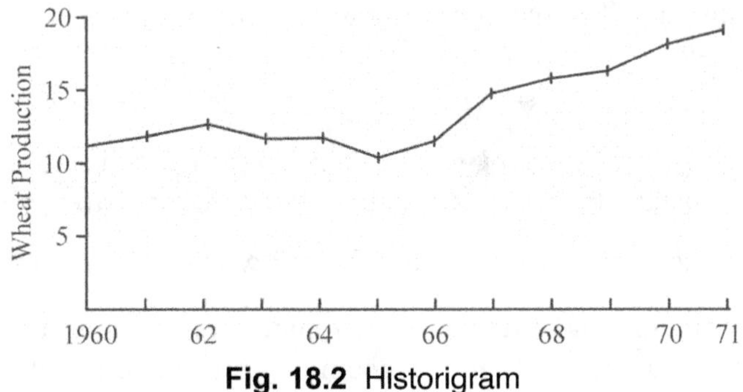

Fig. 18.2 Historigram

The three methods to fit the secular trend are: (a) Trend fitting by observation, (b) Fitting a polynomial by least squares method or fitting any mathematical model, and (c) Method of moving averages. The adoption of these methods depends on the objective of study. If the objective is only to isolate cycles then it would be reasonable to assume that the trend as a straight line drawn in such a way that upper and lower portions of cycles will balance each other. If it is to forecast (or predict) the values for the future it would be desirable to fit a mathematical model of suitable type. Fitting a mathematical model does not remove subject-tiveness in trend since it depends upon the adoption of a particular mathematical model as well as range of time considered in that curve. Similarly for smoothening the seasonal and random fluctuations, method of moving averages would be adopted.

18.2.1.1 Trend fitting by observation

This is the easiest method since fitting is done by mere inspection. If the 'historigram' appears to be of straight line nature, a straight line would be drawn with the help of a scale in such a way that upper and lower phases of cyclical curves would cancel each other. If the 'historigram' appears to be of non-linear form, a smooth hand curve would be drawn closely to the points by removing seasonal variations. This is a subjective method since the trend fitted is based on prefixed notion whether the trend should be a straight line or curve. Further the straight line (or curve) drawn may vary from person to person.

18.2.1.2 Fitting 0f polynomials by least squares method

Let the relation between observed value and the time be of the form $Y = a + bt$ where a and b are called constants. This relation is known as a straight line relation and is also called as polynomial of first degree. In general, $Y = a + bt + ct^2 + \ldots + pt^k$ is known as polynomial of k-th degree.

18.2.1.3 Fitting of first degree polynomial (or straight line)

Fitting of polynomial by least square method is to obtain the values of constants in the polynomial in such a way that the sum of the squares of the deviations between observed and expected values of dependent variable should be minimum. The two normal equations are

$$\Sigma Y = na + b\, \Sigma t$$

$$\Sigma tY = a\, \Sigma t + b\, \Sigma t^2 \qquad \ldots\ldots(18.1)$$

Solving (18.1) 'a' and 'b' would be obtained. In order to simplify the calculations linear transformation could be done from 't' to 'u', $u = \dfrac{(t-A)}{h}$ where A is arbitrary value which would be taken as the middle value when n is odd and average of two middle ones when n is even such that $\Sigma u = \Sigma u^3 = \Sigma u^5 = \ldots = 0$, 'h' is a common divisor and n is the number of observations. The first degree polynomial would become

$$Y = a + b\,(uh + A)$$

$$= (a + bA) + bhu$$

$$Y = a^1 + b^1 u \qquad \ldots\ldots(18.2)$$

where $a^1 = a + bA$ and $b^1 = bh$. The normal equations for this polynomial would be

$$\Sigma Y = na^1 + b^1 \Sigma u$$

$$\Sigma u\, Y = a^1 \Sigma u + b^1 \Sigma u^2 \qquad \ldots\ldots(18.3)$$

Since $\Sigma u = 0$, we have

$$a^1 = \Sigma Y / n \text{ and } b^1 = \Sigma uY / \Sigma u^2$$

Therefore, $b = b^1/h$, $a = a^1 - bA$.

The expected values of Y could be obtained by substituting the values of u in the fitted equation $\hat{y} = a + bu$ and these are known as trend values.

18.2.1.4 Fitting of second degree polynomial

The equation of the second degree polynomial is

$$Y = a + bt + ct^2 \qquad \ldots\ldots(18.4)$$

where a, b and c are constants. The normal equations for estimating the constants are

$$\Sigma y = na + b\Sigma t + c\Sigma t^2$$

$$\Sigma ty = a\Sigma t + b\Sigma t^2 + c\Sigma t^3$$

$$\Sigma t^2 y = a\Sigma t^2 + b\Sigma t^3 + c\Sigma t^4 \qquad \ldots\ldots(18.5)$$

Solving (18.5), the values of a, b and 'c' could be obtained. The expected values of Y are obtained by substituting the values of t in the fitted equation $\hat{Y} = a + bt + ct^2$. In order to simplify the calculations for solving the normal equations, a linear transformation from 't' to another variable 'u', $u = \dfrac{t-A}{h}$ was employed, where A is the arbitrary value which is the middle value when n is odd or average of two middle ones when n is even and h is the common divisor such that $\Sigma u = \Sigma u^3 = \Sigma u^5 = 0$. The polynomial would be written as

$$Y = a + b\,(A + hu) + c\,(A + hu)^2$$

$$= (a + bA + cA^2) + (bh + 2chA)u + ch^2 u^2$$

$$Y = a^1 + b^1 u + c^1 u^2$$

where $a^1 = a + bA + cA^2,\ b^1 = bh + 2chA$ and $c^1 = ch^2$ $\ldots\ldots(18.6)$

The normal equations for (18.6) are

$$\Sigma Y = na^1 + b^1 \Sigma u + c^1 \Sigma u^2$$

$$\Sigma uY = a^1 \Sigma u + b^1 \Sigma u^2 + c^1 \Sigma u^3$$

$$\Sigma u^2 Y = a^1 \Sigma u^2 + b^1 \Sigma u^3 + c^1 \Sigma u^4 \qquad \ldots\ldots(18.7)$$

Since $\Sigma u = \Sigma u^3 = 0$, the normal equations become

$$\Sigma Y = na^1 + c^1 \Sigma u^2$$

$$\Sigma uY = b^1 \Sigma u^2$$

$$\Sigma u^2 Y = a^1 \Sigma u^2 + c^1 \Sigma u^4$$

Hence $b^1 = \dfrac{\Sigma uY}{\Sigma u^2}$

By solving the first and third equations a^1 and c^1 could be obtained. Therefore

$$c = c^1 / h^2,\ b = \frac{b^1 - 2chA}{h}\ \text{ and } a = a^1 - bA - cA^2$$

The expected values (trend values) of Y could be obtained by substituting the values of u in the fitted equation

$$\hat{Y} = a^1 + b^1 u + c^1 u^2$$

EXAMPLE: Fit the second degree polynomial for the following time series data on Bajra production for the years from 1960-61 to 1972-72 in India and obtain the trend values and ratios to trend values.

Year	Prod. (m.t.)	Year	Prod. (m.t.)
1960	3.23	1966	4.67
1961	3.56	1967	5.19
1962	3.89	1968	3.80
1963	3.73	1969	5.33
1964	4.45	1970	8.03
1965	3.75	1971	5.36

Figures rounded off to second decimal place.

Source: 'Agricultural situation in India' journal. Agricultural year 1960-61 is taken as 1960 and so on.

$$Y = a^1 + b^1 u + c^1 u^2$$

The normal equations for fitting the above equation are

$$\Sigma Y = na^1 + b^1 \Sigma u + c^1 \Sigma u^2$$

$$\Sigma uY = a^1 \Sigma u + b^1 \Sigma u^2 + c^1 \Sigma u^3$$

$$\Sigma u^2 Y = a^1 \Sigma u^2 + b^1 \Sigma u^3 + c^1 \Sigma u^4$$

Since $\Sigma u = \Sigma u^3 = 0$, the normal equations become

$$\Sigma Y = na^1 + c1\Sigma u^2$$

$$\Sigma uY = b^1 \Sigma u^2$$

$$\Sigma u^2 Y = a^1 \Sigma u^2 + c^1 \Sigma u^4$$

Hence
$$b^1 = \frac{\Sigma uY}{\Sigma u^2} = \frac{77.23}{572} = 0.1350$$

$$54.99 = 12a^1 + 57c^1$$

$$2705.39 = 572a^1 + 48620c^1$$

Solving these two equations, we have $c^1 = 0.0040$, $a^1 = 4.3902$

$$\hat{Y} = 4.3902 + 0.1350u + .0040u^2$$

Transforming back from u to t, we have A = 6. 5

$$c = c^1/(0.5)^2 = \frac{0.004}{0.25} = 0.016, \quad b = \frac{b^1 - 2ch\,A}{h}$$

$$= \frac{0.1350 - 2 \times .016 \times 0.5 \times 6.5}{0.5} = 0.062$$

$$a = a^1 - bA - cA^2$$

$$= 4.3902 - 0.062 \times 6.5 - 0.016(6.5)^2 = 3.3112.$$

The fitted second degree polynomial is

$$\hat{Y} = 3.3112 + .062t + 0.016t^2$$

TABLE 18.1

Year	Production (m.t) (Y)	't' (S. No .)	$u = \dfrac{t-A}{0.5}$	u^2	uY	u^4	u^2y	Trend value (T)	Ratio to trend
1960	3.23	1	−11	121	−35.53	14641	390.83	3.39	95.28
1961	3.56	2	−9	81	−32.04	6561	280.36	3.50	101.71
1962	3.89	3	−7	49	−27.23	2401	190.61	3.64	106.87
1963	3.73	4	−5	25	−18.65	625	93.25	3.82	97.64
1964	4.45	5	−3	9	−13.35	81	40.05	4.02	110.70
1965	3.75	6	−1	1	−3.75	1	3.75	4.26	88.03
1966	4.67	7	1	1	4.67	1	4.67	4.53	103.09
1967	5.19	8	3	9	15.57	81	46.71	4.83	107.45
1968	3.80	9	5	25	19.00	625	95.00	5.17	73.50
1969	5.33	10	7	49	37.31	2401	261.17	5.53	96.38
1970	8.03	11	9	81	72.27	6561	650.43	5.93	135.41
1971	5.36	12	11	121	58.96	14641	648.56	6.36	84.28
	54.99			572	77.23	48620	2705.39		

Here A = 6.5 h = 0.5

18.2.1.5 Fitting of polynomials of higher degree

The higher degree polynomials would be fitted on the similar lines of fitting first or second degree polynomials. For example the fitting of third degree polynomial $Y = a + bt + ct^2 + dt^3$ would be done by solving the following normal equations for obtaining a, b, c and d as

$$\Sigma Y = na + b\Sigma t + c\Sigma t^2 + d\Sigma t^3$$

$$\Sigma t^2 Y = a\Sigma t + b\Sigma t^2 + c\Sigma t^3 + d\Sigma t^4$$

$$\Sigma t^2 Y = a\Sigma t^2 + b\Sigma t^3 + c\Sigma t^4 + d\Sigma t^5$$

$$\Sigma t^3 Y = a\Sigma t^3 + b\Sigma t^4 + c\Sigma t^5 + d\Sigma t^6 \qquad .....(18.8)$$

where n is the number of observations.

The important point to be considered is that which degree of polynomial would be appropriate for fitting the trend. The approximate procedure is to compute the first differences, second differences, third differences etc. and if the first differences, are constant, the first degree polynomial would be better fit, the second differences are constant the second degree polynomial would be better fit and so on. The second method for finding the suitable degree of polynomial for trend is by variate –differences method.

18.2.1.6 Variate differences method

If the series consists of polynomial element and random element, the polynomial element can be eliminated with the help of successive differencing. Let E_t be the random element at the time, t and Δ^s be the s–th differencing, then

$$\Delta^s E_t = E_{t+s} \binom{s}{1} E_{t+s-1} + \binom{s}{2} E_{t+s-2} +...+ (-1)^s E_t$$

assuming that $\quad E(\Delta^s E_t) = 0$

and $\qquad V(\Delta^s E_t) = \binom{2s}{s} v$

and the estimate of V is given by $'\hat{V}'$ where

$$\hat{V} = \frac{V_2(\Delta^s E_t)}{\binom{2s}{s}} \text{ and } V_2(\Delta^s E_t) = \frac{\text{sum of square of } \Delta^s E_t}{(\text{No.of observation} - s)}$$

$$.....(18.9)$$

If the value of $\hat{v}$ comes out to be approximately constant at say, r-th degree and onwards, then r-th degree polynomial would be taken as a suitable fit. The values of V remains stationary at a particular degree of differencing is a somewhat subjective statement, the variance of the difference between sum of squares at s-th and (s + 1)-th degree is provided here for the normal approximation for large N and s > 6 as

$$V \text{ (difference)} = \frac{(3s+1)}{2(2s+1)^3} \sqrt{\frac{(2\pi s)}{(N-s-)}} \left\{ \left(\frac{(s.s)_s}{(N-s)\binom{2s}{s}} \right) \right\}^2$$

$$.....(18.10)$$

where N is the number of observations and (S.S)s is the sum of squares at the s-th degree. If the reduction in variance is of not much significance from s-th level onwards then the trend line may be taken at s-th degree polynomial.

18.2.1.7 Fitting of exponential curve

If the data indicate a constant ratio of change instead of constant amount of change the exponential curve of the type

$$Y = ab^t$$

is appropriate for fitting the trend. The data plotted on semi-log paper (t on the X-axis and log Y on the Y-axis) showed straight line relationship the exponential curve of the type mentioned above is suitable. For fitting the exponential function the least squares method would be adopted.

$$Y = ab^t \qquad .....(18.11)$$

$$\text{Log } Y = \log a + t \log b$$

The normal equations are

$$\Sigma \log Y = n \log a + \log b \, \Sigma t$$

$$\Sigma t \log Y = \log a \, \Sigma t + \log b \, \Sigma t^2 \qquad .....(18.12)$$

Solving equations (18.12) 'log a' and 'log b' could be obtained. By taking anti log of log a and log b, a and b could be obtained. If the data plotted on semi-log paper show a parabolic curve, a second degree exponential curve of the type $Y = ab^t c^{t^2}$ would be an appropriate trend curve. The procedure of fitting this curve runs on the similar lines as given here. In general the first and second degree exponential curves would be fitted whenever the first and second differences of the logarithms respectively are constant.

18.2.1.8 Fitting of modified exponential curve

If the data show the amount of growth declines by a constant percentage the curve asymptotically approaches to certain upper limit called asymptote then the trend curve may be taken as modified exponential curve. If it is a decreasing series the curve indicates the constant amount of decrease in the decreasing series. The equation of the modified exponential curve is given as

$$Y = a + bc^t \qquad .....(18.13)$$

where a is called asymptote and b, c arc constants. The approximate fitting of this curve is illustrated here with hypothetical data.

TABLE 18.2

t	Y	Partial total	Y increment	Per cent of preceding increment
0	52			
1	68	120	16	
			11	68.75
2	79	165	7	63.64
3	86		5	71.43
4	91	185	3	60.00
5	94			

Here 'a' be the asymptote, 'b' be the distance between trend value Y when t = 0 and the asymptote and c be the ratio between successive first differences. Since there are three constants, three equations are required for fitting the function. The total observations of dependent variable, Y are divided into three equal parts such that T_1 be the total of the first part, T_2 be the total of second part and T_3 be the total of the third part assuming that each part consists of n observations then

$$c = \left(\frac{T_3 - T_2}{T_2 - T_1} \right)^{1/n}$$

$$b = \left(T_2 - T_1 \right) \frac{(C-1)}{(C^n - 1)^2} \text{ and}$$

$$a = 1/n \left[T_1 - b \frac{(c^n - 1)}{c - 1} \right] \text{ or } 1/n \left[\frac{T_1 T_3 - T_2^{\,2}}{T_1 + T_3 - 2T_2} \right] \qquad \dots\dots(18.14)$$

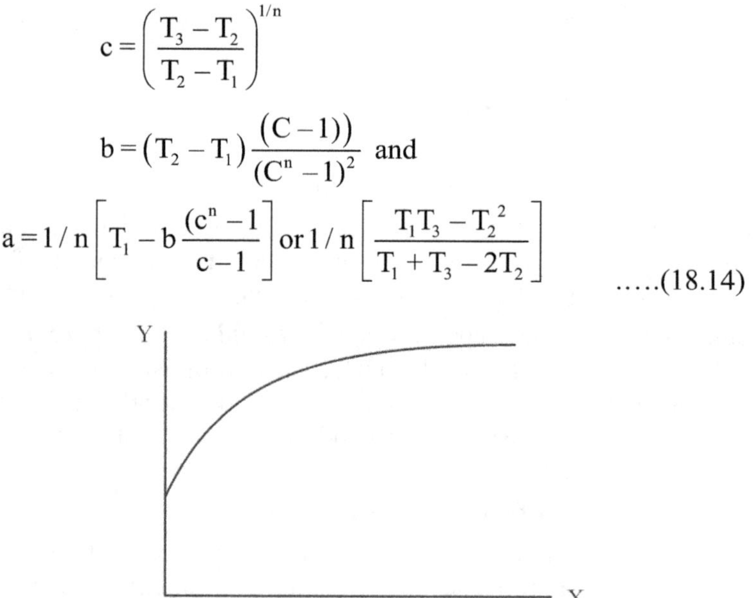

Fig. 18.3 Modified exponential curve.

By substituting the values for T_1, T_2 and T_3 from Table 18.2 as $T_1 = 120$, $T_2 = 165$ and $T_3 = 185$ the values of a, b and c are given as

$$a = \frac{1}{2}\left[\frac{(120\times185-(165)^2}{120+185-2\times165}\right] = -100.5$$

$$b = (165-120)\times\frac{(0.67-1)}{\left[(0.67)^2-1\right]^2} = -49.5$$

$$c = \left[\frac{185-165}{165-120}\right]^{\frac{1}{2}} = 0.67$$

18.2.1.9 Fitting of Gompertz curve

This curve gives a trend in which the logarithms of the increasing values decline by a constant percentage. The Gompertz curve would be used if the approximate trend when plotted on a semi-logarithmic paper resembles a modified exponential curve. The model of the Gompertz curve is given by

$$Y = ab^{c^t} \qquad\qquad .....(18.15)$$

where a, b and c are constants. By taking logarithms for this equation, we have

$$\log Y = \log a + c^t \log b = \log a + (\log b) c^t$$

This curve has two asymptotes and the lower asymptote will be zero. The curve is similar to modified exponential curve after logarithmic transformation is effected. This curve is useful in the study of growth of industries.

The approximate estimates of a, b and c are obtained on the similar lines as in the case of modified exponential curve by considering log Y values in place of Y values. Let T_1 be the total of logarithms of first n years data, T_2 be the total of logarithms of second n years data and T_3 for the third n years data where each part was assumed to be consisting of n observations. Then, we have

$$c^n = \frac{T_3 - T_2}{T_2 - T_1}$$

$$\log b = (T_1 - T_2)\frac{(C-1)}{(c^n-1)^2} \qquad\qquad .....(18.6)$$

$$\log a = 1/n\left[T_1 - \frac{(c^n-1)}{(c-1)}\log b\right]$$

If c >1, there will be no upper asymptote and if c < 1 there will be upper and lower asymptotes with lower asymptote being zero.

18.2.1.10 Fitting of Logistic curve

If the first differences of the reciprocals of Y values decline by a constant percentage, the logistic curve would be appropriate and is given by the formula

$$1/Y = a + bc^t \qquad\qquad .....(18.17)$$

where a. b and c are constants, This curve is useful in the study of population growth, growth of drosophila and yeast. The fitting of this curve is done on the similar lines of modified exponential by using totals of reciprocals of Y values (T_1, T_2 and T_3) when the total number of observations were divided into three equal parts.

CURVE ESTIMATION
(LINEAR, QUADRATIC, EXPONENTIAL CURVES)

SPSS DATA ANALYSIS PROCEDURE

Enter X (independent, protein content) and Y (independent, gain in wt.) variables and feed the data for X and Y column wise and CLICK ANALYSE FROM THE TOOL BAR and then take option REGRESSION AND THEN OPT for 'CURVE ESTIMATION'. In the display table OPT for LINEAR, QUDRATIC AND OTHER CURVES depending upon the requirement of analysis. THEN CLICK O.K . We get the display table with results. The result table contains R^2 value and regression coefficient values for Linear and Quadratic curves and also F- values and probabilities for drawing conclusions. The plotted curves are also given in the same display table.

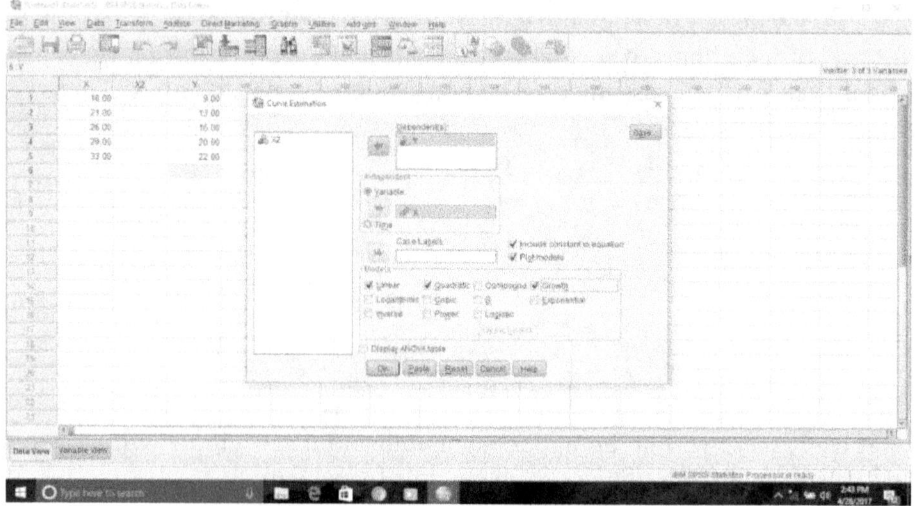

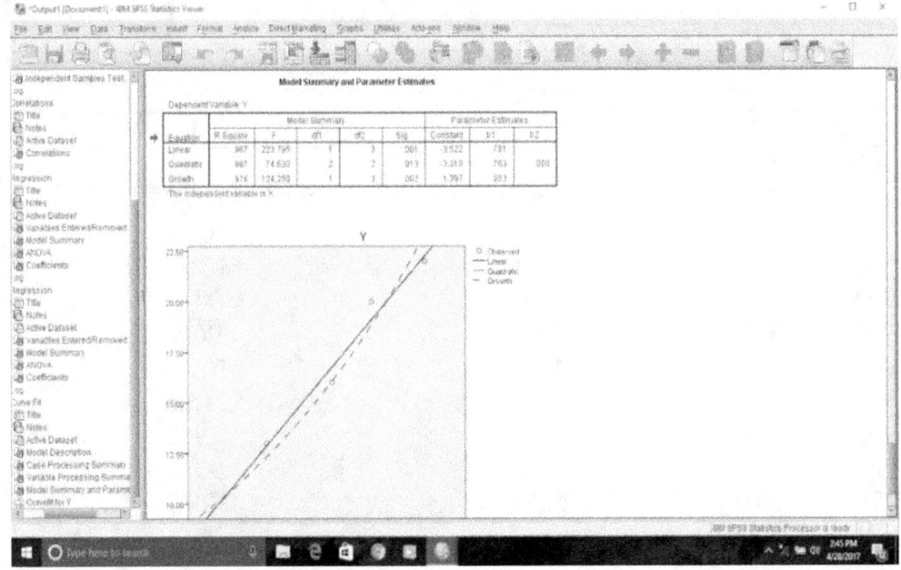

18.2.1.11 Method of moving averages

This is one of the simplest and effective method of fitting the trend of the time series data. In this method, the arithmetic means (or Geometric means) arc calculated successively by taking a specified number of values, say p, at a time and then adding the next value and dropping the initial value until all the values are exhausted. The averages would be written against the middle values of the corresponding observations when p is odd and if it is even, the averages would be written against the middle of the two middle ones, where p is called the period of moving averages. Let Y_t be the t-th observational value in the time series for t = 1, 2, ..., N and let p be the period of moving average then the moving averages are given as

$$m_1 = 1/p (Y_1 + Y_2 + ... + Yp),$$
$$m_2 = 1/p (Y_2 + Y_3 + ... + Y_{p+1}), \text{ etc.} \qquad .....(18.18)$$

It may be noted that one need not compute each moving average afresh since m_2 can be obtained from m_1 m_3 from m_2, etc. as follows.

$$m_2 = m_1 + 1/p (Y_{p+1} - Y_1), m_3 = m_2 + 1/p (Y_{p+2} - Y_2), \text{ etc.} \qquad .....(18.19)$$

EXAMPLE: The following are the data on Bajra production for the years from 1960-61 to 1971-72 in India. The secular trend with the help of 3-yearly and 4-yearly moving averages and also ratios to trend values are presented in Tables 18.3 and 18.4 respectively.

TABLE 18.3

Year	Production	3 – yearly moving average	Ratio to trend $\dfrac{col.(2)}{col.(3)} \times 100$
1	2	3	4
1960	3.23		
1961	3.56	3.56	100.0
1962	3.89	3.73	104.3
1963	3.73	3.69	101.7
1964	3.45	3.64	94.8
1965	3.75	3.89	96.4
1966	4.47	4.47	100.0
1967	5.19	4.49	115.6
1968	3.80	4.77	79.7
1969	5.33	5.72	93.2
1970	8.03	6.24	128.7
1971	5.36		

Production figures were rounded off to second decimal place.

TABLE 18.4

Year	Production	4–yearly moving average	2– yearly moving average for col.(3)	Ratios to trend $\dfrac{col.(2)}{col.(4)} \times 100$
1	2	3	4	5
1960	3.23			
1961	3.56			
1962	3.89	3.60	3.76	103.5
1963	3.73	3.91	3.94	94.7
1964	4.45	3.96	4.03	110.4
1965	3.75	4.10	4.29	87.4
1966	4.47	4.47	4.39	101.8
1967	5.19	4.30	4.50	115.3
1968	3.80	4.70	5.15	73.8
1969	5.33	5.59	5.61	95.0
1970	8.03	5.63		
1971	5.36			

Source: 'Agricultural situation in India' journal. The agricultural year 1960-61 is taken as 1960 and so on.

Properties of moving averages: (i) If the time series data is of exponential type, the Geometric means are used instead of Arithmetic means in the moving averages method. Geometric means can be computed by taking antilogarithm of the arithmetic means of logarithms of the observations. (ii) If the relation between Y and 't' is of linear type (say).Y = a + bt and the time series data is assumed to be of equi-spaced time intervals the moving averages would also lie on the fitted straight line. If the relation between Y and 't' is more than first degree polynomial, the moving averages may not lie on the fitted polynomial. In this case weighted moving averages might be more appropriate. (iii) If the moving averages represent the trend perfectly except, a single cyclical movement of p years is superimposed on it, then the period of moving averages would be taken as p so as to balance the upper values with the lower values. If the number of cyclical movements superimposed on the trend is more, the period of moving average would be arrived at by trial and error.

18.2.2   Seasonal Variation

It is a variation in observed values due to seasons. For example, prices of foodgrains might be lowest at the harvest time and gradually increase up to the time of harvest: sale of ice creams would be highest in the summer season compared to other seasons, sale of umbrellas would be high in rainy and summer seasons compared to winter, bank deposits would be highest in the first week and lowest in the last week of the month, sale of commodities in a super bazar would be highest in the first week of month and lowest in the last week of month, etc. Season here can be a week, month or quarter but less than a year. To evaluate seasonal variation, seasonal indices are calculated by the following method.

18.2.2.1 Method

Supposing that the time series data are of monthly or quarterlywise, the deviation of the monthly (or quarterly) average from yearly average is known as seasonal effect. Here the monthly (or quarterly average) would be obtained by averaging the monthly (or quarterly) observations over number of years. Seasonal index is the percentage of seasonal average to their mean and this can be written as

$$\text{Seasonal index} = \frac{\text{Seasonal Average}}{\text{average of seasonal averages}} \times 100$$

$$.....(18.20)$$

EXAMPLE: Compute the seasonal indices for the following time series data on wholesale monthly prices of wheat per quintal.

TABLE 18.5

Year	June	July	August	September	October	November	December
1961	43.00	44.00	43.12	43.00	45.00	46.00	49.00
1962	44.38	48.00	44.50	42.00	46.00	42.00	44.00
1963	44.00	45.00	46.50	46.75	48.00	48.50	48.95
1964	48.00	48.50	50.00	58.25	58.50	58.50	67.50
1965	58.00	66.70	61.00	53.00	55.70	59.60	57.00
1966	72.80	73.75	73.00	71.30	84.20	86.10	96.30
1967	75.00	74.50	76.25	77.25	80.20	86.50	79.00
1968	79.20	84.00	86.70	89.00	96.00	98.00	100.00
1969	75.00	97.00	97.00	98.20	99.70	100.25	102.70
Total	539.38	581.45	578.07	578.75	613.30	652.45	644.45
Seasonal average	59.93	64.61	64.23	64.31	68.14	69.49	71.61
Seasonal Index	90.57	97.61	97.07	97.19	102.98	105.02	108.22

Year	Jan	February	March	April	May
1961	46.50	45.75	46.00	47.15	45.60
1962	49.00	50.00	48.00	48.50	43.00
1963	45.00	44.00	46.00	43.00	44.20
1964	67.85	61.50	59.00	46.25	48.50
1965	62.00	66.50	68.50	53.00	55.00
1966	60.75	68.80	69.35	61.70	66.95
1967	110.50	109.25	110.75	89.70	88.50
1968	80.00	80.60	80.00	82.50	79.75
1969	102.00	99.00	88.00	90.00	87.60
Total	623.60	625.40	615.60	561.80	559.10
Seasonal average	69.29	69.49	68.40	62.42	62.12, 66.17
Seasonal Index	104.72	105.02	103.37	94.33	93.88

From the above table it can be observed that the seasonal variation was very high in December and very low in June.

18.2.3 Cyclical Variation and Random Fluctuations

In order to isolate cyclical variation the following residual method would be adopted. Let Y be the observed value where Y = T*S*C*R Assuming that Y = T× S × C × R, we have

$$\frac{T \times S \times C \times R}{S} \times 100 = T \times C \times R \times 100$$

$$\frac{T \times C \times R}{T} \times 100 = C \times R \times 100 \qquad\qquad .....(18.21)$$

The cyclical variation can be isolated from data of C × R by smoothening 'random fluctuations' through short term moving averages of appropriate period. In isolating cyclical variation at least monthlywise time series should be considered. The random fluctuations also would be isolated in a similar manner by dividing C × R by C. The behaviour of random fluctuations can be better understood by drawing frequency curve. If the random fluctuations are truly appearing in 'random' then the shape of frequency curve would be of normal curve type. In practice the curve may be slightly different from normal since even after isolating random (or irregular) fluctuations there might be other factors which influence the time series for a longer period than a single month in a monthly wise data.

18.3 INDEX NUMBERS

Index numbers are the measuring devices of comparison within the variable or between the variables over a period of time. For example, it measures the change of price of rice from 1966 to 1976 in percentage. In other words it measures the price of rice in 1976 assuming the price in 1966 as 100. If the index number of price of rice in 1976 is 120 in comparison to 100 in 1966, the price rose by 20 per Cent in a decade. Index numbers arc the scaling devices of the magnitude of the variable at one point of time in comparison to the magnitude of the same variable at another point of time or for comparison of two variables at the same point of time. These are also known as 'Economic barometers'. We shall list out here some of the uses of index numbers.

18.3.1   Uses

The index numbers are useful in situations such as (i) *change of price of a commodity (or commodities) over a period of time.* The knowledge of price movement is essential in order to detect the causes effecting the price changes. For example, if the index numbers of prices of rice are going up, then the causes such as inflation, less production due to drought and floods, less supply of inputs like fertilizers, manures, seeds and irrigation. unremunerative prices, increase in population, etc. may be looked into so that necessary remedial steps may be taken up by the administrative machinery. On the other hand if the index numbers of prices of rice are not increasing in relation to other inputs like fertilizers, irrigation charges, seeds, etc. the farmer would prefer some cash crop to paddy. Hence index numbers are very useful for policy makers as well as for consumers and producers.

(ii) Change of quantities over a period of time: The changes in quantities of import and export of different commodities will be helpful in order to know the foreign exchange position of the country. For example, the index numbers decline for export commodities like jute, textiles, coffee, tea, machine, tools, railway wagons, leather goods, etc. necessitating the review of export policy like quality of goods exported, world market position, incentives to farmers, etc.

(iii) Change of total commodity values over a period of time: The change in cost of living index is of vital importance to wage earners since wages should commensurate with their minimum standard of living. That is why, Govt. of India linked up the dearness allowance of an employee with the cost of living index. For example, if the cost of living index rises by (say) 10 per cent, dearness allowance also would be raised by some percentage of the basic pay so that the same standard of living would be maintained. The cost of living index would also effect the taxation policy followed by the Government. The net rise in per capita income could be determined by dividing the actual per capita income with the cost of living index number. The national income divided by the price index number would eliminate the changed value of money.

(iv) Change in educational efficiency over a period of time: The change in level of attendance of school going children in primary education, change in percentage of passes in secondary education. change in average expenditure per pupil in professional colleges for attaining a degree, change in ratio of educated employed to unemployed over a period of time can be evaluated with the help of index numbers.

18.3.2

Let $\dfrac{Pn}{Po} \times 100$ be the price relative where P_n, p_o, are the prices of a commodity at n-th year and base year respectively. The base year can be any year previous to the n-th year. The base year would be a normal year in which extremeties like drought, war, floods, etc. would not occur. Sometimes base year value will be taken as the average of some normal years instead of one year.

18.3.3   Price Index Number

The price index number is

$$P_{n,,o} = (\Sigma p_n / \Sigma p_o) \times 100 \qquad\qquad .....(18.22)$$

where P_n, P_o are the prices of a particular commodity at n-th and base years respectively for n > 0. The summations in (18.22) indicate the summing over all the commodities involved. Here the yearwise data on prices are assumed to be available. This index number is also known as simple aggregative index number.

This price Index number will be influenced by a commodity which has relatively very high price though its importance may be less in the daily life. For example, Diamonds would inflate the price index number due to very high price but their importance in daily life is negligible compared to other essential commodities.

18.3.4   Laspeyre's Price Index Number

In this index number, the prices are weighted with quantities produced (or consumed) in the base year for each of the commodities. Laspeyre's price index number is

$$L_{n,o} = \frac{\sum p_n q_0}{\sum p_0 q_0} \times 100 \qquad\qquad .....(18.23)$$

where P_n, P_o are the prices of each commodity at n-th year and base year respectively, q_0 be the quantity of the same commodity produced (or consumed) in the base year and 'Σ' extending over all commodities. This index number measures the changing value of goods produced (or consumed) at the base year. All the weighted price index numbers including Laspeyre's price index number are known as cost of living index numbers. This index number may show upward bias since it measures the percentage change in values of goods produced in base year.

18.3.5   Paache's Price Index Number

The quantities produced in the n-th year are given as weights to the prices in Paache's price index number,

$$P_{n,0} = \frac{\sum p_n q_n}{\sum p_0 q_n} \times 100 \qquad\qquad .....(18.24)$$

where P_n, P_o are the prices of each commodity at n-th year and base year respectively and q_n is the quantity produced (or consumed) of the same commodity in the n-th year. This index number gives downward bias since it measures the percentage in value of goods produced (or consumed) in the n-th year.

18.3.6   Marshall-Edgeworth Price Index Number

Here the weights .are taken as the arithmetic mean of the quantities produced (or consumed) at the base and given years for the prices.

$$Mn,0 = \frac{\sum p_n (q_0 + q_n)}{\sum p_n (q_0 + q_n)} \times 100 \qquad \qquad .....(18.25)$$

This index number has no known bias towards any direction since it uses both the quantities produced (or consumed) at the base and given years.

The weights may also be taken as the average of all the quantities of all the years for which the index numbers are to be computed. However, this method is not used much in practice since weights have to be revised every time. The weights may also be taken as the arithmetic mean of the quantities of all the typical years. This is practicable since a list of quantities can be used for computing the weights. If this list of quantities becomes outdated a new list can be prepared.

18.3.7   Keynes Method

Here the weights are to be the highest common factor of the quantities for either base and given years or for all the years. Since the common factor of quantities differs in general for each commodity this index number is expected to utilise the quantities of the crosswise section of the years.

18.3.8   Fisher's Ideal Index Number

This index number is the Geometric mean of the Laspeyre and Paaches' price index numbers and which is given as

$$F_{n,0} = \sqrt{\frac{\sum p_n q_0}{\sum p_0 q_0} \frac{\sum p_n q_n}{\sum p_0 q_n} \times 100^2} \qquad \qquad .....(18.26)$$

This is called 'ideal index number' since it satisfies two properties. 'Time reversal test' and 'Factor reversal test'.

18.3.8.1 Time reversal test

The index number obtained by interchanging the time should be the reciprocal of the given index number. Let $P_{n,o}$ be the price index number and $P_{o,n}$ be the index number obtained by interchanging time then we have

$$P_{0,n} = \frac{1}{P_{n,0}} \quad \text{or} \quad p_{0,n.} \ p_{n,0} = 1$$

Since the index number represents the changing value of prices (or production) the formula should' be able to represent. the changing value of prices (or production) backwards also i.e., from the given year to base year. For example if the index number shows 100 per cent rise in the prices of bajra from 1960 to1975 the same index number should also indicate 50 per cent decrease in prices from 1975 to 1960.

$$F_{n,0} = \sqrt{\frac{\sum p_n q_0}{\sum p_0 q_0} \times \frac{\sum p_n q_n}{\sum p_0 q_n}}$$

Interchanging '0' and 'n' years in $F_{n,o}$, we have

$$F_{0,n} = \sqrt{\frac{\sum p_0 q_n}{\sum p_n q_n} \times \frac{\sum p_0 q_0}{\sum p_n q_0}}$$

$F_{n,o} \times F_{o,n} = I$. Hence Fisher's ideal index number satisfies 'time reversal test'.

18.3.8.2 Factor reversal test

The index number obtained by interchanging factors 'p' and 'q' in the original index number and multiplied by the given index number should be equal to the value index number. Let $P_{n,o}$ be the price index number and $p^1_{n,o}$ be the index number obtained by interchanging the factors in $P_{n,o}$, we have

$$P_{n.o}, P^1_{n,o} = V_{n,o}$$

where $V_{n,o}$ is the value index number and is given as

$$V_{n,o} = \frac{\sum p_n q_n}{\sum p_0 q_0} \times 100$$

Fisher's ideal index number satisfies 'factor reversal test' as follows

$$F_{n,0} = \sqrt{\frac{\sum p_n q_0}{\sum p_0 q_0} \cdot \frac{\sum p_n q_n}{\sum p_0 q_n}}$$

Interchanging factors p and q in $F_{n,0}$ we have

$$F^1 n,0 \sqrt{\frac{\sum q_n p_0}{\sum q_0 p_0} \cdot \frac{\sum q_n p_n}{\sum q_0 p_n}}$$

$$F_{n,0} \cdot F^1 n,0 = \frac{\sum p_n q_n}{\sum p_0 q_0} = V_{n,0.}$$

Hence Fisher's ideal index number satisfies 'factor reversal test'. It can be seen that 'Laspeyre' and 'Paaches' index numbers do not satisfy both 'time' and 'factor' reversal tests.

The advantages and disadvantages of Fisher's ideal number are listed here.

Advantages: (1) It satisfies both 'time' and 'factor' reversal tests and perhaps no other index number satisfies. (2) Since Laspeyre's index number shows upward bias and Paache's index number downward bias the Fisher's index number which is the geometric mean of these two would lie in between them.

TABLE 18.6

Food grains	Prices/ quintal(Rs.)			Quantities (million tons)		
	$1961(p_0)$	$1970(p_{n-1})$	$1971(p_n)$	$1961(q_0)$	$1970(q_{n-1})$	$1971(q_n)$
Paddy	35	60	62	40.7	43.0	43.2
Wheat	70	102	110	45.4	60.2	60.5
Jowar	42	75	80	4.2	5.5	5.9
Bajra	40	78	82	3.5	4.0	4.4
Maize	45	80	85	2.8	3.2	3.3

TABLE 18.7

Food grains	q_0p_{n-1}	q_0p_n	q_0p_0	$q_{n-1}p_{n-1}$	q_np_n	$q_{n-1}p_0$	q_np_0
Paddy	2442.0	2523.4	1424.5	2580.0	2678.4	1505.0	1512.0
Wheat	4630.8	4994.0	3178.0	6140.4	6655.0	4214.0	4235.0
Jowar	315.0	336.0	176.4	412.5	472.0	231.0	247.8
Bajra	273.0	287.0	140.0	312.0	360.8	160.0	176.0
Maize	224.0	238.0	126.0	256.0	280.5	144.0	148.5
	7884.8	8378.4	5044.9	9700.9	10446.7	6254.0	6319.3

Disadvantages: (1) It is difficult to interpret what exactly it measures. (2) The geometric mean of the 'upward' and 'downward' bias index numbers does not necessarily give correct one. (3) This index number requires quantities of both base and given years which may not be feasible due to 'time' and 'cost' involved. Further, index numbers of any two years are not comparable due to change of quantities in the denominator.

But this is not the case with the Laspeyre's index number where the denominator is constant throughout.

EXAMPLE: The following table gives the prices and quantities (hypothetical) of production of different food grains in India for the years 1961, 1970 and 1971. Taking 1961 as base year, compute Laspeyre's, Paascbe's, Marshall-Edgeworth and Fisher's index numbers for the years 1970 and 1971.

Laspeyre's price index no for the year 1970,

$$L_{n-1,0} = \frac{\sum q_0 p_{n-1}}{\sum q_0 p_0} \times 100 = \frac{788.48}{5044.9} \times 100 = 156.29\%$$

$$L_{n,0} = \frac{\sum q_0 p_n}{\sum q_0 p_0} \times 100 = \frac{8378.4}{5044.9} \times 100 = 166.08\%$$

Paasche's price index No. for the year 1970. $F_{n-1,0}$

$$p_{n-1,0} = \frac{\sum q_{n-1} p_{n-1}}{\sum q_{n-1} p_0} \times 100 = \frac{9700.9}{6254.0} \times 100 = 155.12\%$$

$$p_{n,0} = \frac{\sum q_0 p_n}{\sum q_n p_0} \times 100 = \frac{10446.7}{6319.8} \times 100 = 165.30$$

Fisher's ideal index no. for the year 1970, $F_{n-1,o}$

$$= \sqrt{\frac{\sum q_0 p_{n-1}}{\sum q_0 p_0} \times \frac{\sum q_{n-1} p_{n-1}}{\sum q_{n-1} p_0}} = \sqrt{156.29 \times 155.12} = 155.70\%$$

$$F_{n,0} = \sqrt{\frac{\sum q_0 p_n}{\sum q_0 p_0} \cdot \frac{\sum q_n p_n}{\sum q_n p_0}} = \sqrt{166.08 \times 165.30} = 165.69\%$$

For computing the Marshall-Edgeworth Index number, the following table is formed.

Food grains	$(q_{n-1} + q_0)$	$P_0(q_{n-1} + q_0)$	$(q_n + q_0)$	$P_n(q_n + q_0)$	$P_n(q_n + q_0)$	$p_{n-1}(q_{n-1} + q_0)$
Paddy	83.7	2929.5	83.9	5201.8	2936.5	5022.0
Wheat	105.6	7392.0	105.9	11649.0	7413.0	10771.2
Jowar	9.7	407.4	10.1	808.0	424.2	727.5
Bajra	7.5	300.0	7.9	647.8	316.0	585.0
Maize	6.0	270.0	6.1	518.5	274.5	480.0
		11298.9		18825.1	11364.2	17585.7

Marshall-Edgeworth Price index no. for the year 1970,

$$M_{n-1,0} = \frac{\sum p_{n-1}(q_{n-1} + q_0)}{\sum p_0(q_{n-1} + q_0)} \times 100 = \frac{17585.7}{11298.9} \times 100 = 155.64\%$$

$$M_{n,0} = \frac{\sum p_n(q_n + q_0)}{\sum p_0(q_n + q_0)} \times 100 = \frac{18825.1}{1136.2} \times 100 = 165.65\%$$

18.3.9 Geometric Mean of Price Relatives

If geometric mean is used instead of arithmetic mean for the index number, we have

$$P_{n,0} = \left(\Pi \frac{p_n}{p_0} \right)^{1/k}$$

where 'Π' stands for the product of k ratios. If w_i is the weight to be used for i-th commodity and $W = \Sigma w_i$ in the geo-metric mean of price relatives, the formula becomes

$$P_{n,0} = \left[\prod_{i=1}^{k} \left(\frac{P_n}{P_o} \right)^{w_i} \right]^{1/W}$$

These index numbers satisfy 'time' reversal test. The index number of wholesale prices published by Economic and Statistical Advisor, Ministry of Agriculture, Govt. of India is the weighted geometric mean of the price relatives.

Similarly quantity index numbers could be computed by interchanging prices and quantities in Paasche, Laspeyre index numbers. For further reading please refer to Croxton and Cowden (1966).

18.3.10 Important Points in the Construction of Index Number

(i) Objective of constructing Index number: The objective for which the index numbers are to be constructed should be clear. For example, to

measure the cost of living of salaried class employees, the level of employees (skilled or semi-skilled), the region to be covered, the food articles to be included, etc. should be specified.

(ii) Items to be included: Representative as well as sufficient number of items are to be included according to their importance to cover the entire object of study. The number of items should not be too small. Cost and time factors should be considered whenever large number of items are available. For example, the number of items for the cost of living index number discussed above could be: (1) food, (2) clothing, (3) fuel, (4) rent and electricity charges, (5) education, (6) entertainment, and (7) miscellaneous. Again sub-items are to be selected within each item on the basis of representative character. For example, within the item of food the sub-items could be: (1) cereals like rice, wheat, bajra and jowar, (2) Milk, sugar, coffee and tea, (3) fruits, vegetable, meat and eggs, (4) dal and oil, (5) salt and spices. Similarly in constructing price index number of fertilizers the items like: (1) chemical manures, (2) organic manures, etc. are to be included. Again within item (1) chemical manures the sub-items like (1) urea, (2) Ammonium sulphate, (3) mixtures like 28:28:0, 14:35:14, (4) micro nutrients, etc. are to be included.

(iii) Sources of data: The retail and wholesale prices at representative markets in different states for important agricultural commodities are being published by Economic and Statistical Advisor to Ministry of Agriculture and Irrigation. Government of India in 'Agricultural situation in India'. Monthly retail and wholesale prices of other consumable articles like Dal, oil, sugar, jaggery, spices, etc. are being published in bulletins by the State Governments Directorates of Economics and Statistics. Weekly retail and wholesale prices of different commodities are being published in daily newspapers like 'Financial Times', 'Economic Times', 'Price', etc. For example, in constructing consumers price index number for steel mill workers, the retail prices in the market where the workers colonies were located should be considered. In passing it may be noted that a care should be taken in handling secondary data as available in journals, etc. The data should be reliable, appropriate and representative otherwise the index numbers may present misleading picture.

(iv) Selection of base period: The base period may be taken as a representative (or normal) year. Since yearly data is effected by a cyclical variation, an average of some years would be appropriate as base period. For example, 1957-1959 can be taken as base period for agricultural commodities. Base period would be shifted to a more recent period due to variation in prices, growth of population, technological development, currency depreciation, change in food habits, change in quality of goods, etc.

to make comparisons more realistic. However, it is advisable to use same base period for all types of commodities.

(v) Suitable weighting: Though different methods of weights are described earlier in this section, one has to choose them keeping the items such as objective, time, availability of data and cost involved in collecting the data in mind.

18.4 INTERPOLATION

It is an operation of estimating the value of a function f(X) for any intermediary value of X given the values of f(X) for different values of X. It also helps in determining the missing value of f(X).

The assumption is that the value to be interpolated would not be an abnormal one and it follows the same pattern of other values of function.

18.4.1   Finite Differences

Suppose the values of X and f (X) are given at equal intervals of length 'h' as

$$X : a \quad a+h \quad a+2h \ldots a+nh$$

$$f(X): f(a) \quad f(a+h) \quad f(a+2h) \ldots f(a+nh)$$

then $\Delta f(a) = f(a+h) - f(a)$. $\Delta f(a+h) = f(a+2h) - f(a+h)$, etc. are called first differences, and

$\Delta^2 f(a) = \Delta f(a+h) - \Delta f(a)$, $\Delta^2 f(a+h) = \Delta f(a+2h) - \Delta f(a+h)$, etc. are called second differences and so on.

The table of differences for 6 values of X shown in Table 18.9

TABLE 18.9

| | | | | *Difference* | | |
X	$f(X) = Y$	1st	2nd	3rd	4th	5th
a	Y_0	ΔY_0	$\Delta^2 Y_0$			
a + h	Y_1	ΔY_1	$\Delta^2 Y_1$	$\Delta^3 Y_0$		
a + 2h	Y_2	ΔY_2	$\Delta^2 Y_2$	$\Delta^3 Y_1$	$\Delta^4 Y_0$	
a + 3h	Y_3	ΔY_3	$\Delta^2 Y_3$	$\Delta^3 Y_2$	$\Delta^4 Y_1$	$\Delta^5 Y_0$
a + 4h	Y_4	ΔY_4				
a + 5h	Y_5					

18.4.2   Newton's Formula or Newton-Gregory's Formula

In order to apply this formula the values of X must be equidistant. The formula is

$$f(x) = Y_0 + p\Delta Y_0 + \frac{p(p-1)}{2!}\Delta^2 Y_0 + \frac{p(p-1)(p-2)}{3!}\Delta^3 Y_0 + \ldots$$

where $p = \dfrac{X-a}{h}$ and X is the value for which f(X) is to be interpolated, 'a' is the first value of X and 'h' is the difference between any two consecutive values of X and is also known as interval of differencing.

EXAMPLE: The values of the probability integral

$$f(x) = \frac{2}{\sqrt{2\pi}} \int_0^x e^{-x2}\, dx$$

for certain equidistant values of x are given in Table 18. 10. Find the value of f (X) when X = 0.5238 and also when X = 0.5635

The table of differences is given in Table 18.10 itself.

TABLE 18.10

X	$f(x) = Y$	ΔY	$\Delta^2 Y$	$\Delta^3 Y$
0.51	0.5292437	.0086550		
0.52	0.5378987	.0085654	– .0000896	
0.53	0.5464641	.0084751	– .0000903	–.000007
0.54	0.5549392	.0083841	–.0000910	–.000007
0.55	0.5633233	.0082924	–.0000917	–.000007
0.56	0.5716157	.0082001	–.0000923	–.000006
0.57	0.5798158			

In Table 18.10, the differences are calculated up to 3rd difference only since the differences have come out to be constant at this stage.

$$x = 0.5238,\, h = 0.01,\, p = \frac{X - X_0}{h} = 1.38$$

$$f(x) = Y_0 + p\Delta Y_0 + \frac{p\,(p-1)}{2\,!} \Delta^2 Y_0 + \frac{p\,(p-1)\,(p-2)}{3\,!} \Delta^3 Y_0$$

$$= 0.5292437 + 1.38 \times .0086550 + (1.38)\,\frac{(1.38 - 1)}{2\,!}$$

$$(-.0000896) + \frac{(1.38)\,(1.38-1)\,(1.38-2)}{3!}\,(-0000007) = 0.5411642$$

For finding the value of f (X) for X = 0.5635 it is convenient to obtain the value of f(X) by using Newton's backward difference formula for n = 6

$$f(x) = Y_6 + u\Delta Y_5 + \frac{u(u+1)}{2\ !}\Delta^2 Y_4 + \frac{u\ (u+1)\ (u+2)}{3\ !})\Delta^3 Y_3$$

$$X = 0.5635,\ u = \frac{x - x_n}{h} = \frac{0.5635 - 0.57}{0.01} = -0.65$$

$$f(X) = 0.5798158 - 0.65\ (0.0082001) + \frac{(-0.65)\ (-0.65+1)}{2\ !}$$

$$(-.0000923) + \frac{(-0.65)\ (-0.65+1)(-0.65+2)}{3\ !}\ (-.\ 0000006) = 0.\ 5744962$$

18.4.3   Lagrange's Interpolation Formula

This formula is applicable when the values of f(X) are not given at equidistant values of X. If there are n values of f(X) and f(X) is assumed to be the polynomial in X of (n − 1)-th degree. f(X) = a₁ (X − X₂)(X − X₃)…(X − Xₙ) + a₂(X − X₁)(X − X₃)…(X − Xₙ) + …+ aₙ(X − X₁) (X − X₂) … (X − Xₙ₋₁) where there are n terms each of degree (n − 1) in X.

The Lagrange's formula for finding the value of f(X) for given value of X is

$$f(X) = Y_1\ \frac{(x - x_2)\ (x - x_3)...(x - x_n)}{(x_1 - x_2)\ (x_1 - x_3)...(x_1 - x_n)}$$

$$+ Y_2\ \frac{(x - x_1)\ (x - x_3)...(x - x_n)}{(x_2 - x_1)\ (x_2 - x_3)...(x_2 - x_n)}$$

$$+ ... + Y_n\ \frac{(x - x_1)(x - x_2)...(x - x_{n-1})}{(x_n - x_1)\ (x_n - x_2)...(x_n - x_{n-1})}$$

EXAMPLE: Compute the value of f(8) by using Lagrange's formula given that

f(2) = 10. f(3) = 14, f(5) = 18, f(10) = 26

$$f(8) = 10\frac{(8-3)(8-5)(8-10)}{(2-3)(2-5)(2-10)} + 14\frac{(8-2)(8-5)(8-10)}{(3-2)(3-5)(3-10)} + 18\frac{(8-2)(8-3)(8-10)}{(5-2)(5-3)(5-10)}$$

$$+ 26\frac{(8-2)(8-3)(8-5)}{(10-2)(10-3)(10-5)} = 20.86$$

EXERCISES

1. Draw the 'historigram' and fit the 2nd degree polynomial for the following time series data on wheat production for the year from 1960-61 to 1971-72 in India and obtain the trend values and ratios to trend values.

Year	1960	1961	1962	1963	1964	1965
Wheat production (m.t)	12.97	13.45	13.66	13.50	13.46	12.57
Year	1966	1967	1968	1969	1970	1971
Wheat* production (m.t)	12.84	15.00	15.96	16.63	18.24	19.61

*figures rounded off to second decimal place.

Source: 'Agricultural situation in India' Journal. 1960-61
Agricultural year is taken as 1960 and so on.

2. Fit the modified exponential curve given the following
$$T_1 = 20, T_2 = 36, T_3 = 58, n = 10$$

3. Obtain the Laspeyre, Paasche and Fisher's ideal price index numbers for the cost of living per month of middle income families given in the following data with the base year as 1960.

S.No.	Name of the Item	Rate in 1960	Quantity consumed in 1960	Rate in 1977	Quantity consumed in 1977
1.	House rent (2 rooms flat)	100	1	200	1
2.	Cloth	Rs. 520/metre	20	10. 50/metre	18
3.	Milk	1.00/litre	3	2.20/litre	2.25
4.	Kerosene	18.50/tin	1	24.40	1.5
5.	Coal	0.80/kg.	40	1.20/kg	60
6.	Eggs	3.60/dozen	2	4.20/doz.	2.5
7.	Dals	2.50/kg	6	4.50/kg	7.5
8.	Matches	0.80/box	6	1.20/box	6
9.	Rice	1.40/kg	50	2.20/box	70
10.	Edible oils	3.50/kg	2.5	7.00/kg	3

4.   Compute the 4-year moving averages for the following time series data on food production in a particular state.

Year	1964	65	66	67	68	69	70	71	72	73	74	75	76
Food Production (lakh tons)	70	72	73	75	71	80	82	79	84	86	90	92	94

CHAPTER 19

NON-PARAMETRIC STATISTICS

19.1 INTRODUCTION

The following are some of the preliminary considerations in applying statistical test

1. Null Hypothesis
2. Selection of statistical test
3. Level of significance
4. Sampling distribution
5. Decision

Since all these topics are covered in the earlier chapters, the power of the test and different stages of measurement are considered here and which are useful in 'selection of statistical test'

19.1.1 Power

Power of a test may be defined as the probability of rejecting null hypothesis, say, H_0 when it is in fact false. Type I error is defined as the rejection of H_0 when it is true and type II error is defined as the acceptance of H_0 when it is false. The probabilities of committing type I and type II errors are denoted by α and β respectively. Now, Power $= 1-\beta$

In order to make a particular statistical test with fewer or less stringent assumptions as powerful as another statistical test, the sample size for the former test should be considerably increased.

If test P with n_p sample size is as powerful as test Q with n_q sample size, then

$$\text{Power-efficiency of test } Q = \frac{n_p}{n_q} \times 100$$

19.1.2 Measurement

Measurement may be defined as assigning numerical value to observation such that the numerical values follow certain mathematical laws in physical and biological sciences. However, this is not always possible especially when we deal with qualitative characters such as intelligence, colour, shape, etc., in behavioural sciences. The measurement can be done in four stages: (i) nominal, (ii) ordinal, (iii) interval, and (iv) ratio.

(i) ***Nominal Scale:*** If the objects (or individuals) are classified by numbers (or symbols), then they are measured at weakest level known as nominal scale. If nomenclature is used to identify the groups then the measurement is done at nominal (or classificatory scaling). For example, the flowers are classified according to colour such as red, black, purple, etc. then the scaling is done with the help of nomenclature. This scaling satisfies equivalence property such as reflexive, symmetrical and transitive.

Ex: If name is given to person or object then is measured at nominal scale

(ii) ***Ordinal Scale:*** Ex: If name is given to person or object then it is measure at nominal scale in the nominal scaling, the objects (or persons) in a group are not much different from each other but they can be compared such as 'more than' or 'less than' then the numerical values denoting this order is called 'ordinal' (or 'ranking') scale. This scaling satisfies relation of ordering besides equivalence property. For example, the estimation of a particular crop can be done as very good, good, average and below average. This measurement is in ordinal scale since very good> good> average > below average.

(iii) ***Interval Scale:*** If the distance between two values in ordinal scale can be measured then the interval scaling is achieved. In this scale, the ratio of any two intervals is independent of the unit of measurement and of the zero point. The unit of measurement and zero point are arbitrary in this case. For example, the production of paddy crop in a particular region can be scaled as 35 bags per acre as very good crop, 25 bags per acre as good, 18 as average and less than 18 as below average. The distance between good and very good crop can be measured as (35–25) = 10. If the number of bags are measured in kilograms, then the values, for very good, good, average and below average crop are given as 2625, 1875, 1350 and below 1350 kg respectively assuming that bag contains 75 kg. Here it may be verified that the ratio of distances between first three numerical values is independent of units of measurement i.e., 7/10, or 525/750 i.e., 7/10.

(iv) The interval scale satisfies not only the property of ordinal scale but also ratio of the differences. The 'nominal' and 'ordinal' scales are qualitative scales, that is, the scales used for qualitative data whereas the interval scale is used for quantitative data. Therefore, the parametric tests like student's test, F-test can be used for interval scale data.

(v)   *Ratio Scale:* An interval scale with true zero point is called a ratio scale. The ratio of two points of a ratio scale is independent of unit of measurement. For example, the height of a plant in inches (or centimetres) will have the same zero point. The ratio between any two lengths is independent of inches (or centimetres) and also the ratio of two inches points is identical to the ratio of the corresponding centimetre points.

Ratio scale satisfies the mathematical properties such as: (1) equivalence, (2) greater than, (3) known ratio of any two intervals, and (4) known ratio of any two scale values.

19.2 PARAMETRIC VS NON-PARAMETRIC TESTS

Parametric tests depend on the conditions of the parameters of the population and they are used for the data of at least interval scale. If the conditions of the parameters are satisfied and the data are of at least interval scale, Parametric tests are the most powerful tests. The inferences drawn from Parametric tests are applicable to restricted population since the assumptions on parameters are stronger.

Non-parametric tests do not depend upon the conditions of the parameters though the weaker assumptions such as observations are independent and continuous are to be satisfied. The non-parametric tests can be used for the data of nominal and ordinal scale. Non-parametric tests are useful in social sciences research since the data in general are of nominal or ordinal nature. Further, it was observed that the power of these tests can be improved and made as efficient as for the Parametric tests by simply increasing the size of the sample. The inferences drawn from Non-parametric tests are applicable to large population since the assumptions are weak.

19.3 ONE-SAMPLE TESTS

In this section the one sample tests are listed.

19.3.1   Binomial test

The Binomial test is based upon Binomial distribution.

Case (i) Small-samples (n <: 25)

Null Hypothesis: $H_o : P = Q = \frac{1}{2}$ $H_1 : P \neq Q$, where H_0 is the null hypothesis and H_1 is the alternative hypothesis.

The probability of obtaining atmost r cases out of n cases is

$$P(r) = \sum_{i=0}^{r} \binom{n}{i} P^i Q^{n-i}$$

Here r is taken as the smaller number out of the two cases in the sample. If H_1 is $P \neq Q$ then the two-tailed test is used. Half the probability of getting as small as r cases in two-tailed test is the probability of getting as small as r cases in one-tailed test.

CONCLUSION: If the observed probability is less than or equal to $\alpha = .01$ then H_0 is rejected at 1 per cent level of significance in favour of H_1, otherwise H_0 is accepted.

EXAMPLE: A group of 20 young farmers were trained in two methods of poultry keeping. Half of the trainees were randomly selected out of 20 farmers to teach method 1 first and to the remaining method 2 first. After a lapse of some months the two groups of farmers were examined for their adoption of the two methods. The two cases are presented here. Examine whether the percentage adoption of 1st learned method is more than the 2nd learned method.

TABLE 19.1

	1st learned method	*2nd learned method*	*Total*
No .of farmers	16	4	20

Null Hypothesis: $H_0:p = Q = \frac{1}{2}$, $H_1 :P > Q$ where P is the probability of adoption of 1st learned method and Q is the probability of adoption of 2nd learned method.

$$P(r \leq 4) = \sum_{i=0}^{4} \binom{20}{i} \left(\frac{1}{2}\right)^i \left(\frac{1}{2}\right)^{20-i}$$

$$= P(0) + P(1) + P(2) \, P(3) + P(4)$$

$$= 0.006$$

Here r is taken as 4 since it is smaller than 16 in Table 9. 1.

CONCLUSION: Since $P(r \leq 4) = 0.006 < \alpha = 0.01$, the null hypothesis is rejected and the alternative hypothesis is accepted at 1 per cent level of significance. That is the percentage adoption of 1st learned method by young farmers is more than the 2nd learned method.

Case (ii) Large samples: (n>25): If P is nearer to 0. 5 and n is large (say) greater than 25, r given in Case (i) of 19. 3.1 is approximately normally distributed with mean nP and standard deviation $\sqrt{nPQ}$. If P is away from 0.5 then n should be large enough in order to make r normally distributed.

Null Hypothesis:

$$H_0: P = Q = \frac{1}{2}; H_1; P > Q$$

$$Z = \frac{|r - nP|}{\sqrt{nPQ}}$$

Z is approximately normally distributed with mean as zero and standard deviation as unity.

Since r is a discrete variable in binomial distribution and in order to make it continuous for normal distribution, a correction is applied. That is, r is assumed to lie between r–.5 and r + .5. If r> nP, then r– 0 5 is taken in computing Z and if r < nP then r + 0. 5 is considered. The test will be

$$Z = \frac{|(r \pm 0.5) - nP|}{\sqrt{nPQ}}$$

CONCLUSION: If the probability that Z (calculated) is less than or equal to α=.05 then H_0 is rejected in favour of H_1. Otherwise H_0 is accepted. This is the case of one-tailed test since H_1:P > Q. If H_1: P ≠ Q it will become two-tailed test.

19.3.2   Chi-square Test

Chi-square test is used to test the agreement between observed frequencies and expected frequencies when the frequencies fall in different classes. For example, if the researcher is interested to test whether the frequencies in different classes are uniformly distributed or whether they are in agreement with given ratio or not, the χ^2-test can be used. The expected frequencies are computed on the basis of null hypothesis to be tested.

Null Hypothesis: Ho: The observed frequencies are in agreement with expected frequencies.

$$\chi^2 = \sum_{i-1}^{K} \frac{(o_i - E_I)^2}{E_i}$$

where O_i = observed frequency of i-th class for i = 1, 2, ... k

E_i = expected frequency of i-th class for i = 1, 2, ..., k

CONCLUSION: If χ^2 (calculated) $\geq$ χ^2 (tabulated) with (k–1) d.f. at chosen level of significance, H_0 is rejected. Otherwise H_0 is accepted.

EXAMPLE: The number of disease infected tobacco plants in 10 different plots of equal size (in area and population) are given below. Test whether the disease infected plants are uniformly distributed over the entire area.

Plot NO.	1	2	3	4	5	6	7	8	9	10
No .of plants infected	8	10	9	12	15	7	5	12	13	9

Null Hypothesis: The disease infected plants are uniformly distributed over the plots.

TABLE 19.2

Plot NO.	1	2	3	4	5	6	7	8	9	10
Observed frequency	8	10	9	12	15	7	5	12	13	9
Expected frequency	10	10	10	10	10	10	10	10	10	10

Expected frequency of each plot is calculated as total No. of plants divided by total number of plots i.e., Expected frequency = 100/10 = 10

$$\chi^2 = \frac{(8-10)^2}{10} + \frac{(10-10)^2}{10} + \ldots + \frac{(9-10)^2}{10} = 8.2$$

CONCLUSION: Here χ^2 (calculated), 8.2<χ^2 (tabulated), 16.919 with (10-1) d.f at *5* per cent level of significance. Hence the null hypothesis is accepted. The number of diseased plants are uniformly distributed over different plots.

19.3.2.1

For application of χ^2–test, it is important to note that expected frequency of any cell should be at least 5, if the expected frequency of a particular cell is less than 5 then the observed frequency of that cell would be added to the adjacent cell to make the expected frequency of the new cell equal to 5.

NON-PARAMETRIC STATISTICS

SPSS DATA ANALYSIS PROCEDURE

ONE –SAMPLE TEST

The data entry will be done in SPSS Excel sheet in the following FORMAT column wise as follows.

X
13
9
14
5
12

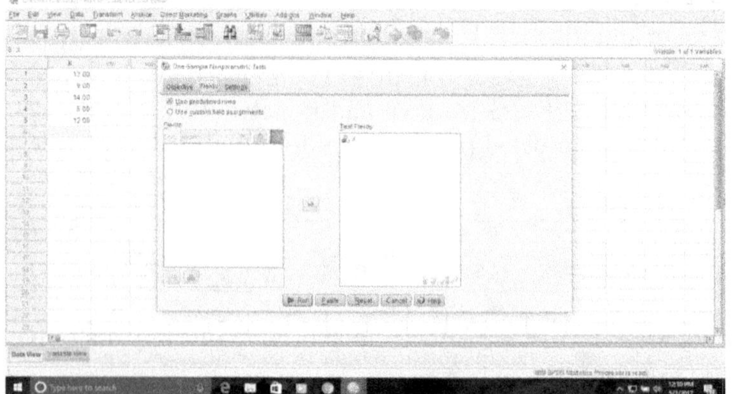

CLICK 'ANALYZE' from TOOL BAR AND then click 'NON-PARAMETRIC STATISTICS' AND opt for 'ONE-SAMPLE '.'SCAN DATA'& Choose 'CUSTOMIZE ANALYSIS' and then 'RUN'.

We get the result of CHI-SQUARE TEST' OF GOODNESS OF FIT and probability value of significance. Here the name of the test conducted is also indicated as there are several one-sample tests are available.

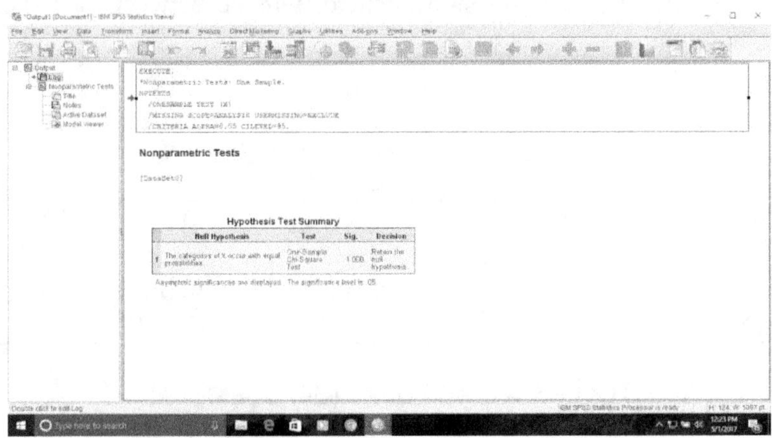

19.3.3 Kolmogorov-Smirnov Test

This is used to test whether the observed cumulative frequencies are in agreement with the theoretical cumulative frequencies. In other words this test is used for testing the goodness of fit. It is based on the maximum difference between observed cumulative frequency and theoretical cumulative frequency.

Null Hypothesis: H_0: The sample comes from a known theoretical distribution.

$$D = \text{Max}\left|T(X) - S(X)\right|$$

where $T(X)$ be the proportion of observations equal to or less than X in theoretical distribution i.e., under null hypothesis, H_0 and $S(X)$ be the proportion of observations equal to or less than X in the observed sample. If m be the number of observations equal to or less than X out of n observations in the sample, then $S(X)=m/n$. The sampling distribution of D under null hypothesis is known. The tabulated values of D for different sample sizes of n at different levels of significance were given by Massey (1951) and Siegel (1956).

CONCLUSION: If D(calculated) > D(tabulated) with size n at 1 per cent level of significance, the null hypothesis is rejected. In other words there is significant difference between observed cumulative frequencies and the theoretical cumulative frequencies. Otherwise, the null hypothesis is accepted.

EXAMPLE: Wheat was graded as 1, 2, 3 and 4 depending upon the colour of the grain. Brown grain was graded as 1, slight brown as 2, slight white as 3 and white as 4, even though the chapati making quality was Same. Here the assumption was that the consumer would prefer white grain in comparison to brown coloured. A sample of 50 consumers were asked to select the grain. The hypothetical data is given below.

Null Hypothesis: There is no significant difference between observed and theoretical cumulative frequencies.

TABLE 19.3

	Grading of wheat			
	1	2	3	4
No. of persons	0	8	20	22
T(X)	1/4	2/4	3/4	4/4
S(X)	0	8/50	28/50	50/50
T(X)-S(X)	1/4	17/50	19/100	0

$$D = \text{Max}\,|T(X) - S(X)| = 17/50 = 0.34$$

CONCLUSION: Here D(calculated) = 0.34> D(tabulated), $1.36/\sqrt{50}$ at 5 per cent level of significance. Hence the null hypothesis is rejected. In other words, the observed frequencies are significantly different from each other. That is, the persons showed significant preference for colour of wheat.

It may be noted that Kolmogorov-Smirnov (K-S) test considers individual cell frequencies whereas its alternative χ^2- test looses information by combining the adjacent cell frequencies when the expected frequency of certain cell is less than 5.Further this test can be used even when the expected frequency of any cell is less than 5 whereas for Chi-square test, the classification of cells have to be modified. Hence (K-S) test is more powerful than the χ^2-test.

19.3.4   Run-test

Repetition of the same symbol followed and preceded by different symbols or by no symbol is called a 'run'. The following is the example of sequence of 8 runs indicated by '+' and '-' ve signs.

– – – + + + + – – + + + – + – + +

1   2   3   4   5 6 7 8

In the above sequence the first three signs comprise one run, the next four would form second run, the next two signs would be third run, etc. In all there are 8 runs in this sequence. Run test is based on the number of runs observed in the sample. In this test, the number of types of symbols (or signs) are assumed to be two only.

In order to arrive at some decision about the population, a sample is to be drawn from the population. If the conclusions are to be valid based on this sample, the sample should be random one. Run-test, therefore, is the test used for testing the randomness of the sample.

Case (i) Small samples ($n_1 \leq 20$, $n_2 \leq 20$): Let n_1 be the number of symbols of first type and n_2 be the number of symbols of second type such that $n = n_1 + n_2$ be the total sample size. Let r be the total number of runs in a sample. The sampling distribution of r was found and the critical values to test the null hypothesis were tabulated in two different tables for different values of n_1 and n_2 at 5 per cent level of significance, [Table VIII of Appendix]. The first table gives the lower limit values of r and the second table refers to the upper limit values of r.

Null Hypothesis: H_0: The observations in the sample are in random order.

The values of r, n1 and n2 would be computed from the sample.

CONCLUSION: If r (calculated)$\leq$ r (tabulated) with n1, n2 at 5 per cent level of significance in the first table, the null hypothesis is rejected. In other words the observations in the given sample are not in random order. It may also be calculated that the observations in the sample are not independent. If r (calculated) > r (tabulated) with n_1, n_2 at 5 per cent level of significance in the second table, the null hypothesis is rejected. If r lies between the two values of Table VIII (a) and VIII (b) the null hypothesis is accepted.

EXAMPLE: In a particular musical chair contest the seating arrangement has to be done at random between 15 males and 15 females in a circular fashion. The following order of seating was observed in the said contest at a particular time. .

M FFF   MM   F   MMM   FF M   FFF   MM F   MMMM   F   MM FFFF

Test whether the above arrangement is in random order.

Null Hypothesis: H_o: The seating arrangement is in random order,

Here n1 = 15: n2 ; n = 30

M FFF MM F MMM   FF M FFF MM   F MMMM F MM   FFFF

1   2   3   4   5   6   7   8   9   10   11   12 13   14

$$r = 14$$

CONCLUSION: Since r (tabulated), 10 < r (calculated), 14 < r (tabulated), 22; the null hypothesis is accepted. Hence the given seating arrangement is in random order.

Case (ii) Large samples (n_1, n_2 > 20): When either of n_1 or n_2 is greater than 20, the small sample test given in case (i) will not hold good.

Null Hypothesis: H_0: The observations in the given sample are in random order.

$$Z = \frac{\left| r - \left(\dfrac{2n_1 n_2}{n_1 + n_2} + 1 \right) \right|}{\sqrt{\dfrac{2n_1 n_2 (2n_1 n_2 - n_1 - n_2)}{(n_1 + n_2)^2 (n_1 n_2 - 1)}}}$$

CONCLUSION: If Z (calculated) > Z (tabulated) at chosen level of significance, the null hypothesis is rejected. In other words, the observations in the given sample are not in random order. Otherwise, the null hypothesis is accepted.

EXAMPLE: A coin is tossed 40 times. The sequence of number of heads and tails obtained are given below. Test whether the coin is unbiased.

T	HH	TTT	HH	T	H	TT	HHH	TTTT	H	T	HH
1	2	3	4	5	6	7	8	9	10	11	12

T	HHH	TT	HH	T	HHH	TTTT	H
13	14	15	16	17	18	19	20

Null, Hypothesis: The coin is unbiased.

Here r = 20, n_1 = number of heads == 20, n_2 = number of tails= 20.

$$Z = \frac{\left| 20 - \left(\dfrac{2 \times 20 \times 20}{40} + 1 \right) \right|}{\sqrt{\dfrac{2 \times 20 \times 20 (2 \times 20 \times 20 - 20 - 20)}{(20 + 20)^2 (20 + 20 - 1)}}} = 0.32$$

CONCLUSION: Z (calculated), 0.32 < Z (tabulated), 1.96 at 5 per cent level of significance. Hence the null hypothesis is accepted. The coin may be unbiased.

19.4 TWO RELATED SAMPLES TESTS

In this section the two sample tests are listed.

19.4.1 Sign Test

This test is used to test the significant difference between two conditions when the observations are in either nominal or ordinal scale. Here the assumption is only that the variable under consideration has a continuous distribution. In small samples the variable under consideration is assumed to follow Binomial distribution and in large samples it is assumed to follow normal distribution. If the difference between scores of each pair in two related samples is positive which is denoted by '+' sign and negative difference by'-' sign. If the difference of scores is found to be zero and that pair of scores will not be counted in finding out the effective size of sample. If there are two pairs whose differences found to be zero, the effective size of sample would be (n – 2) for consideration in sign test where n is the size of sample.

Case (i) Small Samples (n ≤ 25):

Null Hypothesis: H_o: The number of positive signs = The number of negative signs. In other words, the probabilities of occurrence of positive and negative signs are same i.e., $P = Q = \frac{1}{2}$, where P and Q are the probabilities of occurrence of positive and negative signs respectively.

Let 'r' be 'the number of '+' or '-' signs, whichever is small, out of n, then the probability of occurrence of atmost 'r' signs is given by:

$$F(r) = \sum_{x=0}^{r} \binom{n}{x} p^x Q^{n-x}$$

The computed values of F(r) are available in tables [Table D, Siegel (1956)].

CONCLUSION: If F(r) calculated < 0.005 the null hypothesis is rejected at 1 per cent level of significance and the alternative hypothesis (there are more '+' (–) signs than – (+) signs) is rejected in the one tailed test. Otherwise, the null hypothesis is accepted.

EXAMPLE: A national demonstration was conducted to train the farmers in improved methods of cultivation. In order to ascertain the insight of the difference between the two generations, 15 pairs of farmers and their sons were studied before and after the National demonstration was conducted. It is expected that the sons would have more motivation than their fathers in understanding new techniques of cultivation. An agricultural extension expert was asked to evaluate their motivation towards improved techniques of cultivation after the national demonstration was conducted. The scores of the pairs given on 3 point scale, where 1indicates for less motivation 2 for good motivation and 3 for high motivation. The results are given as

TABLE 19.4

Pair			1	2	3	4	5	6	7	8	9	10	11	12	13	14	15
Rating of	Father	..	1	2	3	1	2	3	1	1	2	3	2	1	2	1	2
Motivation	son	..	3	2	1	2	3	3	2	3	3	2	1	3	3	2	3
Difference	..	..	–	0	+	–	–	0	–	–	–	+	+	–	–	–	–

Null Hypothesis: Ho: $P = Q = \dfrac{1}{2}$ or number of +ve signs are equal to number of –ve signs.

n = 13, no. of –ve signs = 10, No. of + ve signs = 3. Therefore, r = 3 since r is the lesser among + ve and – ve signs.

CONCLUSION: Probability at n = 13 and r = 3 is 0.046 which is < 05. Hence the null hypothesis is rejected therefore it can be concluded that there exist difference in motivational pattern of father and son.

Case (ii) Large samples (n > 25):

When n>25, the binomial distribution in case (i) of this test approximates to normal distribution, The 'r' is approximately normally distributed with mean nP and standard deviation $\sqrt{nPQ}$ where P is the probability of occurring positive sign (or negative sign) as the case may be,

Null Hypothesis: H_0: $P = Q = \dfrac{1}{2}$ or number of +ve signs are equal to number of -ve signs.

$$Z = \frac{|r - nP|}{\sqrt{nPQ}}$$

Since $P = Q = \dfrac{1}{2}$, we have

$$Z = \frac{|r - n/2|}{\dfrac{1}{2}\sqrt{n}}$$

CONCLUSION: If Z (calculated) >Z (tabulated), 1.96 at 5 per cent level of significance, the null hypothesis is rejected, otherwise it is accepted.

Note: Since 'r' is a discrete variable whereas normal distribution is used for continuous variable, a correction is used to make the distribution of 'r' a good approximation to normal distribution. If r <n/2; (r +0.5) is used instead of 'r' and if r>n/2; (r − 0.5) is used instead of r in the expression of 'Z' mentioned above.

EXAMPLE: The example in case (i) of 19.4.1 for small sample would be appropriate for large samples also when n>25.

19.4.2   Wilcoxon Test

This test considers the overall direction of the differences between the pairs as well as the magnitude of the individual differences of pairs whereas the sign test given in sub-section 19.4.1 considers only the overall direction of the   differences of pairs. Therefore, 'Wilcoxon test' is considered more powerful than the 'sign test'.

Case (i) Small samples (n < 25): In this method the ranks were given based on the magnitude of the differences of the scores for each pair irrespective of the negative or positive differences. If two or more differences are equal then the ranks will be awarded as in the case of tied observations. For example, let d_i be the difference of scores for i-th pair for i = 1, 2, .., n. Let $d_1 = -2$ and $d_2 = -2$ then the rank for d_1 and d_2, would be $(1+2)/2 = 1.5$ and the next difference would get the rank '3'. After effecting the ranks for each d_i in the above described manner, the ranks for the negative d'_is would be given negative sign rank and the ranks for the positive d_i s would be denoted by positive sign rank. For consulting table the number of effective pairs would be taken by subtracting the number of pairs which give $d_i = 0$ from the total number of pairs.

Null Hypothesis: H_0: Sum of the negative and positive ranks is zero or sum of the negative ranks is equal to the sum of the positive ranks.

The value of 'S' would be computed from the sample where 'S' is the total value of the positive ranks or negative ranks whichever is smaller and n is the effective number of pairs of observations in the sample.

CONCLUSION: If S (calculated) $\leq$ S (tabulated) at chosen level of significance, the null hypothesis is rejected. In other words, there is significant difference between two treatments (or conditions) under consideration. Otherwise the null hypothesis is accepted. The tabulated value is obtained from Table G. (Cf. Siegel (1956))

EXAMPLE: One of the two brothers of each farmer's family was randomly selected and put under training for 'poultry rearing' in Agricultural University and the other one remains at home with traditional knowledge on poultry. After completion of training both the brothers were interviewed on the knowledge of 'poultry rearing'. The scores awarded on the performance of interview are given in Table 19.5. Test the significant difference between the two brothers with respect to knowledge on 'Poultry rearing'.

TABLE 19.5

Pair	T	UT	Difference	Rank	S
1	75	60	15	10	
2	81	68	13	9	
3	70	76	−6	−4	4
4	83	71	12	7.5	
5	65	70	−5	−3	3
6	86	78	8	5	
7	72	74	−2	−1	1
8	92	80	12	7.5	
9	85	74	11	6	
10	65	61	4	2	
Total					8

T = trained, UT = untrained

Null Hypothesis: H_0: Sum of the negative ranks = Sum of the positive ranks.

CONCLUSION: Here S(calculated), 8 = S(tabulated), 8 at 5 per cent level of significance. Hence, the null hypothesis is rejected. The training had significant impact on the knowledge of 'Poultry rearing'.

Case (ii) Large samples (n > 25)

Null Hypothesis: H_0: There is no significant difference between the two treatments:

$$Z = \frac{\left| S - \dfrac{n(n+1)}{4} \right|}{\sqrt{\dfrac{n(n+1)(2n+1)}{24}}}$$

Where n is the number of pairs of observations.

CONCLUSION: If Z(calculated) > Z(tabulated) at chosen level of significance, the null hypothesis is rejected. Otherwise it is accepted.

EXAMPLE: An experiment was conducted on a batch of 32 students on individual timings taken for answering question paper having two choices of answers in each question. An experimenter who was also an experienced teacher himself could predict the answer each student would choose. Some of the predictions of an experimenter might go wrong due to the indifference on the part of the student. It was expected that the student would take more time for the incorrectly predicted cases than the correctly predicted cases. The differences between median time taken for incorrectly predicted answers and correctly predicted answers for each student are given in Table 19.6. Test whether there is any significant difference between correctly and incorrectly predicted answers.

Null Hypothesis H_0: Sum of the positive differences is equal to sum of the negative differences or there is no significant difference between the median times taken for correctly and incorrectly predicted answers.

$$Z = \frac{\left| S - \dfrac{n(n+1)}{4} \right|}{\sqrt{\dfrac{n(n+1)(2n+1)}{24}}}$$

$$Z = \frac{\left| 53 - \dfrac{26(27)}{4} \right|}{\sqrt{\dfrac{26(27)(53)}{24}}} = 3.11$$

where S = 53, sum of the negative ranks since they are small in number and n =(32–6) = 26.

CONCLUSION: Here Z(calculated), 3.11 $\geq$ Z (tabulated), 1.96 at 5 per cent level of significance. Hence, the null hypothesis is rejected. In other words, there is significant difference between median times taken for correctly and incorrectly predicted asnwers.

TABLE 19.6

Student	Difference (d_i)	rank	s	student	Difference (d_i)	rank	s	student	Difference (d_i)	rank	s
1	1	5.5		11	−1	−5.5	5.5	22	5	25.5	
2	0	−		12	−2	−14.0	14.0	23	1	5.5	
3	0	−		13	1	5.5		24	3	19.5	
4	−1	−5.5	5.5	14	1	5.5		25	−2	−14.0	14.0
5	2	14.0		15	3	19.5		26	5	25.5	
6	1	5.5		16	4	23.0		27	4	23.0	
7	−2	−14.0	14.0	17	2	14.0		28	1	5.5	
8	0	−		18	0	−		29	2	14.0	
9	3	19.5		19	1	5.5		30	1	5.5	
10	0	−		20	4	23.0		31	2	14.0	
				21	0	−		32	3	19.5	
		19.5				19.5				14.0	53.0

19.5 TESTS FOR TWO INDEPENDENT SAMPLES

In the preceding sections the tests for two related samples were considered. Though the test for two related samples are more precise than the tests for the two independent samples, the number of situations where these tests are applicable is less. In this section the different tests involving two independent samples are considered. The Student's t-test is applicable only when the sizes of samples are small and observations are Independent which are drawn from normal population. But the tests considered in this section do not require the assumption of normality and independence but the scores should be at least on interval scale.

19.5.1 Fisher's Test

This is used to test the significant difference between two groups with respect to proportions having been classified in a 2×2 contingency table. Further this test is useful when the sizes of the two independent samples are small say R_1, $R_2 < 15$. Since this test is an exact probability test it is more useful in 2×2 contingency tables.

Null Hypothesis H_0: There is no significant difference between the proportions of two classes for the two groups.

TABLE 19.7

Class Group	1	2	
1	a	b	R_1
2	c	d	R_2
	C_1	C_2	N

When the marginal totals R_1, R_2 C_1 and C_2 in Table 19.7 are fixed, the probability of getting observed frequencies a, b, c and d in the different cells in Table 19.7 is given by:

$$p = \frac{\binom{R_1}{a}\binom{R_2}{c}}{\binom{N}{C_1}}$$

The probabilities, p were computed for the different values of R_1, R_2, a and c and tabulated by Finney (1948) and given in Table IX of Appendix.

CONCLUSION: If the probability (calculated) < 0.05 then the null hypothesis is rejected at 5 per cent level of significance. Otherwise it is accepted.

EXAMPLE: The Following data (hypothetical) are based on Study conducted on 17 Farmers to Know whether progressiveness of their outlook depends on education. Test whether progressiveness of farmer differs with respect to education.

Null Hypothesis: Progressive and less progressive farmers have equal proportions in the case of education.

Here $R_1 == 10$; $R_2 = 7$, c=2 and a= 9.

TABLE 19.8

	Educated	Less educated	
Progressive	9	1	10
Less progressive	2	5	7
	11	6	17

The computed value of p = 0.025 (from table).

CONCLUSION: Since p (calculated) o. 025 < 0.05, the null hypothesis is rejected at 5 per cent level of significance. Hence it can be concluded that the proportions of educated persons are not same for progressive and less progressive farmers.

19.5.2   Chi-square Test

Fisher's test is applicable when the sizes of the two samples (R_1 and R_2) are small in 2 × 2 contingency tables but Chi-Square test is applicable even for m × n contingency table where m, n $\geq$ 2 and when the scores are on nominal scale and the sizes of the samples are large. This test was dealt with in detail in Chapter 11 of Part I.

19.5.3   Median Test

In this case the individuals or objects would be classified according to two groups based on above or below common median as shown in the following Table 19.9.

TABLE 19.9

Group \ Common median	above	below	
I	a	b	R_1
II	c	d	R_2
	C_1	C_2	N

Null Hypothesis H_0: There is no significant difference between two groups with respect to number of individuals (or objects) above common median or the two groups come from the population having the same median.

$$\chi^2 = \frac{(ad - bc)^2}{R_1\, R_2\, C_1\, C_2}\, N$$

where a, b, c and d are the observed frequencies. R_1, R_2, C_1, C2 and N are the marginal totals and grand total respectively in Table 19.9.

CONCLUSION: If χ^2 (calculated) $\geq \chi^2$ (tabulated) at chosen level of significance, the null hypothesis is rejected. In other words there is significant difference between medians of both groups.

Note: Yates correction for continuity in χ^2-test is used when N > 30 and the expected frequency of any one cell is not less than 5. If the expected frequency of any one cell is less than 5 or N < 30, Fisher's test given in 19.5.1 is used.

19.5.4 Mann-Whitney test

This test is used to test whether two groups are from the same population when the measurement is on ordinal scale. This is a powerful non-parametric test for two Independent samples corresponding to t-test in parametric case.

Let n_1, n_2, be the sizes of small and large groups respectively out of the two groups to be tested, where $n_1 < 8$ and $n_2 < 8$. Common ranking would be done for all the $(n_1 + n_2)$ scores such that lowest algebraic value (may be negative) would be given 1, next higher value 2, etc.

The scores of both the groups would be arranged in ascending order of magnitude starting from highest negative value, if any. These scores would then be placed in their respective groups. Let W be the number of times the score in the group of size n_2 precedes the score of n_1 size group. W also can be obtained directly with the help of the following formula as

$$W = n_1 n_2 + \frac{n_1(n_1 + 1)}{2} - T_1$$

or
$$W_1 = n_1 n_2 + \frac{n_2(n_2 + 1)}{2} - T_2$$

where T_1, T_2 are the sums of the ranks in the groups of sizes n_1 and n_2 respectively.

Case (i) Small samples: $(n_1, n_2, \leq 20)$:

Null Hypothesis H_0: The two groups belong to the same population.

$$W = n_1 n_2 + \frac{n_1(n_1 + 1)}{2} - T_1$$

If W is larger than the available values in the table [Table X of Appendix] the actual value would be $W_1 = n_1 n_2 - W$. Then W_1 instead of W would be considered for the test.

CONCLUSION: If calculated probability obtained from table for given W, n_1 and n_2 ≤ 0.05, the null hypothesis would be rejected at 5 per cent level of significance. That is, the two groups do not belong to the same population. Otherwise, the null hypothesis is accepted.

EXAMPLE: The following are the ranks of yields obtained on an experiment conducted on 5 'demonstration plots' and 7 neighbouring plots of farmers for standing paddy crop in a district of A. P. Test whether the 'demonstration plots' are superior to farmers plots with respect to yields.

Demonstration 'Plots'	3	4	1	6	5		
Farmers plots	12	2	7	9	10	8	11

Null Hypothesis H_0: There is no significant difference between yields of 'demonstration plots' and 'farmers' plots. The alternative Hypothesis, H_1 is that demonstration plots would be superior to farmers plots.

Here $\qquad T_1 = 19; T_2 = 59, n_1 = 5, n_2 = 7$

$$W = (5)\ (7) + \frac{5(6)}{2} - 19 = 31$$

since W = 31 which is not available in the table, the actual value would be $W_1 = (5)\ (7) + \frac{7(8)}{2} - 59 = 4$.

CONCLUSION: The calculated probability with $W_1 = 4, n_1 = 5, n_2 = 7$ is .015 (from table) < 0.05. Hence the null hypothesis is rejected at 5 per cent level of significance. In other words, there is significant difference between 'demonstration plots' and 'farmers plots'.

Case (ii) Large samples ($n_2 > 20$):

Null Hypothesis: H_0: There is no significant difference between two groups or the two groups belong to the same population.

$$Z = \frac{\left| W - \frac{n_1 n_2}{2} \right|}{\sqrt{\frac{n_1 n_2 (n_1 + n_2 + 1)}{12}}}$$

CONCLUSION: If Z (calculated) > Z (tabulated), at chosen level of significance, the null hypothesis is rejected. Otherwise, the null hypothesis is accepted.

TWO-INDEPENDENT SAMPLES TEST: (MANN-WHTNEY TEST) SPSS DATA ANALYSIS PROCEDURE

The data entry will be done in SPSS Excel sheet in the following FORMAT column wise as follows.

D	F
1	34.6
1	31.5
1	20.8
1	18.6
2	13.5
2	20.6
	12.5

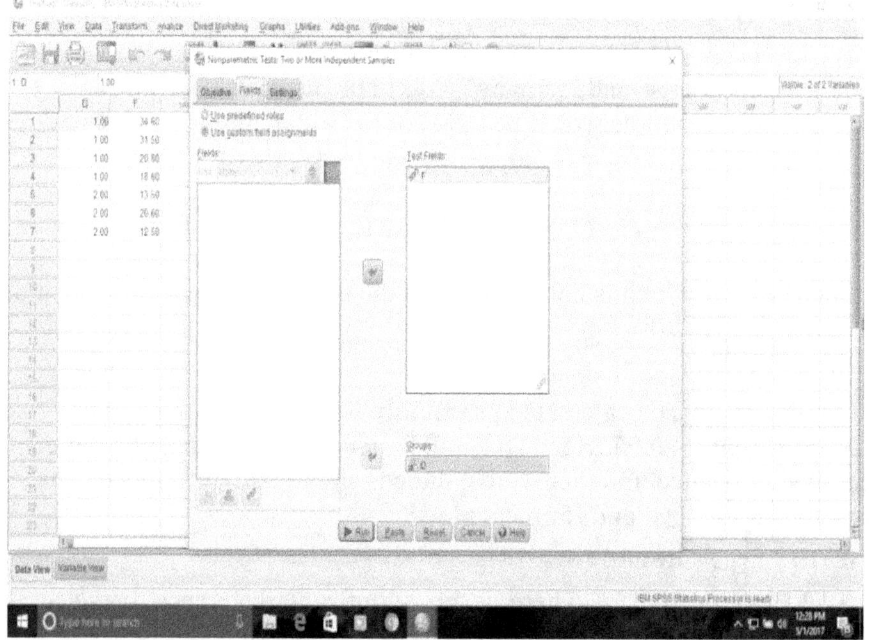

Where 'D' is 'GROUPS' and 'F' is Test FIELDS. The Test variable contain continuous variable values such as 28.5,13.6 etc. and should not only discrete values such as 12.14.,16,17 etc. CLICK 'ANALYZE' from TOOL BAR AND then click 'NON-PARAMETRIC STATISTICS' AND opt for 'INDEPENDENT SAMPLES'& 'SCAN' & 'CUSTOMIZE ANALYSIS'. Enter 'F' Test-fields and 'D' for 'Groups' in the R.H.S enclosures as per indicator. Choose 'CUSTOM' and then 'RUN'.

The Table of results contain the 'NULL HYPOTHESIS', NAME OF THE TEST USED', 'PROBABILITY VALUE' AND 'DECISION' ARRIVED.

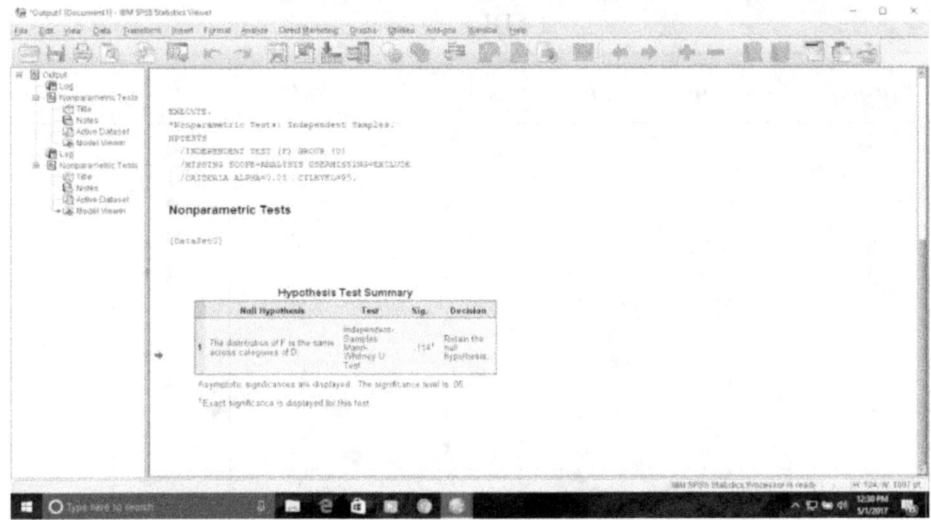

19.6   TESTS BASED ON P RELATED SAMPLES (P > 2)

19.6.1   Cochran's Test

When more than two related samples are under consideration and the observations are measured on nominal scale, Cochran's test is used as a counterpart to the Analysis of variance of one-way classification in the parametric case for testing the equality of proportions having particular character for all the samples.

Since the measurement is of the nominal scale, the scores may be given as S, F where S denotes the 'Success' and 'F' 'Failure' and Y, N where Y denotes .'Yes' and N denotes 'No'. etc. for each of the sample. The different samples and the scores may be arranged in a two-way table as follows:

TABLE 19.10

Sample / Score	1	2	....	k	
1	S	F	....	S	R_1
2	F	S	....	F	R_2
⋮	⋮	⋮		⋮	⋮
n	S	F		S	R_n
	C_1	C_2	.....	C_p	N

where R_i, C_J be the total number of successes in the i-th row and j-th column respectively for i = 1, 2, ... , n and j = 1, 2,... , p, so that

$$\sum_{i=1}^{n} R_i = \sum_{j=1}^{p} C_j = N$$

TABLE 19.11

Farmer	Mashuri	I.R.8	Varieties Jaya	Padma	R_i	R_i^2
1	Y	Y	N	Y	3	9
2	Y	N	N	N	1	1
3	N	N	Y	Y	2	4
4	Y	N	N	Y	2	4
5	Y	Y	Y	N	3	9
6	N	Y	Y	Y	3	9
7	Y	Y	Y	Y	4	16
8	N	N	N	Y	1	1
9	Y	N	Y	Y	3	9
10	Y	Y	Y	N	3	9
11	Y	N	N	Y	2	4
12	N	Y	Y	Y	3	9
13	Y	Y	N	N	2	4
14	Y	Y	Y	N	3	9
15	N	Y	Y	Y	3	9
16	Y	Y	Y	Y	4	16
17	N	N	Y	Y	2	4
18	N	Y	Y	Y	3	9
19	N	Y	Y	N	2	4
20	N	Y	Y	Y	3	9
C_j	11	13	14	14	52	148
C_j^2	121	169	196	196	682	

Null Hypothesis H_0: There is no significant difference between 'p' samples with respect to proportion of successes.

$$\chi_c^2 = \frac{p(P-1)\left[\sum_{j=1}^{p} C_j^2 - (\sum_{j=1}^{p} C_j)^2 / p\right]}{p\sum_{i=1}^{n} R_i - \sum_{i=1}^{n} R_i^2}$$

If n is not too small, χ_c^2 is approximately distributed as χ^2 with (p-1) d.f.

EXAMPLE: 20 farmers were asked for the acceptability of 4 varieties of paddy for cultivation in a particular region for kharif season. Test whether there is any significant difference between varieties with respect to acceptability (Table 19. 11).

Null Hypothesis: The proportion of answers 'Yes' are same for all the 4 varieties

$$\Sigma C_j = 52,\ \Sigma C_i^2 = 682,\ \Sigma R_i = 52,\ \Sigma R_i^2 = 148$$

$$\chi_c^2 = \frac{4\,(4-1)\left[682 - \dfrac{(52)^2}{4}\right]}{4 \times 52 - 148} = 1.2$$

CONCLUSION: χ_c^2 (calculated), $1.2 < \chi^2$ (tabulated), 7.82 with $(4 - 1)$ d.f. at 5 per cent level of significance, the null hypothesis is accepted. There is no significant difference between 4 varieties of paddy with respect to acceptability.

19.6.2 Friedman's Test

This test would be used corresponding to the Analysis of variance of one way classfication in the parametric case when the data are in ordinal scale. The scores in this case would be ranks and they would be arranged in two way table as in Table .19.10 with samples as columns and ranks within each sample as rows.

Null Hypothesis H_0: There is no significant difference between samples based on ranks.

$$\chi_F^2 = \frac{12}{np(p+1)} \sum_{j=1}^{p} C_j^2 - 3n(p+1)$$

where n, p are the total number of rows and columns respectively and Cj be the total of ranks in the j-th column for j=1, 2, .. , p. χ_F^2 is approximately distributed as χ^2 with (p-l) d.f,

CONCLUSION: If χ^2 F (calculated) $> \chi^2$ (tabulated) with (p–l) d.f. at chosen level of significance, the null hypothesis is rejected. Otherwise, it is accepted.

EXAMPLE: In the selection of officers' cadre in the Indian army, three types of tests psychological, physical endurance and map reading were conducted. The following table gives the scores (hypothetical) awarded out of 100 for 15 candidates. Test whether the candidates performance was same in all the three tests.

TABLE 19.12

Test Candidates	I (psychological)	II (Physical)	III (Map Reading)
1	60 (3)	55 (2)	43 (1)
2	50 (2)	62 (3)	49 (1)
3	65 (3)	60 (1)	64 (2)
4	60 (1)	62 (3)	61 (2)
5	72 (3)	65 (1)	70 (2)
6	50 (1)	64 (3)	55 (2)
7	40 (1.5)	50 (3)	40 (1.5)
8	35 (1)	55 (3)	42 (2)
9	58 (2)	60 (3)	35 (1)
10	65 (3)	61 (2)	58 (1)
11	50 (1)	52 (2)	55 (3)
12	45 (2)	50 (3)	40 (1)
13	75 (3)	60 (1)	70 (2)
14	60 (1)	65 (3)	62 (2)
15	78 (3)	64 (1)	65 (2)
C_j	30.5	34.0	25.5

The ranks would be given to the scores (marks) of the three tests appeared by each candidate. The lowest score would be given rank as 1, the next higher one as 2 and the highest as 3. For example, for the first candidate the ranks would be 3, 2 and 1 for the scores 60, 55 and, 43 respectively. These ranks are presented in parentheses beside's scores in Table 19.12 itself.

Null Hypothesis H_0: There is no significant difference between the three tests based on ranks.

$$\Sigma C_j^2 = 2736.50, \; p = 3, \; n = 15.$$

$$\chi_F^2 = \frac{12}{np(p+1)} \sum_{j=1}^{p} C_j^2 - 3n(p+1)$$

$$= \frac{12}{15(3)(4)}(2736.50) - (3)(15)(4) = 2.43$$

CONCLUSION: $\chi^2{}_F$(calculated), $2.43 < \chi^2$ (tabulated), 5.99 at 5 per cent level of significance. Hence the null hypothesis is accepted. That is, there is no significant difference between the three tests with respect to scoring.

More than Two-Related Samples (Fried Mann's Test)

SPSS DATA ANALYSIS PROCEDURE

The data entry will be done in SPSS Excel sheet in the following FORMAT column wise as follows.

X	Y	Z
3	2	1
1.5	3	1.5
2	1	3
1	2.5	2.5
1.5	1.5	3

Here 'X', 'Y' and 'Z' are taken as Test variables.

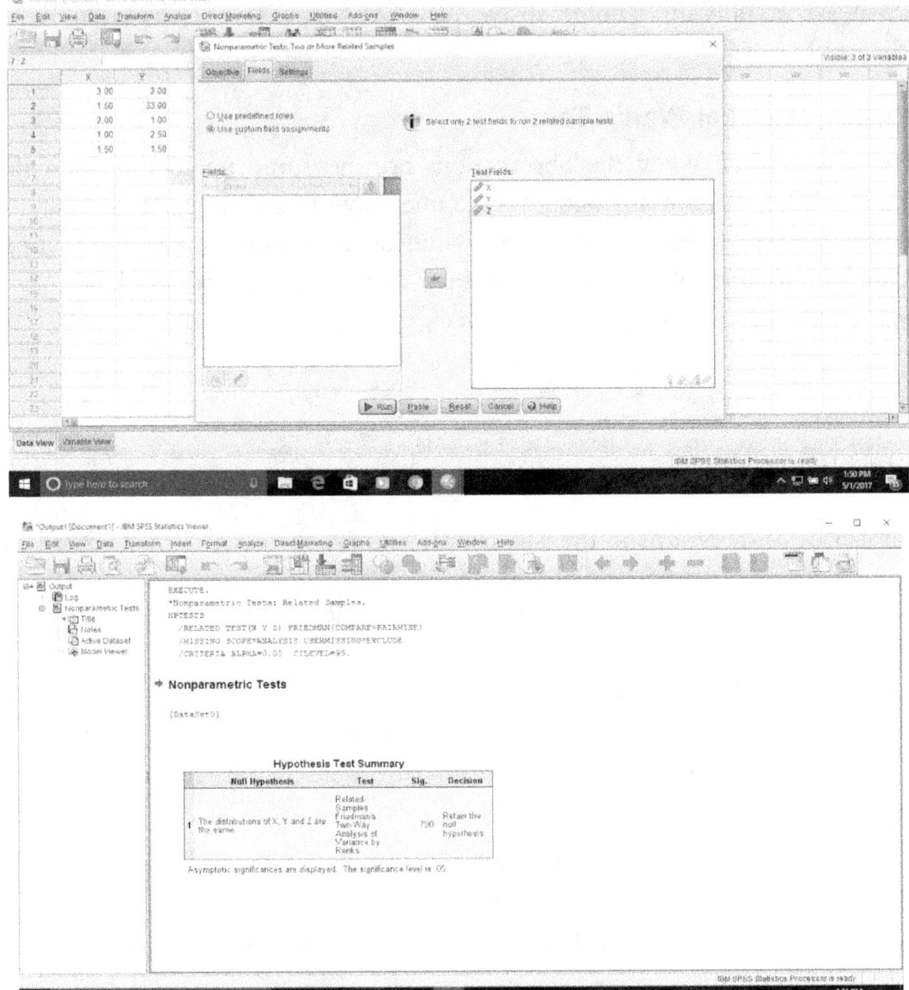

CLICK 'ANALYZE' from TOOL BAR AND then click 'NON-PARAMETRIC STATISTICS' AND opt for 'RELATED SAMPLES'. Enter Test-variables as 'X,'Y' and Z''. in the R.H.S enclosure as per indicator and then CLICK' TESTINGS' on the TOP AND SELECT 'FRIEDN MANN'S TEST AND 'RUN'.

The Table of results contain the 'NULL HYPOTHESIS', NAME OF THE TEST

19.7   TESTS BASED ON P INDEPENDENT SAMPLES (P>2)

19.7.1   Chi-Square Test

When more than two independent samples are involved and the observations are measured only up to nominal scale, the Chi-square test discussed in Chapter 11 is used.

19.7.2   Kruskal-Wallis Test

This test is used when the observations are measured up to ordinal scale (ranks). This is again analogous to the one-way classification of analysis of variance in parametric case. Here all the observations are ranked on one scale. The number of observations in each sample may not be equal.

Null Hypothesis H_0: There is no significant difference between p samples.

$$\chi_w^2 = \frac{12}{n(n+1)} \sum_{j=1}^{p} \frac{C_j^2}{n_j} - 3(n+1)$$

where C_j be the total of ranks in the j-th sample for j =1, 2, .. p; n_j be the number of observations in the j-th sample and n= $\sum_{j=}^{p} n_j$, χ_w^2 is distributed as χ^2 with (p–1) d.f. when the number of observations in each sample is at least 5. If the number of observations in any sample is less than 5, the exact probabilities were computed and presented by Kruskal and Wallis (1952). In case $n_1 > 5$ for every j, we have the following conclusions.

CONCLUSION: If χ_w^2 (calculated) $> \chi^2$ (tabulated)at chosen level of significance, the null hypothesis is rejected. Otherwise, it is accepted.

EXAMPLE: A study was conducted to know the difference between chapati making quality of three varieties of wheat out of which two of them belong to hybrid variety (Mexican dwarf and RS 31-1) and the other one local variety.

TABLE 19.13

S.No.	Local variety	rank	Mexican dwarf	rank	RS. 31-1	Rank
1	80	16	62	5	74	12
2	78	14.5	67	7	82	17
3	92	23	58	3	78	14.5
4	87	20	70	9.5	69	8
5	70	9.5	61	4	83	18
6	91	22	52	1	90	21
7	85	19	54	2	76	13
8			64	6	72	11
		124.0		37.5		114.5

8 samples were observed for each of the hybrid varieties and 7 samples for local variety and the following results (hypothetical) are obtained. Test whether the three varieties of wheat differ in chapati making quality.

Null Hypothesis Ho: All the three varieties are equal in chapati making quality.

Transforming all the scores to ranks such that lowest score gets rank 1, the next higher one gets 2, etc. and presented in Table 19.13 itself.

$$n = 7 + 8 + 8 = 23, p = 3$$

$$\chi^2_w = \frac{12}{23(24)}\left[\frac{(124.0)^2}{7} + \frac{(37.5)^2}{8} + \frac{(114.5)^2}{8}\right] - 3(24) = 15.20$$

CONCLUSION: χ^2_w (calculated), $15.20 > \chi^2$ (tabulated), 5.99 with (3–1) d.f, at 5 per cent level of significance. Hence the null hypothesis is rejected. The three varieties significantly differ in their chapati making quality.

More than Two-Independent Samples (Kruskal –Wallis Test)

SPSS DATA ANALYSIS PROCEDURE

The data entry will be done in SPSS Excel sheet in the following FORMAT column wise as follows.

A	B
1	11
1	8
1	7
1	9.5
1	6
1	9.5
2	4.5

2	4.5
2	10
3	1
3	2.5
3	2.5

Here 'A' is 'GROUPS' VARIABLE' AND 'B' IS TEST VARIABLE.

CLICK 'ANALYZE' from TOOL BAR AND then click 'NON-PARAMETRIC STATISTICS' AND opt for ' INDEPENDENT SAMPLES'. Enter Test-variables as 'B' and 'GROUPS' VARIABLE AS 'A'. in the R.H.S enclosures as per indicator and then CLICK' TESTINGS' AND SELECT 'KRUSKAL WALLIS TEST'& 'RUN'.

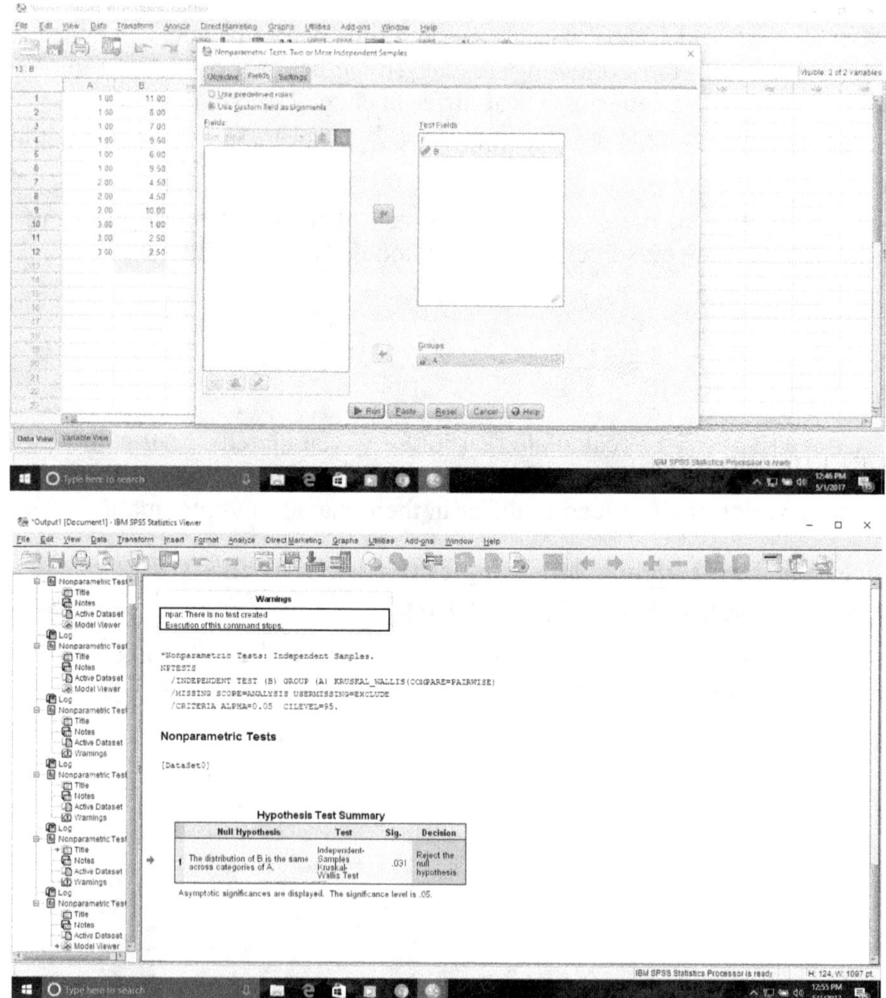

The Table of results contain the 'NULL HYPOTHESIS', NAME OF THE TEST USED', 'PROBABILITY VALUE' AND 'DECISION' ARRIVED.

19.8 COEFFICIENT OF CONCORDANCE

This is an extension of Spearman's rank correlation Coefficient. Spearman gave the Coefficient of correlation between ranks of two variates whereas Kendall developed coefficient of concordance (Kc) for measuring the relationship between the k variates. It measures the extent of relationship (or the degree of associationship) between 'k' variates based on 'n' rankings for each variate. This coefficient will avoid the procedure of computing several Spearman rank correlation coefficients pairwise.

The two way table of ranks for the variates and the observations is given in-Table 19.14.

TABLE 19.14

Variate \ Observations	1	2	..	n
1	6	$(n-1)$	...	2
2	3	5	...	$n-2$
⋮	⋮	⋮		⋮
k	$(n-5)$			1
	C_1	C_2		C_n

$$K_c = \frac{\displaystyle\sum_{j=1}^{n} C_j^2 - \frac{(\Sigma C_j)^2}{n}}{\dfrac{1}{12}k^2 n(n^2 - 1)}$$

where C_J be the j-th column total for $j = 1, 2, \ldots , n$ and n be the number of observations in each variate and k be the number of variates. Kc, always lies between 0 and 1 i.e., $O \le K_c \le 1$.

EXAMPLE: In a certain cattle judging competition, 10 cows were ranked by 4 judges (A, B, C and D) and are presented here. Compute the Kendall's coefficient of concordance between the rankings of 4 judges.

Judge	Cow									
	1	2	3	4	5	6	7	8	9	10
A	5	3	4	1	8	7	6	9	10	2
B	4	6	3	2	7	5	10	8	9	1
C	4	7	5	3	6	9	8	10	2	1
D	3	5	1	4	9	10	7	6	8	2
	16	21	13	10	30	31	31	33	29	6

$$\Sigma C_j^2 = 5754, \ \Sigma C_j = 220$$

$$K_c = \dfrac{5754 - \dfrac{(220)^2}{10}}{\dfrac{1}{2}(4)^2 10(100-1)} = 0.69$$

19.8.1 Test of Significance

Case (i) Large samples (n>7): If the number of observations for each of the variate is greater than 7, a Chi-square approximation is used for testing the significance of coefficient of concordance in the population.

Null Hypothesis H_0: There exists no correlation between k variates based on ranks.

$$\chi^2 = k(n-1)\, K_c$$

CONCLUSION: If χ^2 (calculated) $>$ χ^2 (tabulated) with (n-l) d.f. at chosen level of significance, the null hypothesis is rejected. Otherwise, the null hypothesis is accepted.

EXAMPLE: Test the coefficient of concordance obtained in example given in Section 19.8 for $K_c = 0\cdot69$.

Null Hypothesis H_0: There exists no correlation between rankings of 4 judges in the population.

$$\chi_c = 0.69, \ n = 10, \ k = 4$$

$$\chi^2 = 4\,(10-1)\,(0.69) = 24.84$$

CONCLUSION: Here χ^2 (calculated), 24.84 $>$ χ^2 (tabulated), 16.92 with 9 d.f. at 5 per cent level of significance. Hence the null hypothesis is rejected and there exists significant correlation between the rankings of 4 judges.

Zero Order Correlation Matrix For Ordinal Data For More Than Two Variables (kendall'coefficient/ spearmann's coefficient)

SPSS DATA ANALYSIS PROCEDURE

The data entry will be done in SPSS Excel sheet in the following FORMAT column wise as follows. Supposing that we are ranking 5 subjects by 3 judges.

A	B	C
3	5	2
1	2	4
4	3	1
2	4	5
5	1	3

CLICK 'ANALYZE' from TOOL BAR AND then click 'CORRELATE' AND opt for ''BIVARIATE'. Enter Test-variables as 'A,'B' and C''. in the R.H.S enclosure as per indicator and then 'OPT ' FOR 'KENDAL COEFFICIENT ' OR 'SPEARMANN'S COEFFICIENT 'AS THE CASE MAY BE AND THEN 'O.K.'

We get the table of results containing 'ZERO ORDER CORRELATION MATRIX' AND ALSO SIGNIFICANCE OF THE CORRELATION COEFFICENTS.

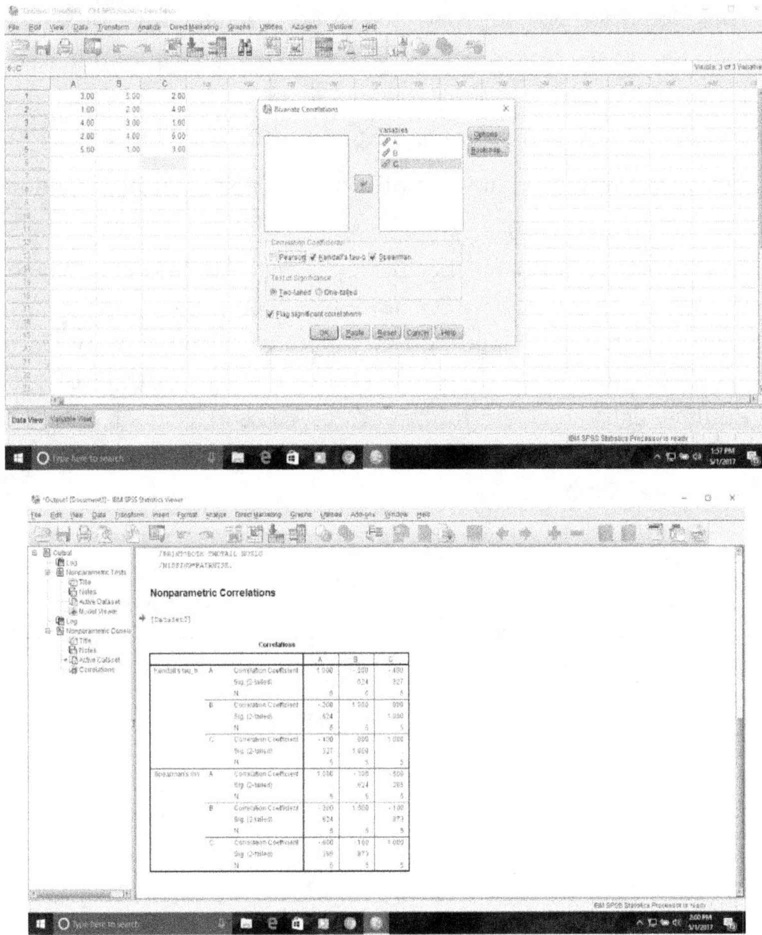

EXERCISES

1.   Tobacco leaves are graded as 1, 2, 3, 4 and 5 depending on the shape of the leaf. Very narrow leaf is graded as 1, Narrow as 2, Medium as 3, broad as 4 and very broad as 5. Here the assumption is that manufacturing firm prefers broad leaves though the quality is same.

A sample of 100 buyers were asked to grade the leaves and the frequency distribution is given below. Test whether there is any preference for broad leaves.

Grade	1	2	3	4	5
No. of buyers	2	10	20	32	36

2. The distribution of rainfall in a rainy season was recorded at a certain research station. If the total rainfall in a rainy season is considered as 'Satisfactory' (S) if it is greater than or equal to 40 cm, otherwise 'Not satisfactory' (N). The distribution for 25 years is given as follows:

S S N N S N S S S N N N sN S S. S N S N N S S S

Test whether the occurrence of 'Satisfactory' rainfall is random or not.

3. In order to evaluate the performance of Churcha Mandals in different villages, an interview was conducted by an Extension expert for members and non-members of Churcha Mandals in each of 20 villages with respect to knowledge on package of practices, Poultry, dairy, etc. and ratings are given as 1 for poor, 2 for average, 3 for good and 4 for very good. Test whether members are better than non-members.

Village	Group I (Members)	Group II (Non-members)	Village	Group I (Members)	Group II (Non-members)
1	3	1	11	3	1
2	4	2	12	4	3
3	3	2	13	3	2
4	2	3	14	2	3
5	3	2	15	3	3
6	3	3	16	4	1
7	3	4	17	3	2
8	2	1	18	2	1
9	3	4	19	3	4
10	4	2	20	2	3

4. Two types of package programmes were offered to 30 farmers in an investigation and were asked to award scores for each type of programme based on its merit and the scores are presented. Test whether there is any significant difference between two types of programmes.

Farmer	Type I	Type II	Farmer	Type I	Type II
1	64	68	16	39	50
2	70	72	17	47	40
3	65	60	18	35	35
4	72	69	19	57	50
5	35	42	20	68	52
6	52	49	21	70	68
7	45	45	22	43	51
8	76	73	23	59	55
9	60	58	24	38	39
10	48	54	25	59	51
11	39	42	26	45	50
12	67	54	27	62	68
13	50	65	28	72	63
14	76	75	29	78	68
15	42	40	30	48	52

5. The following contingency table gives the frequency distribution of farmers with respect to adoption of package of practices and type of communication. Test whether adoption depends on type of communication.

Type of communication	Adoption		Total
	Adopted	Non-adopted	
Mass media	32	10	42
Extension agency	20	14	34
Progressive farmer	16	8	24
	68	32	100

6. The following are the scores awarded for the market centres run by Poultry corporation and also by Private individuals with respect to their performance. Test whether there is any significant difference between two types of centres run by two different organisations.

S.NO.	Market centres (poultry crop.)	Market centres (private)	S.NO.	Market centres (poultry crop.)	Market centres (private)
1	62	50	13	51	46
2	58	47	14	38	27
3	69	59	15	60	65
4	56	72	16	67	58
5	42	64	17	54	59

Contd...

S.NO.	Market centres (poultry crop.)	Market centres (private)	S.NO.	Market centres (poultry crop.)	Market centres (private)
6	78	43	18	46	56
7	64	57	19	56	64
8	50	49	20	48	54
9	64	75	21	78	67
10	50	62	22	40	
11	49	58	23	36	
12	67	74	24	69	

7. In a crop competition, 6 judges gave the following scores for 10 samples brought by 10 farmers of jowar crop. Test whether there is any significant difference between the 10 samples based on the scores given by judges.

Judge	Sample									
	1	2	3	4	5	6	7	8	9	10
1	30	70	45	52	65	41	38	60	43	68
2	35	65	53	60	70	46	39	64	50	72
3	41	74	50	61	67	44	40	67	55	65
4	39	64	42	50	60	52	46	62	58	62
5	42	60	56	65	67	59	43	69	54	69
6	37	64	52	56	68	53	47	65	53	70

8. A study was conducted to determine the difference between Politically oriented farmers, urban based farmers and the Professional farmers with respect to knowledge on Poultry farming and the following scores were recorded.

S.No.	Political oriented farmer	Urban based farmer	Professional farmer
1	65	49	60
2	47	67	53
3	50	39	46
4	56	50	67
5	38	64	58
6		58	49

Test whether there is any significant difference between the three groups of farmers with respect to knowledge on Poultry.

9. In an experiment on 10 cooking qualities of food grain, 4 judges gave the following hypothetical ranks.

Judge	Cooking qualities									
	1	2	3	4	5	6	7	8	9	10
A	5	3	4	1	8	7	6	9	10	2
B	4	3	5	2	10	6	7	8	9	1
C	6	1	7	4	9	2	5	10	8	3
D	3	5	4	1	7	8	9	6	10	2

Compute the coefficient of concordance between the judges.

PART IV

MULTIVARIATE STATISTICAL
METHODS

CHAPTER 20

MULTIVARIATE STATISTICAL METHODS

20.1 MULTIVARIATE NORMAL DISTRIBUTION

The density function of a univariate normal distribution for a random varibale x is

$$\varphi(x) = \frac{1}{\sigma\sqrt{2\pi}}\ e^{-1/2\left(\frac{x-u}{\sigma}\right)^2} \qquad \ldots.(20.1)$$

The density function of a p-vatiate normal distribution is

$$\phi(x_1\ x_2\ \ldots\ x_p) = \frac{1}{\sigma_1, \sigma_2 \ldots \sigma_p (2\pi)^{p/2}} \qquad \ldots.(20.2)$$

$$e^{-1/2}\ \sum_{i=1}^{p}\left(\frac{x_1 - u_i}{\sigma_i}\right)^2$$

If $\quad x^1 = (x_1\ x_2, \ldots\ x_p),\ u^1 = (u_1, u_2, \ldots\ u_p)$

and $\quad \Sigma = \begin{bmatrix} \sigma_1^2 0 \ldots 0 \\ 0\sigma_2^2 0 \ldots 0 \\ 00 \ldots \sigma_p^2 \end{bmatrix}$

The density function of a p-variate normal distribution can be written as

$$\phi(x) = \frac{1}{(2\pi)^{p/2}\ |\Sigma|^{1/2}}\ e^{-1/2}\ (x-u)^1\ \Sigma^{-1}\ (x-u)$$

Where Σ is $p \times p$ symmetric positive definite matrix.

20.1.1 Bivariate Normal Distribution

If $p = 2$ in $\phi(x)$ given above, we have

$$u = \begin{pmatrix} u_1 \\ u_2 \end{pmatrix}, \qquad \Sigma = \begin{pmatrix} \sigma_1^2 & \sigma_1\ \sigma_2 \\ \sigma_1\ \sigma_2 & \sigma_2^2 \end{pmatrix}$$

The density function is

$$\phi(x_1\ x_2) = \frac{1}{2\pi\ \sigma_1\ \sigma_2\sqrt{1-\rho^2}}\ e^{-\frac{1}{2(1-\rho^2)}\left\{\left(\frac{x_1-u_1}{\sigma_1}\right)^2\right.}$$

$$\left.-2\ \rho\left(\frac{x_1-u_1}{\sigma_1}\right)\left(\frac{x_2-u_2}{\sigma_2}\right)+\left(\frac{x_2-u_2}{\sigma_2}\right)^2\right\} \qquad \ldots\ldots (20.3)$$

20.1.2 Standard Brivariate Normal Distribution

If $z_1 = \dfrac{x_1-u_1}{\sigma_1}$, $z_2 = \dfrac{x_2-u_2}{\sigma_2}$, then

$$\phi(z_1\ z_2) = \frac{1}{2\pi\ \sqrt{1-\rho^2}}\ e^{-\frac{1}{2(1-\rho^2)}\left(z_1^2-2\rho\ z_1\ z_2+z_2^2\right)} \qquad \ldots\ldots (20.4)$$

is a density function of standard bivariate normal distribution with mean as zero and variance as unity.

20.1.3 Tests of Hypotheses

Here multivariate case of student's t-test is presented. Student's t-test is applicable in univariate case for testing a sample mean with population mean and between two population means. Hotelling's T^2 is used for testing sample mean with population mean and two population means in multivariate case where more than one character under consideration. For example, in order to test whether there is any significant difference between two groups of farmers i.e., beneficiaries and non-beneficiaries of integrated rural development project (IRDP), several factors have to be taken into consideration such as family size (x_1), working labur force (x_2), productive assets (x_3), non-productive asscets (x_4), toal consumption expenditure (x_5), expenditure on education (x_6), per capita total income (x_7), employment/man unit (x_8) etc. Student's test cannot be used for testing the above two groups of farmers when 8 characters are considered simultaneously. Therefore Hotelling's T^2 is used with the help of multivariate normal distribution.

20.1.4 Case (i) one sample case

Null Hypothesis, Ho: $\underline{u} = \underline{\mu}_0$

where $\underline{u} = \begin{pmatrix} u_1 \\ u_2 \\ \vdots \\ u_p \end{pmatrix}$, $\underline{u}o = \begin{pmatrix} u_{o1} \\ u_{o2} \\ \vdots \\ u_{op} \end{pmatrix}$

$\underline{u0}$ is the given population mean vector

Alternate Hypothesis, $H_1 : \underline{u} \neq \underline{uo}$

Hotelling T^2 – statistic is given by

$$T^2 = n \ (\overline{\underline{x}} - \underline{u})' S^{-1} (\overline{\underline{x}} - \underline{u})... \qquad .....(20.5)$$

$$\text{where} \qquad \underline{x} = \begin{pmatrix} \overline{x_1} \\ \overline{x_2} \\ \\ \overline{x_p} \end{pmatrix} \qquad S = \begin{pmatrix} s_{11} & s_{12} & \cdots & s_{1p} \\ s_{21} & s_{22} & \cdots & s_{2p} \\ & & & \\ s_{p1} & s_{p2} & \cdots & s_{pp} \end{pmatrix}$$

$\overline{\underline{x}}$ and S are the estimates of u and S respectively based on n observations on each of the p-characters and S is the variance - convariance matrix.

For example, $\overline{x}_1$ is the mean of n observations on 1st character and s_{11} is the estimate of variance based on n observations for the 1st character.

In order to test the null hypothesis. F-test is used by following the relationship between F and T^2 which is given as

$$F = \frac{n - p}{p \ (n-1)} T^2 \qquad .....(20.6)$$

which follows F-distribution with (p, n–p) d.f.

CONCLUSION: If F (calculated) value is greater than F (tabulated) value with (p, n–p) d.f. at the chosen level of significance, the null hypothesis is rejected. Otherwise, it is accepted.

If the null hypothesis is rejected, then it can be concluded that there is significant difference between sample means and population means based on p-characters. Otherwise there is no significant difference between sample means and population means of p-characters.

EXAMPLE: In an investigation on the performance of a newly released variety of a paddy crop from a sample of 50 farms, the following results were obtained on the height of the plant (x_1), number of tillers (x_2) and ear length (x_3). Test whether there is any significant difference between sample means obtained from 50 farms and the population means i.e., quoted by the research station at the time of release of the said variety.

$$\underline{x} = \begin{pmatrix} 115 \\ 8.4 \\ 13.8 \end{pmatrix} \qquad \underline{u} = \begin{pmatrix} 108 \\ 9.7 \\ 12.4 \end{pmatrix}$$

$$S = \begin{pmatrix} 38.0 & 11.0 & 16.0 \\ 11.0 & 25.0 & 20.0 \\ 16.0 & 20.0 & 30.0 \end{pmatrix}$$

Null hypothesis: $H_o : \underline{u} = \begin{pmatrix} 108 \\ 9.7 \\ 12.4 \end{pmatrix}$

Alternate hypothesis, $H_1 : \underline{u} \neq \begin{pmatrix} 108 \\ 9.7 \\ 12.4 \end{pmatrix}$

$$T^2 = n \ (\overline{x} - \underline{u}) \ S^{-1} \ (\overline{x} - \underline{u})$$

Here $n = 50$, $(\overline{x} - \underline{u}) = (7, -13, 1.4)$

$$S^{-1} = \begin{pmatrix} .0339 & -.0010 & -.0175 \\ -.0010 & .0857 & -.0566 \\ -.0175 & -.0566 & .0804 \end{pmatrix}$$

S^{-1} can be obtained by following any method of inverting matrix

$$(\overline{x} - \underline{u}) = \begin{pmatrix} 7 \\ -13 \\ 1.4 \end{pmatrix}$$

Therefore, $T^2 = 50 \ (7, -13, 1.4)$

$$\begin{pmatrix} .0339 & -.0010 & -.0175 \\ -.0010 & .0857 & -.0566 \\ -.0175 & -.0566 & .0804 \end{pmatrix} \begin{pmatrix} 7 \\ -13 \\ 1.4 \end{pmatrix}$$

$$= 910.04$$

$$F = \frac{(n-p)}{p(n-1)} \ T^2$$

where $n = 50$, $p = 3$, $T^2 = 910.04$

$$F = \frac{(50-3)}{3(50-1)} \times 910.04 = 290.97$$

CONCLUSION: Here F (calculated) value, 290.97 > F (tabulated) value, 2.81 with (3, 47) d.f. at 5 percent level of significance. Therefore, the null hypothesis is rejected. That is, there is significant difference between the results based on sample and quoted figures based on the three characters.

20.1.5   Confidence Limits for the Population Mean Vector

Just as in the case of student's   t-test, here also a confidence region can be constructed for the population mean vector $\underline{u}$. In one sample case, 100 $(1-\infty)\%$ confidence region is given by the vectors satisfying the following inequality.

$$n(\underline{\bar{x}} - \underline{u})' \, S^{-1}(\underline{\bar{x}} - \underline{u}) \le \frac{(n-1)}{n-p} F\,\alpha;p,n-p$$

where "α" is the specified level of significance.

EXAMPLE: The lower and upper limits for mean of each character in $\underline{u}$ for the example given in section 20.1.4 above are obtained as follows.

Ist character (height of the plant)

$$n(\underline{\bar{x}} - u_1)\, S_{11}^{-1}\, (\bar{x}_1 - u) \le \frac{(n-1)}{(n-p)} F\,\alpha;\, p,\, n-p$$

Here $n = 50,\ \bar{x}_1 = 115,\ S^{-1} = 1/38,\ p = 3$

$F\,\alpha;\, p,\, n - p = F\ 5;\ 3,\ (50{-}3) = 2.81$

substituting the above numerical values, we have

$$50\,(115 - u_1)\ 1/38\ (115{-}u_1) \le \frac{(50-1)3}{50-3} \times 2.81$$

$$(115 - u_1)^2 \le \frac{38}{50} \times \frac{(50-1)3}{47} \times 2.81 = 6.68$$

$$(115 - u_1) \le \pm \sqrt{6.68}$$

i.e., the upper and lower bounds for u_1 are

$$(115 + 2.58) \text{ and } (115 - 2.58)$$

i.e., 117.58 and 112.42 respectively.

Similarly the upper and lower bounds for the population means of 2nd and 3rd characters are obtained as follows.

2nd character (number of tillers)

$$8.4 \pm (1.25 \times 2.81) = 8.4 \pm 3.51$$

i.e., 11.91 and 4.89 are the upper and lower bounds respectively for the mean of number of tillers.

3rd character (ear length):

$$13.8 \pm (1.37 \times 2.81) = 13.8 \pm 3.85$$

i.e., 17.65 and 9.95 are the upper and lower bounds respectively for the mean ear length.

Taking all the three characters into consideration the upper and lower bounds vectors are

$$\begin{pmatrix} 117.58 \\ 11.91 \\ 17.65 \end{pmatrix} \text{ and } \begin{pmatrix} 112.42 \\ 4.89 \\ 9.95 \end{pmatrix} \text{ respectively}$$

20.1.6   Two Sample Case

In order to test the significant difference between two sample means based on p–characters, Hotelling–T^2 statistic is used analogus to usual student's t–test. For example, two samples are to be tested from two groups of varieties of jowar based on height of plant (x_1) yield (x_2), number of tillers (x_3) etc. Hotelling–T^2 can be used. The method of analysis is given in Table 20.1.

TABLE 20.1

S.No.		Sample 1	Sample 2
1.	Size of sample	n_1	n_2
2.	Mean	$\left(\overline{x}_{11}, \overline{x}_{12}....\overline{x}_{1p}\right)$	$\left(\overline{x}_{21}, \overline{x}_{22}....\overline{x}_{2p}\right)$
3.	Matrix of sum of squares and products	S_1	S_2
4.	Pooled covariance matrix	$S = \dfrac{(S_1 + S_2)}{n_1 + n_2 - 2}$	

Null hypothesis: $H_0 : \underline{\mu}_1 = \underline{\mu}_2 \quad H_1 : \underline{\mu}_1 \neq \underline{\mu}_2$

Where
$$\underline{u}_1 = \begin{pmatrix} u_{11} \\ u_{12} \\ .. \\ .. \\ u_{1p} \end{pmatrix} \quad \underline{u}_2 = \begin{pmatrix} u_{21} \\ u_{22} \\ .. \\ .. \\ u_{2p} \end{pmatrix}$$

Then Hotelling $- T^2$ statistic is given as

$$T^2 = \frac{n_1 n_2}{n_1 + n_2}\left(\overline{\underline{x}}_1 - \overline{\underline{x}}_2\right)' S^{-1}\left(\overline{\underline{x}}_1 - \overline{\underline{x}}_2\right) \qquad .....(20.7)$$

$$F = \frac{n_1 + n_2 - p - 1}{p\left(n_1 + n_2 - 2\right)} T^2$$

CONCLUSION: If F (calculated) value > F (tabulated) value with $(p, n_1 + n_2 - p - 1)$ d.f. at the chosen level significance, the null hypothesis is rejected. Otherwise, the null hypothesis is accepted.

EXAMPLE: 50 varieties of jowar crop were classified as 'high yielding' and 'not high yielding' based on 3 characters of the crop such as yield, number of days to 50 percent flowering, fertilizer consumption and the results are presented in the following Table 20.2. Test whether there is any significant difference between the two groups.

TABLE 20.2

S.No.	Character	Group		$\bar{x}_1 - \bar{x}_2$
		I (High yielding)	II (Nothigh yielding)	
1.	Sample size	20	30	
2.	Yield	35	26	9
3.	Mean number of days to 50 percent flowering	40	52	−12
4.	Mean fertilizer consumption	220	160	60

$$S = \begin{pmatrix} 18 & 9 & 6 \\ 9 & 12 & 4 \\ 6 & 4 & 29 \end{pmatrix}$$

The inverted matrix S^{-1} is given as

$$S^{-1} = \begin{pmatrix} .0915 & -.0653 & -.0099 \\ -.0653 & .1340 & -.0050 \\ -.0099 & -.0050 & .0372 \end{pmatrix}$$

Null hypothesis: $H_0 : \underline{u}_1 = \underline{u}_2$, $H_1 : \underline{u}_1 \neq \underline{u}_2$

$$T^2 = \frac{n_1 \, n_2}{n_1 + n_2} (\bar{x}_1 - \bar{x}_2)^1 \, S^{-1} (\bar{x}_1 - \bar{x}_2)$$

$$T^2 = \frac{20 \times 30}{20 + 30} (9, -12, 60) \begin{pmatrix} .0915 & -.0653 & -.0099 \\ -.0653 & .1340 & -.0050 \\ -.0099 & -.0050 & .0372 \end{pmatrix} \begin{pmatrix} 9 \\ -12 \\ 60 \end{pmatrix}$$

$$T^2 = 2054.88$$

$$F = \frac{n_1 + n_2 - p - 1}{p \, (n_1 + n_2 - 2)} T^2$$

$$= \frac{20+30-3-1}{3\,(20+30-2)} \times 2054.88 = 656.42$$

CONCLUSION: Here F (calculated) value, 656.42 > F (tabulated) value with (3, 46) d.f. at 5 percent level of significance i.e., 2.81. Therefore, the null hypothesis is rejected. In otherwords, there is significant difference between high yielding and not high yielding varieties based on 3 characters.

20.1.7   Confidence Limits for the Differences of Two Population Means

Let $\underline{\delta} = \underline{u}_1 - \underline{u}_2$ be the difference of two population means and $l^1\,\underline{\delta}$ be the linear function of the difference of population means, then the confidence limits for $l^1\,\underline{\delta}$ is given by

$$l^1(\bar{\underline{x}}_1 - \bar{\underline{x}}_2) - \sqrt{l^1\,s\underline{l}\,\frac{n_1+n_2}{n_1\,n_2}}\,T$$

$$\underline{l}^1\underline{\delta} \leq l^1(\bar{\underline{x}}_1 - \bar{\underline{x}}_2) + \sqrt{l^1\,s\underline{l}\,\frac{n_1+n_2}{n_1\,n_2}}\,T \qquad\qquad ......(20.8)$$

$$\text{where}\quad T = \sqrt{\frac{n_1+n_2-2}{n_1+n_2-p-1}}\,F$$

and F is the tabulaed value of F with p, $n_1 + n_2 - p - 1$ d.f.

EXAMPLE: In the example given in section 20.1.6 the lower and upper bounds for the difference between two population means of high yielding and not high yielding varieties of jowar are obtained as follows:

for yield 1 : $\bar{x}_1 - \bar{x}_2 = 9 \; l^1(\bar{x}_1 - \bar{x}_2) = 9$

since $\qquad l_1 = 1, l_2 = -1$ in the above function

$$\underline{l}^1\,S\underline{l} = 18 \quad T = \sqrt{\frac{p\,(n_1+n_2-2)}{n_1+n_2-p-1}}\,F$$

$$T = \sqrt{\frac{3\,(50-2)}{50-3-1}} \times 2.81 = 2.97$$

Therefore, the lower and upper limits for yield are

$$9 - \sqrt{18 \times \frac{50}{20 \times 30}}\,2.97 = 5.38$$

$$\text{and}\qquad 9 + \sqrt{18 \times \frac{50}{20 \times 30}}\,2.97 = 12.64$$

Similarly for other characters such as number of days to 50 per cent flowring and fertilizer consumptions are give as –14.97, –9.03 and 55.39, 64.61 respectively. Therefore, the lower and upper bound vectors are

$$\begin{bmatrix} 5.38 \\ -14.97 \\ 55.39 \end{bmatrix} \text{ and } \begin{bmatrix} 12.64 \\ 9.03 \\ 64.61 \end{bmatrix} \text{ respectively}$$

20.1.8   P Related Sample Case

When the observations are collected on the same experimental unit at different time intervals under different conditions, an extension of paired t-test in the multivariate case is used for testing the significant difference between p conditions. For example, in cotton crop several pickings on each plant will be effected for yield in order to test the significant difference between different pickings of cotton taking into consideration of different plots, the multivariate use of paired t-test can be used. Similarly p drugs can be tested on n patients.

Null hypothesis, $H_0 = u_1 = u_2 \ .... = u_p$

$$H_1 : u_1 \neq u_2 \neq ... \neq u_p$$

This can be written as

$$H_0 : \begin{bmatrix} u_1 - u_2 \\ u_2 - u_3 \\ \vdots \\ u_{p-1} - u_p \end{bmatrix} = \begin{bmatrix} 0 \\ 0 \\ \vdots \\ 0 \end{bmatrix} \qquad H_1 : \begin{bmatrix} u_1 - u_2 \\ u_2 - u_3 \\ \vdots \\ u_{p-1} - u_p \end{bmatrix} \neq \begin{bmatrix} 0 \\ 0 \\ \vdots \\ 0 \end{bmatrix}$$

The sample observations can be written in the Table 20.3 as follows

TABLE 20.3

Condition	1	Sampling unit	n	Mean
1	X_{11}	X_{12}	X_{1n}	$\overline{X}_1$
2	X_{21}	X_{22}	X_{2n}	$\overline{X}_2$
$\vdots$				
p	X_{p1}	X_{p2}	X_{pn}	$\overline{X}_p$

We reduce the above p conditions into (p – 1) conditions by using the following transformation

$$Z_j = C X_j$$

Where C is a (p – 1) x p matrix with

$$C J_{p',1} = O_{p-1,1}$$

Where $J_{p',1}$ is a $p \times 1$ column vector with $+1$ everywhere and $O_{p-1,1}$ is a $(p - 1)$ dimensional zero vector.

$$\underline{Z} = \left(\overline{Z}_1, \overline{Z}_2 \dots \overline{Z}_{p-1} \right)$$

S_z be the $(p - 1) \times (p - 1)$ covariance matrix of $\overline{z}_j$ values.

The Hotelling $- T^2$ for testing the null hypothesis is given as

$$T^2 = n\ \overline{Z}\ S_z^{-1}\ \overline{Z}$$

and $\qquad F = \dfrac{n - p + 1}{(p - 1)(n - 1)} T^2$ $\qquad\qquad$ ....(20.9)

CONCLUSION: If F (calculated) value > F (tabulated) value with $(p - 1)$, $(n - p + 1)$ d.f. at chosen level of significance, the null hypothesis is rejected. Otherwise, it is accepted.

EXAMPLE: An experiment was conducted on particular variety of cotton in order to test the significant difference between three pickings based on yields for 10 plots. The yields, Z-values and S_z matirx are presented in the Table 20.4. Test the significant difference between the three pickings.

TABLE 20.4

Plot

Picking	1	2	3	4	5	6	7	8	9	10
1	14.1	12.2	13.7	15.4	16.5	17.8	12.6	18.0	19.0	14.5
2	9.6	8.7	11.2	12.3	14.8	15.1	9.8	14.7	15.2	10.4
3	7.6	5.3	6.8	7.4	8.6	9.1	6.4	7.3	8.6	5.9

The Z-values are presented in Table 20.5

TABLE 20.5

Picking	1	2	3	4	5	6	7	8	9	10	$\overline{Z}_i$
Z_1	4.5	3.5	2.5	3.1	1.7	2.7	2.8	3.3	3.9	4.1	3.21
Z_2	8.5	10.3	11.5	12.5	14.1	14.7	9.6	18.1	17.1	13.1	12.93

where $\qquad z_1 = x_{1i} - x_{2i}$ and $z_2 = x_{1i} + x_{2i} - 2x_{3i}$

For example $z_1 = 14.1 - 9.6 = 4.5$ and

$\qquad z_2 = 14.1 + 9.6 - 2 \times 7.6 = 8.5$

S_z matrix of order 2×2 is computed from z-values given in Table 20.5 and presented here

$$S_z = \begin{bmatrix} 0.69 & -0.34 \\ -0.34 & 9.9 \end{bmatrix}$$

The inverted matrix S_z^{-1} for S_z is obtained as

$$S_z^{-1} = \begin{bmatrix} 1.4742 & 0.0506 \\ 0.0506 & 0.1027 \end{bmatrix}$$

H_0: There is no significant difference between 1st and 2nd, 2nd and 3rd pickings of yield cotton.

Hotelling $-T^2$ formula is

$$T^2 = n\,\overline{Z}'\,S_z^{-1}\,\overline{Z}$$

$$T^2 = 10\,(3.21,\ 12.93) \begin{bmatrix} 1.4742 & 0.0506 \\ 0.0506 & 0.1027 \end{bmatrix} \begin{bmatrix} 3.2 \\ 12.93 \end{bmatrix}$$

$$= 365.60$$

$$F = \frac{n-p+1}{(p-1)(n-1)}\,T^2 = \frac{(10-3+1)}{(3-1)(10-1)}(365.60) = 162.49$$

CONCLUSION: Here F (calculated) value i.e., 162.49 > F (tabulated) value with (2, 8) d.f. at 5 per cent level of significance i.e., 4.46. Therefore, the null hypothesis is rejected. Hence, it can be concluded that there is significant difference between the three pickings of yield cotton.

20.2 CLASSIFICATION BY LINEAR DISCRIMINANT FUNCTION

Sometimes there is a need to assign an observation to one of several populations. For example, a botanist may wish to classify a new speciman as one of recognized species of a flower in plant taxonomy. Similarly a farmer may have to be identified as, progressive or non-progressive farmer for the purpose of extending benefit by the funding agency.

If an individual (object) was charactrised by a single character then this individual was classified into a group based on the characteristic value. If this characteristic value is greater than some pre determined value than the individual was classified into that group otherwise in the second group. For example, the per hectare yield of a particular variety of paddy exceeds, say, 30 quintals/ha then the variety may be classified as high yielding variety otherwise not. If an individual (object) was characterised by more than one character a suitable linear function was necessary for classifying into one of the two groups by taking into consideration of the measurements of all the characters. That linear function is called linear discriminant function.

Let n_1 and n_2 random samples of observation vectors have been drawn independently from respective p-dimensional multinormal populations with mean vectors $\underline{u}_1$ and $\underline{u}_2$ and a common covariance matrix Σ.

A linear function (index) has to be constructed which discriminates between the populations by some measure of maximal separation. If $\bar{\underline{x}}_1$ and $\bar{\underline{x}}_2$ are the sample mean vectors and S is the pooled estimate of Σ, the coefficient vector $\underline{l}$ of the index $\underline{l}^1 \underline{x}$ has to be determined which gives the greatest squared critical ratio.

$$t^2(\underline{l}) = \frac{\underline{l}^1(\bar{\underline{x}}_1 - \bar{\underline{x}}_2)^2 \dfrac{n_1 n_2}{(n_1 + n_2)}}{\underline{l}^1 S \underline{l}} \qquad \text{..... (20.10)}$$

or which maximizes the absolute difference $\underline{l}^1(\bar{\underline{x}}_1 - \bar{\underline{x}}_2)$ in the average values of the index for the two groups subject to the constraint $\underline{l}^1 S \underline{l} = 1$.

The coefficient vector $\underline{l}$ is given by the homogeneous system of equations.

$$\frac{n_1 n_2}{n_1 + n_2}(\bar{\underline{x}}_1 - \bar{\underline{x}}_2)(\bar{\underline{x}}_1 - \bar{\underline{x}}_2) - \lambda S \underline{l} = 0 \qquad \text{.....(20.11)}$$

where $\quad \lambda = \max t^2(\underline{l})$

$$\frac{n_1 n_2}{n_1 + n_2}(\bar{\underline{x}}_1 - \bar{\underline{x}}_2)^1 S^{-1}(\bar{\underline{x}}_1 - \bar{\underline{x}}_2)$$

The rank of the coefficient matirx is $(p - 1)$, so that the system has only the single solution

$$\underline{l} = S^{-1}(\bar{\underline{x}}_1 - \bar{\underline{x}}_2)$$

The linear discriminant function is

$$Y = \underline{l}' \bar{\underline{x}}$$

$$= (\bar{\underline{x}}_1 - \bar{\underline{x}}_2)^1 S^{-1} \underline{x}$$

If the variances of the different components (characters) are nearly equal, the elements at $\underline{l}$ give the relative importance of the contribution of each response to the T^2 statistic, otherwise the multiplication of the i-th element by the standard deviation of its character will give comparable discriminant coefficients.

20.2.1 Classification for Two Groups

First the means of the two samples using discriminant function are given as

$$\bar{y}_1 = (\bar{\underline{x}}_1 - \bar{\underline{x}}_2)^1 S^{-1} \bar{\underline{x}}_1$$

$$\bar{y}_2 = (\bar{\underline{x}}_1 - \bar{\underline{x}}_2)^1 S^{-1} \bar{\underline{x}}_2$$

The mid-point of the above two means is

$$1/2\left(\overline{\underline{x}}_1 - \overline{\underline{x}}_2\right)' S^{-1}\left(\overline{\underline{x}}_1 - \overline{\underline{x}}_2\right)$$

Now, the individual with observation is assigned to the population I if

$$\left(\overline{\underline{x}}_1 - \overline{\underline{x}}_2\right)' S^{-1}\underline{x} > 1/2\left(\overline{\underline{x}}_1 - \overline{\underline{x}}_2\right)S^{-1}\left(\overline{\underline{x}}_1 + \overline{\underline{x}}_2\right) \qquad .....(20.12)$$

and to populations II if

$$\left(\overline{\underline{x}}_1 - \overline{\underline{x}}_2\right)' S^{-1} \leq \frac{1}{2}\left(\overline{\underline{x}}_1 - \overline{\underline{x}}_2\right)S^{-1}\left(\overline{\underline{x}}_1 + \overline{\underline{x}}_2\right) \qquad .....(20.13)$$

That is, the individuals are assigned to the group with the closer discriminant mean score.

Let

$$W = \underline{x}^{-1} S^{-1}\left(\overline{x}_1 - \overline{x}_2\right) - 1/2\left(\overline{x}_1 - \overline{x}_2\right)' S^{-1}\left(\overline{x}_1 + \overline{x}_2\right) \qquad .....(20.14)$$

$\underline{x}$ is assigned to population I if $W > 0$ and otherwise to population II. 'W' is called the Wald-Anderson classification statistic.

20.2.2 Classification for Several Groups

Here the methods of classification of an unknown observation to one of K populations $(K > 2)$ is considered. The independent groups are described by multinormal random variables with mean vectors, u_1, u_2, ... u_k and common covriance matrix Σ.

$$\overline{\underline{x}}_j \text{ and } S = \frac{1}{n-k}\sum_{j=1}^{k} A_j \text{ are the estimates of } \underline{\mu}_j \text{ and } \Sigma \text{ respectively.}$$

A_j is the sums of squares and products matrix of the j-th group. If $\underline{x}$ is the new observation to be classified then the linear discriminant function is

$$W_{ij} = \underline{x}^{-1} S^{-1}\left(\overline{x}_i - \overline{x}_j\right) - 1/2\left(\overline{x}_i + \overline{x}_j\right)' s^{-1}\left(\overline{x}_i - \overline{x}_j\right) \qquad .....(20.15)$$

$\underline{x}$ is to be assigned to population 'c' if $W_{ij} > 0$ for all $j \neq i$. $W_{ij} = W_{ji}$ and that any $(k-1)$ linearly independent W_{ij}'s form a basis for the complete set of the statistics if $(k-1) \leq p$. If $P < (k-1)$, the space of the W_{ij} will have rank p, and the classification rule can be defined in terms of p observations.

EXAMPLE:

Let $K = 3$ and $P \geq 2$. The distinct discriminant functions are

$$W_{12} = \underline{x}'S^{-1}\left(\overline{\underline{x}}_1 - \overline{\underline{x}}_2\right) - 1/2\left(\overline{\underline{x}}_1 + \overline{\underline{x}}_2\right)' s^{-1}\left(\overline{\underline{x}}_1 - \overline{\underline{x}}_2\right)$$

$$W_{13} = \underline{x}' S^{-1}\left(\overline{\underline{x}}_1 - \overline{\underline{x}}_3\right) - 1/2\left(\overline{\underline{x}}_1 + \overline{\underline{x}}_3\right)' S^{-1}\left(\overline{\underline{x}}_1 - \overline{\underline{x}}_3\right) \qquad .....(20.16)$$

$$W_{23} = \underline{x}' S^{-1}\left(\overline{\underline{x}}_2 - \overline{\underline{x}}_3\right) - 1/2\left(\overline{\underline{x}}_2 + \overline{\underline{x}}_3\right)' S^{-1}\left(\overline{\underline{x}}_2 - \overline{\underline{x}}_3\right)$$

Since $W_{23} = W_{13} - W_{12}$, W_{12} and W_{13} are only needed.

Classify $\underline{x}$ as from

(i)   Population I if $W_{12} > 0$ and $W_{13} > 0$. Since in W_{13} function also $W_{13} > 0$, if $\underline{x}$ is to be classified from population I.

(ii)   Population II if $W_{12} < 0$ and $W_{13} > W_{12}$ since $W_{23} > 0$. i.e., $W_{13} - W_{12} > 0$.

(iii) Population III if $W_{13} < 0$ and $W_{12} > W_{13}$ i.e., $W_{23} < 0$.

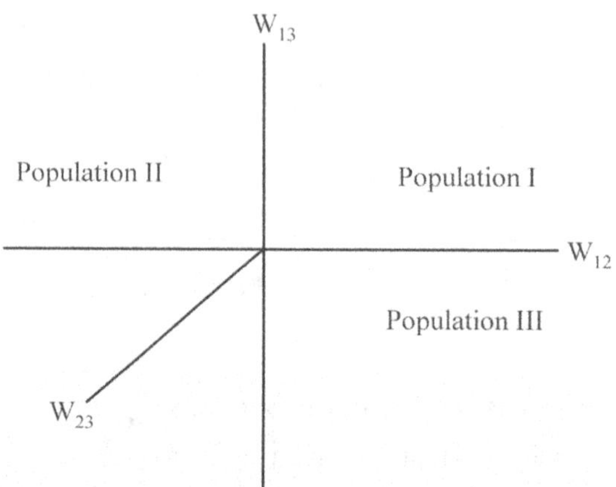

Fig. 20.1

EXAMPLE: Let $k = 3$ and $p = 1$, the three populations could be labeled so that $\overline{x}_1, < \overline{x}_2 < \overline{x}_3$. Classify X as from

(i)   Population I if $x < 1/2 \left(\underline{\overline{x}}_1 + \underline{\overline{x}}_2 \right)$

(ii)   Population II if $1/2 \left(\underline{\overline{x}}_1 + \underline{\overline{x}}_2 \right) \le x \le \left(\underline{\overline{x}}_2 + \underline{\overline{x}}_3 \right)$

(iii) Population III if $x > 1/2 \left(\underline{\overline{x}}_2 + \underline{\overline{x}}_3 \right)$

20.3 PRINCIPAL COMPONENT ANALYSIS

Suppose that the random variables x_1, x_2, ...x_p have a multivariate distribution with mean $\underline{u}$ and covariance matrix Σ.

The rank of Σ is $r \le p$ and the q largest characteristic roots

$\lambda_1 > \lambda_2 > ..... > \lambda_q$ of Σ are all distinct.

From this population, a sample of n independent observations has been drawn. The observations can be presented in the following n × p matrix form

$$X = \begin{bmatrix} x_{11} & x_{12} & \cdots & x_{1p} \\ x_{21} & x_{22} & \cdots & x_{2p} \\ . & .. & & . \\ . & . & & . \\ x_{n1} & x_{n2} & \cdots & x_{np} \end{bmatrix} \qquad .....(20.17)$$

Let S be the estimate of Σ. The information for the principal component analysis will be contained in S. If the observations recorded for different characters are in widely different units i.e., age in years, weight in kgs, biochemical measurements, linear function of original quantities would have no relevance and in place of it standardized variates and correlation matrix would be more meaningful. If the characters are measured in reasonably related units, covariance matrix would be more appealing. For example, i-th principal component explains the i-th largest portion of the total variance. The first principal component of the sample values of the characters x_1, x_2 ..... x_p is the linear function

$$y_1 = l_{11} x_1 + l_{21} x_2 + ... l_{p1}, x_p \qquad .....(20.18)$$

whose coefficients l_{i1}, are the elements of the characteristic vector associated with greatest characteristic root λ_1 of the sample covariance matrix S. The characteristic root λ_1 is interpretable as the sample variance, of y_1. The coefficients l_{11}, l_{21}, ... l_{p1} are chosen such that $\underline{l'}_1 \underline{l}_1 = 1$ where $l'_1 = (l_{11}, l_{21} ....$ $l_{p1})$. In the extreme case of x of rank 'l' the first principal component would explain all the variation in the multivariate system. If 85 percent of the variation in a system of p characters could be accounted for by a simple weighted average of the response values, it would appear that almost all the variation could be expressed along a single continuum rather than p-dimensional space. In such case, the coefficients of p characters would indicate the relative importance of each original character in the new derived component.

The second principal component is the linear function

$$y_2 = l_{12} x_1 + l_{22} x_2 + ... + l_{p2} x_p \qquad .....(20.19)$$

whose coefficients have been chosen subject to the constraints

that $\qquad \underline{l'_2} \, \underline{l}_2 = 1$ and $\underline{l'_1} \, \underline{l}_2 = 0$

The first constraint is to assure the uniqueness of the cofficients, while the second requires that l_1 and l_2 to be orthogonal which ensures that the variances of the successive components sum to the total variance of the responses (characters).

The j-th principal component of the sample of p-variate observations is the linear function

$$y_j = l_{1j}\, x_1 + l_{2j}\, x_2 + ... + l_{pj}\, x_j \qquad .....(20.20)$$

whose coefficients are the elements of the characteristic vector of the sample covariance matrix S corresponding to the j-th largest characteristic root λ_j.

$$\lambda_1 + \lambda_2 + ... + \lambda_p = \text{tr } S. \qquad .....(20.21)$$

The importance of j-th component is measured by

$$\frac{\lambda}{\text{trS}}$$

The correlation coefficient between i-th character and j-th

$$\text{component is } \frac{l_{ij}\ \sqrt{\lambda_j}}{s_i} \qquad .....(20.22)$$

where s_i is the estimate of the variance of i-th character and it is the i-th diagonal element of matrix 'S'.

20.3.1

Reddy (1991) used principal component analysis for grouping 20 districts in Andhra Pradesh with respect to area and productivity of paddy, jowar, sugarcane and groundnut. He considered data for 33 years from 1956-57 to 1988-89 and divided this period into two sub periods as 1956-59 to 1965-66 as sub period I (old technology) and 1966-67 to 1988-89 as sub period II (new technology). He obtained first two principal component vectors for each of the sub periods as well as total period and classified the districts into groups based on the coefficients of principal component vector.

EXAMPLE: The following is the correlation matrix obtained from the measurements of average ear length (x_1), average number of grains per ear head (x_2), average number of panickles (x_3) and average density per 100 grains (x_4) on 120 paddy hills. Obtain the four principal components by following the iterative procedure.

$$R = \begin{bmatrix} 1.00 & 0.85 & .075 & 0.82 \\ 0.85 & 1.00 & 0.91 & 0.65 \\ 0.75 & 0.91 & 1.00 & 0.74 \\ 0.82 & 0.65 & 0.74 & 1.00 \end{bmatrix} \qquad .....(20.23)$$

The second power of the matrix is

$$R^2 = \begin{bmatrix} 2.9574 & 2.9155 & 2.8803 & 2.7475 \\ 2.9155 & 2.9731 & 2.9385 & 2.6704 \\ 2.8803 & 2.9385 & 2.9382 & 2.6865 \\ 2.7475 & 2.6704 & 2.6865 & 2.6425 \end{bmatrix} \qquad .....(20.24)$$

A reasonable initial vector for the iterative process is chosen as $\underline{l}_0^1 =$

$[1,1,1,1]$

Thus $\underline{l}_0^1 R^2 = [11.5007, 11.4975, 11.4435, 10.7469]$

The standardized vector is obtained by dividing the elements of the $\underline{l}_0^1 R^2$ by highest element i.e., 11.5007 and the first solution is proportional to this standardized vector i.e.,

$$[1, 0.99972, 0.99503, 0.93446] \qquad\qquad .....(20.25)$$

$$10^{-1} R^4 = \begin{bmatrix} 3.3091 & 3.3091 & 3.2929 & 3.0909 \\ 3.3091 & 3.3105 & 3.2942 & 3.0901 \\ 3.2929 & 3.2942 & 3.2781 & 3.0753 \\ 3.0909 & 3.0901 & 3.0753 & 2.8880 \end{bmatrix} \qquad .....(20.26)$$

$$\underline{l}_0^1 \, 10^{-1} R^4 = (13.0020, 13.0039, 12.9405, 12.1443)$$

The standardized vector is obtained by dividing with 13.0039 which is the highest element i.e.,

$$(0.9999, 1, 0.9951, 0.9339) \qquad\qquad .....(20.27)$$

Since the vectors in eqs. (20.25) and (20.26) are equal up to second decimal place, the iteration process can be stopped.

The vector of direction cosines is obtained as

$$\frac{x_i}{\sqrt{x_i^2}} \qquad \text{for } i = 1, 2, ..., 4$$

where x_i is the i-th element in eq. (20.27). From eq. (20.27) the vector of direction cosines is obtained as

$$\underline{l}_1' = [0.5088, 0.5089, 0.5064, 0.4752]$$

The characteristic vector must satisfy the linear equation

$.5088 \,(1-\lambda_1) + 0.5089 \, r_{12} + 0.5064 \, r_{13} + 0.4752 \, r_{14} = 0$

i.e., $0.5088 \,(1-\lambda_1) + 0.5089 \times 0.85 + 0.5064 \times 0.75 + 0.4752 \times$

$0.82 = 0 \qquad \lambda_1 = 3.36$

The first characteristic root $\lambda_1 = 3.36$. The contribution of 1st characteristic root to its total variance is calculated as $3.36 / 4 \times 100 = 84$ percent where 4 is trS in the original correlation matrix R.

The residual matrix R_1 required in the extraction of the second principal component is found by substracting the following matrix from matrix R to obtain R_1

$$\lambda_1 \underline{l} \ \underline{l}'_1 = \begin{bmatrix} 0.8698 & 0.8700 & 0.8657 & 0.8124 \\ & 0.8702 & 0.8659 & 0.8125 \\ & & 0.8616 & 0.8086 \\ & & & 0.7587 \end{bmatrix} \qquad .....(20.28)$$

$$R_1 = R - \lambda_1 \underline{l} \ \underline{l}'_1 = \begin{bmatrix} 0.1302 & -0.0200 & -0.1157 & 0.0076 \\ & 0.1298 & 0.0441 & -0.1625 \\ & & 0.1384 & -0.0686 \\ & & & 0.2413 \end{bmatrix} \qquad .....(20.29)$$

The standardized vectors will be obtained for the reduced matrix R_1 by taking another initial vector. The same procedure will be followed for obtaining 2nd characteristic vector and characteristic root. These are given as

$$\underline{l}'_2 = [0.2263, -0.5231, -0.3837, 0.7266] \qquad .....(20.30)$$

and the characteristic root. $\lambda_2 = 0.40$

The contribution of this characteristic root to total variance = $0.40/4 \times 100 = 10$ percent.

The 3rd and 4th characteristic vectors and roots are obtained from the residual matrices R_2 and R_3 and are given as follows.

$$\underline{l}'_3 = [0.6936, 0.2232, -0.5812, -0.3623] \qquad .....(20.31)$$

$$\lambda_3 = 0.22$$

The contribution of this characteristic root to total variance = $\dfrac{0.22}{4} \times 100 = 5.5$ percent.

$$\underline{l}'_4 = [0.4569, -0.6463, 0.5085, 0.3390] \qquad .....(20.32)$$

$$\lambda_4 = 0.02$$

It can be seen that major contribution to total variance is from the 1st characteristic root i.e., 84 percent.

20.4 FACTOR ANALYSIS

In order to study the mutual relationship between the variables, a statistical technique is needed. Though principal component analysis fulfills this to some extent but it is only a transformation but not change in fundamental model. Further, in principal component analysis, the different components are not invariant under changes in the scales of the observations and also when to stop the process of extracting components so that sufficient variance has been achieved is not well defined and also there is no provision for sampling variation of the individual observations.

The main features of factor analysis are that it greatly facilitates identification of key traits from the mosaic of overlapping relationship and is capable of achieving scientific parismony by reducing a set of large number of variables for a convenient size of underlying factors (often called dimensions axes or vectors) which cannot be easily accomplished by any other analytical technique including the multiple regression analysis.

Through factor analysis some of the deficiencies in principal components analysis can be met by way of representing each observation as a linear function of a small number of unobservable common factor variables and a single latent variate. The common factors provide the covariances among the observable reponses, while the specific terms contribute only to the variances of their particular responses.

20.4.1   Model for Factor Analysis

Let $X_1, X_2, ..., X_p$ be p random variables which can be represented as follows

$$X_1 = \lambda_{11} Y_1 + \lambda_{12} Y_2 + \lambda_{1m} Y_m + e_1$$
$$\vdots \qquad \vdots \qquad \vdots \qquad \vdots \qquad\qquad .....(20.33)$$
$$X_p = \lambda_{pl} Y_1 + \lambda_{12} Y_2 + \lambda_{pm} Y_m + e_p$$

Where

$Y_j = j^{th}$ common factor variable.

λ_{ij} = Coefficient of jth common factor with i-th random variable.

e_i = Effect of i-th specific random variable

In the factor analysis λ_{ij} is called the loading of the i-th random (response) variable on the j-th common factor.

In matric notation, equation (20.33) can be written as

$$\underline{X} = \Lambda\, \underline{Y} + \underline{E} \qquad\qquad .....(20.34)$$

Where 　 $\underline{X}' = [X_1, X_2, ..., X_p]$

$\underline{Y}' = [Y_1, Y_2, ..., Y_m]$

$\underline{E}' = [e_1, e_2, ..., e_p]$

$$\Lambda = \begin{pmatrix} \lambda_{11} & \lambda_{12} & \cdots & \lambda_{1m} \\ \cdot & \cdot & & \cdot \\ \cdot & \cdot & & \cdot \\ \lambda_{pl} & \lambda_{p2} & .., & \lambda_{pm} \end{pmatrix} \qquad\qquad .....(20.35)$$

$\underline{Y}'$ is assumed to be independently and normally distributed with mean as 0 and variance – covariance matrix I. Similarly the elements $\underline{E}'$ are normally and independently distributed with mean as zero and var $(\underline{E}) = \psi$ where

$$\psi = \begin{pmatrix} \psi_1 & 0 & .....0 \\ 0 & \psi_2 & .....0 \\ 0 & 0 & \psi_p \end{pmatrix} \qquad .....(20.36)$$

The variance of the i-th response can be written as

$$\sigma_i^2 = \lambda_{i1}^2 + \lambda_{i2}^2 + ... + \lambda_{im}^2 + \Psi_i \qquad .....(20.37)$$

and the covariance of i-th and j-th response is

$$\sigma_{ij} = \lambda_{i1} \lambda_{j1} + \lambda_{i2} \lambda_{j2} + ... + \lambda_{jm} \lambda_{jm} \qquad ..... (20.38)$$

The relations (20.37) and (20.38) can be written in matrix form as

$$= \wedge \wedge' + \underline{\Psi} \qquad .....(20.39)$$

The diagonal elements can be written as

$$\sigma_i^2 - \Psi_i = \sum_{j=1}^{m} \lambda_{ij}^2 \qquad .....(20.40)$$

and which are called the communalities of the responses. The equation (20.40) can be described as the variance of i-th response and is the sum of the squares of common factor loadings and i-th specific variance (error variance).

20.4.2   Principal–Axes Method

The principal–axes method extracts the maximum amount of variance (i.e., the sum of squares of factor loadings is maximized on each factor) and gives the smallest possible residuals. The correlation matrix is converted into the smallest number of orthogonal factors by this method. This method also gives the unique solution for a given correlation matrix.

20.4.3   Iterative Method

The iterative method for obtaining factor loadings by principal axes method is given as follows in different steps.

1. Write the sum of each column of the correlation matrix (R) in the row labelled "S_1".

2. Divide each value in the row labelled 'S_1' by the highest value in that row and enter the results in the next row labelled "u_1".

3. Obtain the sum of the cross products of S_1 with each row of the correlation matrix and enter the results in the next row named as "S_2". For example, the first entry of row S_2 is the sum of the cross products of row 1 and row S_1; the second entry is the sum of the cross products of row 2 and row S_1 and so on.

4. Divide each value in row S_2 by the highest value in that row and enter the results in the next row named as "u_2".

5. Now compare the values in rows "u_1" and "u_2". If they agree to the desired degrees of accuracy (say upto four decimals) the iteration will be stopped and the factor loading may be computed. If the desired accuracy is not achieved, the iteraction is continued by squaring the correlation matrix "R".

6. Obtain R^2 matrix by multiplying R with itself.

7. Total the columns of matrix R^2 and enter the values in the row named as "S_2". These values should be the same as values in row "S_2" in the matrix R. Copy the values from row u_2 of matrix R into the next row in matrix R^2.

8. Sum the cross products of row S_2 with each row of matrix R^2 and enter the values in row S_3 of matix R^2. These values in S_3 will be same as the column totals of matrix R^4.

9. Divide each value in row S_3 by the highest value in that row and enter into the next row named as u_3.

10. Compare the values in rows u_2 and u_3 to the desired degree of accuracy and if they tally the iteration procedure is terminated. If not, the process has to be continued with R^4, R^8, etc., until two successive rows of u will agree to the desired degree of accuracy.

Computation of first factor loadings

1. Sum the cross products of each row of original correlation matrix R with final row of u, when the iteration procedure was terminated. The values are entered as row $V_{i1} = R\,u$

2. The loading for variable 'i' on the first factor is

$$\lambda_{i1} = \frac{V_{i1}}{\sqrt{\sum u V_{i1}}}$$

Where $\sum u V_{i1}$ is the sum of the cross products of the final row 'u' and row 'V_{i1}'.

Second the succeeding factors

1. The matrix of residuals $R^{(1)}$ is obtained from the original correlation matrix R as follows

$$r_{ij}^{(1)} = r_{ij} - \lambda_{i1}\lambda_{j1}$$

Where $r_{ij}^{(1)}$ denotes the first residual and λ_{i1} and λ_{j1} are the first factor loadings for i-th and j-th variables.

2. The process that was used for extracting loadings of 1st factor is repeated here for obtaining loadings of 2nd factor.

3. Compute $\qquad V_{i2} = r^{(1)} u$

Multiply each row of the first residual matrix by u and enter the results in the next row labeled v_{i2}.

4. The loadings of the second factor can be computed by the following formula as

$$\lambda_{i2} = \frac{V_{i2}}{\sqrt{\sum V_{i2} u}}$$

5. The next residual matrix is obtained in the same way as obtained for the its residual matrix for obtaining factor loadings of the 3rd factor. This process is continued till the values of the elements in the residual matrix is negligible.

6. It may be noted that in the principal axes method the sum of the cross products of any pair of factor loadings is zero.

20.4.4

Murthy (1986) used factor analysis for identifying the variables responsible for the performance of the Karnataka State Co-operative Marketing Federation Ltd. He found that the key physical indicators of the federation in the first dimension were Number of employees. Number of direct recruits, membership and Number of branches/depots.

The other key variables in the first dimension in respect of financial indicators of the Federation were : Establishment expenses, Total operating expenses, working capital, Total expenses and net working capital constituted the financial indicators of expenses.

Total sales, fertilizer sales, total purchases, fertilizer purchases, purchases of other commodities and sales of other commodities constituted the financial indicators pertaining to business transaction.

Total share capital, Government's share capital, share capital exclusing Government's share, Owned funds, Depreciation fund, Statutory reserve fund, Bad and doubtful reserve fund and other funds are the financial indicators belonging to share capital funds.

Networth, Fixed assets, Net value of fixed assets and long term investments accounted for the financial strength of the federation.

Total amount borrowed and interest on borrowings were the financial indicators reflecting borrowings.

He found that the above variables were positively associated among themselves while they were negatively associated with Net income (–0.097) and subsidy received from the Government (–0.305).

The first dimension explained 51.46 per cent of the total variation in the correlation matrix. He felt that from the point of overall performance of the Federation, Physical; and Financial indicators reflected the growth and strength of the Federation.

EXAMPLE: The example given in principal components analysis is considered here for illustration purpose. The correlation matrix is given as

$$R = \begin{pmatrix} 1 & 0.85 & 0.75 & 0.82 \\ 0.85 & 1 & 0.91 & 0.65 \\ 0.75 & 0.91 & 1 & 0.74 \\ 0.82 & 0.65 & 0.74 & 1 \end{pmatrix}$$

TABLE 20.6

Variable	1	2	3	4
1	1	0.85	0.75	0.82
2	0.85	1	0.91	0.65
3	0.75	0.91	1	0.74
4	0.82	0.65	0.74	1
S_1	3.42	3.41	3.40	3.21
u_1	1	0.9971	0.9942	0.9386
S_2	11.5007	11.4975	11.4435	10.7469
U_2	1	0.9997	0.9950	0.9345

Since the rows u_1 and u_2 in the Table 20.6 are tallying only upto 2nd decimal place, we proceed for computing R^2 matrix.

TABLE 20.7 (R^2)

	1	2	3	4
1	2.9574	2.9155	2.8803	2.7475
2	2.9155	2.9731	2.9385	2.6704
3	2.8803	2.9385	2.9382	2.6865
4	2.7475	2.6704	2.6865	2.6425
S_2	11.5007	11.4975	11.4435	10.7469
u_2	1	0.997	0.9950	0.9345
S_3	130.02034	130.03875	129.4057	121.44274
u_3	0.9999	1	0.9951	0.9339

Since u_2 and u_3 rows in Table 20.7 are agreeing only upto 3rd decimal, we proceed for obtaining R^4 matrix from R^2 matrix.

TABLE 20.8 (R^4).

Variable	1	2	3	4
1	33.09123	33.09106	32.92945	30.90920
2	33.09106	33.10528	32.94190	30.90051
3	32.92945	32.94190	32.78121	30.75314
4	30.90920	30.90051	30.75314	28.87989
S_3	130.02090	130.03875	129.4057	121.44274
u_3	0.9999	1	0.9951	0.9339
S_4	16620.628	16623.011	16542.061	155323.990
u_4	0.9999	1	0.9951	0.9339
V_{i1}	3.362023	3.362491	3.346111	3.140192
λ_{i1}	0.93	0.93	0.93	0.87

Since the rows u_3 and u_4 in Table 20.8 are tallying upto 4th decimal place, the iteration process is stopped and the factor loadings for the 1st factor can be computed as follows.

$$V_{i1} = R\, u_4, \quad \Sigma\ V_{i1}\, u_4 = 12.986517$$

$$\lambda_{i1} = \frac{V_{i1}}{\Sigma V_{i1}\ u_4}\sqrt{\Sigma V_{i1} u_{14}} = 3.603681$$

The factor loadings for the 1st factor are presented in Table 20.8 itself. The first residual correlation matrix $R^{(1)}$ is computed and presented in Table 20.9 for obtaining loadings of the 2nd factor. The elements of $R^{(1)}$ matrix are obtained as follows.

$$r_{ij}^{(1)} = r_{ij} - \lambda_{i1}\lambda_{j1}$$

for example
$$r_{11}^{(1)} = r_{11} - \lambda_{11}\lambda_{11}$$
$$= 1 - (.93)(.93) = 0.14$$
$$r_{12}^{(1)} * = r_{12} - \lambda_{11}\lambda_{21}$$
$$= 0.85 - (.93)(.93) = .01$$

Since the rows u_1 and u_2 in Table 20.9 are not tallying even upto 1st decimal place $R^{(1)2}$ matrix is computed and presented in Table 20.10.

TABLE 20.9 ($R^{(1)}$)

Variable	1	2	3	4
1	0.14	−0.01	−0.11	0.01
2	−.01	0.14	0.05	−0.16
3	−0.11	0.05	0.14	−0.07
4	0.01	−0.16	−0.07	0.24
S_1	0.03	0.02	0.01	0.02
u_1	1	0.67	0.33	0.67
S_2	0.0031	−0.0002	−0.0023	0.0012
u_2	1	−0.0645	−.7419	.3871

TABLE 20.10 ($R^{(1)2}$)

Variable	1	2	3	4
1	0.0319	−0.0099	−0.032	0.0131
2	−0.0099	0.0478	0.0263	−0.0644
3	−0.032	0.0263	0.0391	−0.0357
4	0.0131	−0.0644	−0.0357	0.0882
S_2	0.0031	−0.0002	−0.0023	0.0012
u_2	1	−0.0645	−0.7419	0.3871
S_3	0.00019	−0.00018	−0.00024	0.00024
u_3	0.7917	−0.7500	−1	1

Since u_2 and u_3 rows are not tallying in the Table 20.10 the iteration process is continued by computing $R^{(1)4}$.

TABLE 20.11 $(R^{(1)4})$

	1	2	3	4
1	0.002311	−0.002474	−0.00300	0.003353
2	−0.002474	0.007222	0.004901	−0.009827
3	−0.00300	0.004901	0.004519	−0.006657
4	0.003353	−0.009827	−0.006657	0.013373
S_3	0.00019	0.00018	−0.00024	0.00024
u_3	0.7917	−0.7500	−1	1
S_4	0.0000015	−0.0000026	−0.0000022	0.0000036
u_4	0.4164	−0.7222	−0.6111	1

Since u_3 and u_4 rows in Table 20.11 are not tallying upto 4 decimals the iteration is continued upto total 21 times with the help of computer and the factor loadings for the 2nd factor are obtained as

$$\lambda_{i2} = (0.14 - 0.33 - 0.24 - 0.46)$$

For finding out the factor loadings of the third factor, we obtain the residual matrix $R^{(2)}$ by using λ_{i2} and Table 20.9

After performing 6 iterations for the residual matrix $R^{(2)}$ with the help of computer, the factor loadings for the 3rd factor are obtained as follows

$$\lambda_{i3} = (0.32 - 0.10 - 0.27 - 0.17)$$

After obtaining residual matrix $R^{(3)}$ the factor loadings for the 4th factor obtained after 2 iterations with the help of computer are as follows.

$$\lambda_{i4} = (0.07 \quad -0.10 - 0.08 - 0.05)$$

Finally the factor loadings of all the four factor are presented in Table 20.12

TABLE 20.12 FACTOR LOADIGS

Variable	1	2	3	4	Total
1	0.93	0.14	0.32	0.07	
2	0.93	−0.33	0.10	−0.10	
3	0.93	−0.24	−0.27	0.08	
4	0.87	0.46	−0.17	−0.05	
Percentage to total variance	83.79	9.94	5.36	0.66	99.75

20.5 CANONICAL CORRELATIONS

In this section the correlations between two sets of variables which are drawn from multi dimensional population are considered.

Suppose there are r + s varieties (x_1', x_2')

where $\underline{X_1'} = (x_1,\ x_2,\ ...\ x_r),\ \underline{X_2'} = (x_1,\ x_2\ ...\ x_s)$

which are from multi dimensional population with covariance matrix, Σ and which is partitioned as

$$\Sigma = \begin{bmatrix} \Sigma_{11} & \Sigma_{12} \\ \Sigma_{21} & \Sigma_{22} \end{bmatrix} \qquad .....(20.41)$$

assuming that the elements of Σ are finite, (Σ) is of full rank (r + s) and the first p < min (r, s) characteristic roots of $\Sigma_{11}^{-1} \Sigma_{12} \Sigma_{22}^{-1} \Sigma_{12}'$ are distinct given that $(\Sigma_{21} = \Sigma_{12}')$.

From the multidimensional population, a sample of n observations have been drawn and the sample covariance matrix S is also partitioned as follows.

$$S = \begin{bmatrix} S_{11} & S_{12} \\ S_{12}' & S_{22} \end{bmatrix} \qquad .....(20.42)$$

To study the correlations between and within sets of (r + s) variates, there will be rs between sets and r(r–1)/2 + s(s–1)/2 within set correlations.

In order to reduce these correlations, the linear combinations of 1st set and linear combinations of 2nd set of variables will be formed and the correlations between these linear transformations will be computed and is given as follows.

$$l_1 = \underline{a_1'},\ x_1 \qquad m_1 = \underline{b_1'},\ x_2$$
$$l_2 = \underline{a_2'},\ x_1 \qquad m_2 = \underline{b_2'},\ x_2$$
$$\vdots \qquad\qquad \vdots \qquad\qquad .....(20.43)$$
$$l_q = \underline{a_q'},\ x_1 \qquad m_q = \underline{b_q'},\ x_2$$

Such that the sample correlation between l_1 and m_1 is greatest, the sample correlation between l_2 and m_2 is greatest among all linear combinations uncorrelated with l_1 and m_1 and so on, for all q = min (r, s) possible pairs.

If we introduce the new conditions

$$\underline{ai}\, S_{11}\, \underline{aj} = 0$$
$$\underline{bi}\, S_{22}\, \underline{bj} = 0$$
$$\underline{ai}\, S_{12}\, \underline{bj} = 0 \qquad .....(20.44)$$
$$\underline{aj}\, S_{12}\, \underline{bi} = 0$$

for $i \neq j$, the coefficients of i-th pair are given by the homogeneous linear equations

$$(S_{12} S_{12}^{-1} S_{12}' - c_i S_{11})\, \underline{a}_i = 0$$

$$(S_{12}' S_{11}^{-1} S_{12} - c_i S_{22})\, \underline{b}_i = 0 \qquad\qquad .....(20.45)$$

where ci is the i-th largest root of the determinantal equations

$$(S_{12} S_{22}^{-1} S_{12}' - \lambda S_{11}) = 0 \ \text{ or}$$

$$(S_{12}' S_{11}^{-1} S_{12} - \lambda S_{22}) = 0 \qquad\qquad .....(20.46)$$

and c_i is the squared product moment correlation of i-th linear transformations.

$$c_i = r_2\, l_i\, m_i \qquad\qquad .....(20.47)$$

$$= \frac{\left(\underline{a}_i'\, S_{12}\, \underline{b}_i\right)^2}{\left(\underline{a}_i'\, S_{11}\, \underline{a}_i'\right)\left(\underline{b}_i'\, S_{22}\, b_i\right)} \qquad\qquad .....(20.48)$$

we started with sample covariance matrix (20.42) and transformed original variables into l_i, m_i ($i = 1, 2, ... q$) with correlation matrix

$$\begin{bmatrix} 1 & 0 & \cdots & 1 & \sqrt{c_1} & \cdots & 0 \\ 0 & 1 & \cdots & 0 & 0 & \sqrt{c_2}\,.. & 0 \\ 0 & 0 & \cdots & 1 & 0 & ...... & \sqrt{c_q} \\ \sqrt{c_1} & 0 & \cdots & 0 & 1 & 0.. & \cdots & 0 \\ 0 & \sqrt{c_2} & \cdots & 0 & 0 & 1.. & \cdots & 0 \\ 0 & 0 & \cdots & \sqrt{c_q} & 0 & 0.. & \cdots & 1 \end{bmatrix}$$

Now it can be seen that all the correlations between the sets of original variables have been transformed into q canonical correlations.

In otherwords the equation (20.46) can also be written as

$$\left| S_{11}^{-1} S_{12}' S_{22}^{-1} S_{12}' - \lambda \right| = 0$$

$$\left| S_{22}^{-1} S_{12}' S_{11}^{-1} S_{12} - \lambda \right| = 0 \qquad\qquad .....(20.49)$$

The characteristic roots of one of the equation (20.49) would give the squares of the canonical correlations.

20.5.1

Kumari (1992) used canonical roots for classifying the districts of Andhra Pradesh with the help of indicators (i), percentage area (to the total cropped area) under the crop (in the district), (ii) percentage irrigated area under the crop (iii) crop yield ($kgha^{-1}$) iv) harvest price of the crop (Rs Qtl^{-1}) and

(v) total rainfall during the crop period (mm). She used two way analysis of variance technique with W as matrix of error sum of squares by considering 'districts' as one factor and 'years' as another factor. B is the matrix of sum of squares for beteween districts and C is the matrix of sum of squares between 'years'. L_j be the latent vector corresponding to the latent root λ_j of $(w^{-1} B)$ which satisfy the matrix equation.

$$[B - \lambda_j \, w] \, L_j = 0$$

or $\qquad (w^{-1} B - \lambda_j I) L_j = 0 \qquad\qquad .....(20.50)$

The vectors $(\underline{L}_1, \underline{L}_2, \, ... \, \underline{L}_r)$, $(r < p)$, where p is the number of indicators, are the canonical vectors and can be used to estimate the plane of the true treatment means; r (r = min (p, k) where k is the number of districts (treatments) is the dimension of the plane spanned by the true treatment means.

EXAMPLE: Reddy (1992) conducted an experiment on uniformity trial on cotton crop at Student's farm, College of Agriculture, APAU, R' Nagar, Hyderabad. The field was divided into 20 equal parts and a random sample of 5 plants were selected from each plot and the number of branches/plant (x_1), number of leaves/plant (x_2), height of the plant (x_3) were recorded for each plant at different stages of crop growth. The correlation coefficients between the above three parameters were worked out at 15 days and 105 days (harvesting stage) of the growth of the crop and presented in the matrix form as follows.

| | | *Days of growth* | | | | | |
| | | *15* | | | *105* | | |
Days of Growth		x_1	x_1	x_3	x_1	x_2	x_3
15	x_1	1.0000	0.4151	0.5040	0.7201	0.5500	0.5099
	x_2	0.4151	1.0000	0.3950	0.2102	0.5408	0.1584
	x_1	0.5040	0.3950	1.0000	0.5805	0.6090	0.7078
105	x_1	0.7201	0.2102	0.5805	1.0000	0.6658	0.6614
	x_2	0.5500	0.5408	0.6090	0.6658	1.0000	0.5357
	x_3	0.5099	0.1584	0.7078	0.6614	0.5357	1.0000

The above matrix can be partitioned as

$$R_{11} = \begin{bmatrix} 1.0000 & 0.4151 & 0.5040 \\ 1.4151 & 1.0000 & 0.3950 \\ 0.5040 & 0.3950 & 1.0000 \end{bmatrix}$$

$$R_{12} = \begin{bmatrix} 0.7201 & 0.5500 & 0.5099 \\ 0.2102 & 0.5408 & 0.1584 \\ 0.5805 & 0.6090 & 0.7078 \end{bmatrix}$$

$$R_{21} = R'_{12}$$

$$R_{22} = \begin{bmatrix} 1.000 & 0.6658 & 0.6614 \\ 0.6658 & 1.0000 & 0.5357 \\ 0.6614 & 0.5357 & 1.0000 \end{bmatrix}$$

$$R_{11}^{-1} = \begin{bmatrix} 1.4478 & -0.3706 & -0.5833 \\ -0.3706 & 1.2797 & -0.3187 \\ -0.5833 & -0.3187 & 1.4199 \end{bmatrix}$$

$$R_{22}^{-1} = \begin{bmatrix} 2.3448 & -1.0243 & -1.0021 \\ -1.0243 & 1.8500 & -0.3135 \\ -1.0021 & -0.3135 & 1.8308 \end{bmatrix}$$

Therefore

$$R_{11}^{-1} R_{12} R_{22}^{-1} R_{11}^{1} = \begin{bmatrix} 0.4249 & 0.0153 & 0.2317 \\ -0.0854 & 0.2757 & -0.0247 \\ 0.2776 & 0.1398 & 0.4684 \end{bmatrix}$$

using computer, the 1st character root of the above matrix is 0.6803 and its contribution to total variance is 58.25 percent. 2nd characteristic root is 0.3227 and its contribution to toal variance is 27.63 percent and 3rd characteristic root is 0.1650 and its contribution to total variance is 14.12 percent.

Therefore, the matrix with canomical correlations is

$$\begin{bmatrix} 1 & 0 & 0 & 0.8248 & 0 & 0 \\ 0 & 1 & 0 & 0 & 0.5681 & 0 \\ 0 & 0 & 1 & 0 & 0 & 0.4062 \\ 0.8248 & 0 & 0 & 1 & 0 & 0 \\ 0 & 0.5681 & 0 & 0 & 1 & 0 \\ 0 & 0 & 0.4062 & 0 & 0 & 1 \end{bmatrix}$$

PART V
ECONOMETRICS

CHAPTER 21

TWO-VARIABLE LINEAR MODEL

21.1 TWO-VARIABLE LINEAR MODEL

The two – variable linear model is given as

$$Yi = \alpha + \beta X_i + u_i \text{ for } i = 1, 2, \ldots, n$$

Where Y is dependent variable and X is independent variable or explanatory variable which influences the Y variable. u is called disturbance term or error term which may take positive or negative values. The assumptions for the u variable are

$$E(u_i) = 0 \quad \text{for} \quad \text{all } i$$

$$E(u_i u_j) = 0 \quad \text{for} \quad i \neq j, i, j = 1, 2, \ldots n$$

$$= \sigma_u^2 \quad \text{for} \quad i = j, i, j = 1, 2, \ldots n$$

α, β and σ_u^2 are unknown parameters. We would like to estimate these parameters using sample observations on X and Y. Also we want to test hypotheses about these parameters.

21.1.1 Least – squares estimators

We denote the sample observations of X and Y as

$$X_1, X_2 \ldots, X_n$$
$$Y_1, Y_2, \ldots, Y_n$$

then

$$\overline{X} = \frac{1}{n}\sum_{i=1}^{n} X_i, \quad \overline{Y} = \frac{1}{n}\sum_{i=1}^{n} Y_i$$

$\overline{X}$ and $\overline{Y}$ are the Arithmetic means of X and Y respectively.

The two – variable linear model with estimated values is given as

$$\hat{Y} = \hat{\alpha} + \hat{\beta}X \quad\quad\quad \ldots(21.1)$$

where $\hat{\alpha}$ and $\hat{\beta}$ are the estimates of α and β respectively.

$e_i = Y_i - \hat{Y}_i$ = residual or deviation value of Y_i from its estimated value.

$$\sum e_i^2 = f(\hat{\alpha}, \hat{\beta}) = \text{function of } \hat{\alpha}, \hat{\beta}.$$

The principle of least squares states that $\hat{\alpha}, \hat{\beta}$ should be obtained such that $\sum e_i^2$ should be as small as possible.

$$\sum_{i=1}^{n} e_i^2 = \sum_{i=1}^{n} (Y_i - \hat{Y}_i)^2$$

$$= \sum_{i=1}^{n} (Y_i - \hat{\alpha} - \hat{\beta} X_i)^2$$

So that differentiating partially, we have

$$\frac{\partial}{\partial \hat{\alpha}} \sum e_i^2 = -2 \sum_{i=1}^{n} (Y_i - \hat{\alpha} - \hat{\beta} X_i) = 0$$

$$\frac{\partial}{\partial \hat{\beta}} \sum e_i^2 = -2 \sum_{i=1}^{n} X_i (Y_i - \hat{\alpha} - \hat{\beta} X_i) = 0$$

Simplifying the above two equations, we get the standard form of normal equations as

$$\sum_{i=1}^{n} Y_i = n\hat{\alpha} + \hat{\beta} \sum_{i=1}^{n} X_i \qquad \qquad .....(21.2)$$

$$\sum_{i=1}^{n} X_i Y_i = \hat{\alpha} \sum_{i=1}^{n} X_i + \hat{\beta} \sum_{i=1}^{n} X_i^2$$

The equations (21.2) can be solved for obtaining the value of $\hat{\alpha}$ and $\hat{\beta}$. We can also find the estimates of α and β by using the following method. From equation (21.1), we can have

$$\overline{Y} = \hat{\alpha} + \hat{\beta} \overline{X} \qquad \qquad .....(21.3)$$

Subtract (21.3) from (21.1), we have

$$\hat{Y} - \overline{Y} = \hat{\beta} (X - \overline{X})$$

Let $\qquad \qquad x_i = (X_i - \overline{X}), \; y_i = (Y_i - \overline{Y}), \; \hat{y}_i = (\hat{Y}_i - \overline{Y})$

Then the above equation can be written as

$$\hat{y} = \hat{\beta} x_i \qquad \qquad .....(21.4)$$

The residual, e_i may be written as

$$e_i = y_i - \hat{y}_i = y_i - \hat{\beta} x_i$$

$$\sum_{i=1}^{n} e_i^2 = \sum_{i=1}^{n} (y_i - \hat{\beta} x_i)^2$$

Minimizing this expression, we get

$$\hat{\beta} = \frac{\displaystyle\sum_{i=1}^{n} x_i y_i}{\displaystyle\sum_{i=1}^{n} x_i^2} \qquad \qquad ....(21.5)$$

then

$$\hat{\alpha} = \overline{Y} - \hat{\beta}\overline{X} \qquad\qquad .....(21.6)$$

$$E(\hat{\beta}) = \beta, \qquad E(\hat{\alpha}) = \alpha$$

i.e., Expected values of $\hat{\beta}$ and $\hat{\alpha}$ are β and α respectively

$$V(\hat{\beta}) = \frac{\sigma_u^2}{\sum\limits_{i=1}^{n} x_i^2} \qquad\qquad .....(21.7)$$

$$V(\hat{\alpha}) = \frac{\sum\limits_{i=1}^{n} X_i^2}{n\sum\limits_{i=1}^{n} x_i^2} \qquad\qquad .....(21.8)$$

$$\text{Cov}\ (\hat{\alpha},\hat{\beta}) = E\left[(\hat{\alpha}, -\alpha)(\hat{\beta} - \beta)\right]$$

$$= \frac{-\overline{X}}{\sum\limits_{i=1}^{n} x_i^2} \qquad\qquad .....(21.9)$$

21.1.2   Maximum Likelihood estimators

The likelihood function for the sample is

$$L = \frac{1}{\left(\sigma_u^2 2\pi\right)^{n/2}}\ \text{Exp}\left[-\frac{1}{2\sigma_u^2}\sum_{i=1}^{n}(Y_i - \alpha - \beta X_i)^2\right]$$

$$\text{Log } L = -\frac{n}{2}\log 2\pi - \frac{n}{2}\log \sigma_u^2 - \frac{1}{2\sigma_u^2}\sum_{i=1}^{n}(Y_i - \alpha - \beta X_i^2)$$

Differentiating partially with respect to

α, β and σ_u^2, we have

$$\frac{\partial}{\partial\alpha}(\log L) = \frac{1}{\sigma_u^2}\sum(Y_i - \alpha - \beta X_i) = 0$$

$$\frac{\partial}{\partial\beta}(\log L) = \frac{1}{\sigma_u^2}\sum X_i(Y_i - \alpha - \beta X_i) = 0$$

$$\frac{\partial}{\partial\sigma_u^2}(\log L) = \frac{-n}{2\sigma_u^2} + \frac{1}{2\sigma_u^4}\sum(Y_i - \alpha - \beta X_i)^2 = 0$$

The first two equations reduces to

$$\sum Y_i = n\tilde{\alpha} + \tilde{\beta}\,\Sigma X_i$$

$$\sum X_i Y_i = \tilde{\alpha}\Sigma X_i + \tilde{\beta}\,\Sigma X_i^2$$

The third equation gives

$$\tilde{\sigma}_u^2 = \frac{1}{n}\sum_{i=1}^{n}(Y_i - \tilde{\alpha} - \tilde{\beta}X_i)^2 \qquad .....(21.10)$$

Since $\tilde{\alpha}$ and $\tilde{\beta}$ are identical to the least squares estimators $\hat{\alpha}$ and $\hat{\beta}$ and they are linear functions of u_i which have a multivariate normal distribution.

$$\frac{n\tilde{\sigma}_u^2}{\sigma_u^2} \quad \text{has a } \chi^2 - \text{distribution with } (n-2)$$

degrees of freedom and this statistic is distributed independently of $\hat{\alpha}$ and $\tilde{\beta}^2$

Test of significance of α:

$$t = \frac{(\hat{\alpha}-\alpha)}{\sqrt{\mathrm{Var}(\hat{\alpha})}} = \frac{(\hat{\alpha}-\alpha)\sqrt{n\Sigma x^2}}{\hat{\sigma}_u\sqrt{\Sigma X^2}} \qquad .....(21.11)$$

95% confidence limits of α:

$$\hat{\alpha} \pm t_{0.025}\frac{\hat{\sigma}_u\sqrt{\Sigma X^2}}{\sqrt{n\Sigma x^2}} \qquad .....(21.12)$$

Test of significance of β:

$$t = \frac{(\hat{\beta}-\beta)\sqrt{\Sigma x^2}}{\hat{\sigma}_u} = \frac{(\hat{\beta}-\beta)}{\sqrt{\mathrm{Var}(\hat{\beta})}} \qquad .....(21.13)$$

Confidence limits of β:

$$\hat{\beta} \pm t_{0.025}\frac{\hat{\sigma}_u}{\sqrt{\Sigma x_i^2}} \qquad .....(21.14)$$

21.1.3   Correlation Coefficient

Correlation is the relation between two variables X and Y. Correlation Coefficient is the amount of Correlation between two variables X and Y.

$$r = \frac{\Sigma x_i y_i}{n\, S_X S_Y} \qquad .....(21.15)$$

where $\qquad S_X = \sqrt{\dfrac{\Sigma x_i^2}{n}} \qquad$ and $\qquad S_Y = \sqrt{\dfrac{\Sigma y_i^2}{n}}$

$$r = \frac{\Sigma x_i y_i}{\sqrt{\Sigma x_i^2 \Sigma y_i^2}} \qquad \ldots\ldots(21.16)$$

$$\hat{\beta} = \frac{\Sigma xy}{\Sigma x^2} = r\frac{S_Y}{S_X} \qquad \ldots\ldots(21.17)$$

where $\hat{\beta}$ is the regression Coefficient of y on x and (21.17) gives the relationship between regression coefficient and Correlation coefficient and S_X and S_Y are the standard deviations of X and Y respectively.

21.1.4   Analysis of Variance

$$y_i = \hat{y}_i + e_i$$

$$\Sigma y_i^2 = \Sigma \hat{y}_i^2 + \Sigma e_i^2 + 2\Sigma \hat{y}_i e_i$$

But $\quad \Sigma \hat{y}_i e_i = \hat{\beta}\Sigma e_i x_i$

since $\quad \hat{\beta} x_i = \hat{y}_i \quad$ using $\quad$ (21.4)

$$= \hat{\beta}\Sigma x_i (y_i - \hat{\beta}x_i)$$

$$= \hat{\beta}\Sigma x_i y_i - \hat{\beta}\Sigma x_i^2$$

$$= \hat{\beta}\hat{\beta}\Sigma x_i^2 - \hat{\beta}\Sigma x_i^2 = 0$$

Thus $\quad \Sigma y_i^2 = \Sigma \hat{y}_i^2 + \Sigma e_i^2$

Hence $\quad \Sigma y_i^2 = $ Total sum of square

$\quad\quad \Sigma \hat{y}_i^2 = $ Regression sum of squares

$\quad\quad \Sigma e_i^2 = $ Error or Disturbance sum of squares.

$$\frac{\Sigma \hat{y}_i^2}{\Sigma y_i^2} = \frac{(\Sigma \hat{\beta}x_i)^2}{\Sigma y_i^2} = \frac{\hat{\beta}^2 \Sigma x_i^2}{\Sigma y_i^2} = r^2$$

$$r^2 = 1 - \frac{\Sigma e_i^2}{\Sigma y_i^2} \qquad \ldots\ldots(21.18)$$

$$r^2 = 1 \text{ when } \Sigma e_i^2 = 0$$

Thus the limits of r are –1 and 1.

Analysis of Variance Table

Source	d.f	S.S	M.S	F_{cal}
Regression	1	$Q_i = \Sigma \hat{y}_i^2$	$Q_i / 1$	$\dfrac{R.M.S}{E.M.S}$
Residual (Error)	$n-2$	$Q_2 = \Sigma e_i^2$	$Q_2 / (n-2)$	
Total	$n-1$	$Q_i + Q_2$		

where $\quad \Sigma \hat{y}_i^2 = \hat{\beta} \Sigma xy = Q_1 =$ explained sum of squares

$$Q_2 = \Sigma y^2 - Q_1 = \text{un explained sum of squares}$$

$$R.M.S. = Q_1 / 1 \, , \, E.M.S = Q_2 / (n-2)$$

Test of significance of r:

$$t = \frac{r \sqrt{n-2}}{\sqrt{1-r^2}} = \frac{\hat{\beta} S_x \sqrt{n-2}}{S_y \sqrt{1-r^2}} \qquad .....(21.19)$$

$$= \frac{\hat{\beta} \sqrt{\Sigma x^2}}{\sqrt{\dfrac{\Sigma y^2 (1-r^2)}{(n-2)}}} = \frac{\hat{\beta} \sqrt{\Sigma x^2}}{\sigma_u}$$

where $\quad \hat{\sigma}_u = \dfrac{\Sigma e_i^2}{n-2} = \dfrac{\Sigma y^2 (1-r^2)}{n-2} \qquad .....(21.20)$

since $\quad r^2 = 1 - \dfrac{\Sigma e_i^2}{\Sigma y_i^2}$

This is exactly same as testing of $\beta = 0$.

Thus correlation coefficient, regression coefficient and analysis of Variance testing methods are same for testing the hypothesis of no relationship between Variables X and Y.

21.1.5   Prediction

Predicting the mean value of Y for a given value of X say X_0 which may or may not lie within the range of sample observations X_1 to X_n. Our prediction may be point or interval estimation.

$$\hat{Y} = \hat{\alpha} + \hat{\beta} X_0$$

where $\hat{\alpha}$ and $\hat{\beta}$ are the least squares estimators of α and β respectively.

Confidence Interval: The upper and lower limits of $E(Y_0 / X_0)$ are

$$(\hat{\alpha} + \hat{\beta} X_0) \pm t_{n-2} \sqrt{\frac{1}{n} + \frac{(X_0 - \overline{X})^2}{\Sigma x_i^2}} \qquad .....(21.21)$$

where Y_0 is the mean value of Y for a given value of X_0.

$$t_{(n-2)} = \text{tabulated of t with } (n-2) \text{ degrees of freedom.}$$

21.2 THE GENERAL LINEAR MODEL

$$Y_i = \beta_1 + \beta_2 X_{2i} + \beta_3 X_{3i} + \ldots + \beta_k X_{ki} + u_i \qquad \ldots(21.22)$$

The β coefficients and parameters of the u distribution are unknown, and our problem is to obtain estimates of these unknowns.

In matrix notation for n = equation is

$$\underline{Y} = X\underline{\beta} + \underline{u} \qquad \ldots(21.23)$$

where

$$\underline{Y} = \begin{bmatrix} Y_1 \\ Y_2 \\ Y_n \end{bmatrix}, \quad X = \begin{bmatrix} 1 & X_{21} & \cdots & X_{k1} \\ 1 & X_{22} & & X_{k2} \\ 1 & \vdots & & \vdots \\ 1 & X_{2n} & & X_{kn} \end{bmatrix}$$

$$\underline{\beta} = \begin{bmatrix} \beta_1 \\ \beta_2 \\ \beta_k \end{bmatrix}, \quad \underline{u} = \begin{bmatrix} u_1 \\ u_2 \\ \vdots \\ u_n \end{bmatrix} \qquad \ldots(21.24)$$

The important assumption is

$$E(\underline{u}) = 0$$

$$E(\underline{u}\underline{u}^1) = \sigma^2 I_n$$

X is a set of fixed numbers

X has rank $k < n$

The first assumption states that $E(u_i) = 0$ for all i i.e. the u_i are random variables with zero expectation.

$$E(\underline{u}\,\underline{u}^1) = \begin{bmatrix} E(u_1^2) & E(u_1 u_2) & \cdots & E(u_1 u_n) \\ E(u_2 u_1) & E(u_2^2) & \cdots & E(u_2 u_n) \\ \vdots & & & \\ E(u_n u_1) & E(u_n u_2) & \cdots & E(u_n^2) \end{bmatrix}$$

$$= \begin{bmatrix} \sigma^2 & 0 & 0 \\ 0 & \sigma^2 & 0 \\ 0 & 0 & \sigma^2 \end{bmatrix}$$

The terms on the main diagonal show that $E(u_i^2) = \sigma^2$ for all i, i.e., u_i have constant variance σ^2, which property referred to a 'homoscedsticity.'

Least squares Estimators:

$$\hat{\underline{\beta}} = \left[\hat{\beta}_1, \hat{\beta}_2, \ldots, \hat{\beta}_k\right]$$

$$\underline{Y} = X\hat{\underline{\beta}} + \underline{e}$$

where $\underline{e}$ denotes is the column vector of n residuals $(\underline{Y} = X\hat{\beta})$

$$\sum_{i=1}^{n} e_i^2 = \underline{e}^1\underline{e} = (\underline{Y} - X\hat{\beta})'(\underline{Y} - X\hat{\beta})$$

$$= \underline{Y}^1\underline{Y} - 2\hat{\underline{\beta}}'X'\underline{Y} + \hat{\underline{\beta}}'X'X\hat{\underline{\beta}}$$

since $\quad \hat{\beta}'X'Y = Y'X\hat{\underline{\beta}} \quad = $ scalar.

$$\frac{\partial}{\partial\hat{\underline{\beta}}}(\underline{e}'\underline{e}) = -2X'Y + 2X'X\hat{\underline{\beta}} = 0$$

$$X'X\hat{\underline{\beta}} = X'\underline{Y}$$

$$\hat{\underline{\beta}} = (X'X)^{-1}X'\underline{Y} \qquad\qquad \ldots\ldots(21.25)$$

To establish the mean and variance of $\hat{\underline{\beta}}$ we substitute (21.23) into (21.25)

$$\hat{\underline{\beta}} = (X'X)^{-1}X'\left[X\underline{\beta} + \underline{u}\right]$$

$$= \underline{\beta} + (X'X)^{-1}X'\underline{u} \qquad\qquad \ldots\ldots(21.26)$$

$$E(\hat{\underline{\beta}}) = E(\beta) + E\left[(X'X)^{-1}X'\underline{u}\right]$$

$$= \underline{\beta} + 0 \quad \because E(u) = 0$$

Now consider

$$E\left[(\hat{\underline{\beta}} - \underline{\beta})(\hat{\underline{\beta}} - \underline{\beta})'\right]$$

$$= \begin{bmatrix} E(\hat{\beta}_1 - \beta_1)^2 & E(\hat{\beta}_1 - \beta_1)(\hat{\beta}_2 - \beta_2) & \cdots & E(\hat{\beta}_1 - \beta_1)(\hat{\beta}_k - \beta_k) \\ E(\hat{\beta}_2 - \beta_2)(\hat{\beta}_1 - \beta_1) & E(\hat{\beta}_2 - \beta_2)^2 & \cdots & E(\hat{\beta}_2 - \beta_2)(\hat{\beta}_k - \beta_k) \\ \vdots & & & \\ E(\hat{\beta}_k - \beta_k)(\hat{\beta}_1 - \beta_1) & E(\hat{\beta}_k - \beta_k)(\hat{\beta}_2 - \beta_2) & \cdots & E(\hat{\beta}_k - \beta_k)^2 \end{bmatrix}$$

$$\ldots\ldots(21.27)$$

$$E(\hat{\beta}_i - \beta_i)^2 \text{ is the variance of } \hat{\beta}_i$$

$$E(\hat{\beta}_i - \beta_i)(\hat{\beta}_j - \beta_j) \text{ is the covariance } \hat{\beta}_i \text{ and } \hat{\beta}_j.$$

$$(\hat{\underline{\beta}} - \underline{\beta}) = (X'X)^{-1}X'\underline{u} \qquad \qquad \dots. (21.28)$$

Thus $$\text{Var } (\hat{\underline{\beta}}) = E\left[(\hat{\underline{\beta}} - \underline{\beta})(\hat{\underline{\beta}} - \underline{\beta})'\right]$$

$$= E\left[(X'X)^{-1}X'\underline{u}\underline{u}'X(X'X)^{-1}\right]$$

$$= (X'X)^{-1}X'E(\underline{u}\underline{u}')X(X'X)^{-1}$$

$$= \sigma^2(X'X)^{-1} \qquad \qquad \dots..(21.29)$$

Since σ^2 is a scalar can be moved in front of X' .

21.3 AUTO CORRELATION

One of the Crucial assumptions of the linear model is that serial independence of the disturbance term implied in

$$E(UU') = \sigma^2 I \qquad \qquad \dots..(21.30)$$

which gives

$$E(\, u_t u_{t+s}) = 0 \text{ for all t and for all s} \neq 0$$

It implies that succession disturbances are drawn independently of previous values, though of course finite samples would show non – zero correlations between successive values, if it were possible to measure the disturbances.

There are, however, circumstances in which the assumptions of a serially independent disturbance term may not be very plausible. For example, one may make an incorrect specification of the form of relationship between the variables. Suppose we specify a linear relation between Y and X when the true relation is say quadratic. If there is any serial correlations in X values then we shall have serial correlation in the composite disturbance term. If the omitted variables tend to move in phase, then there is a real possibility of an auto correlated disturbance term. A disturbance term may also contain a component of measurement error in the explained variable. This too may be a source of sireal correlation in the composite disturbance.

In general, autocorrelation in the disturbance term may occur due to incorrect relationship, omitted variables & measurement errors.

Let $$Yt = \alpha + \beta x_t + u_t \qquad \qquad \dots..(21.31)$$

where we assume that the disturbance u_t follows

first – order auto – regressive scheme

$$u_t = \rho u_{t-1} + E_t \qquad\qquad .....(21.32)$$

were $|\rho| < 1$ and $\in_t$ satisfies the assumptions

$$E(\in_t) = 0$$

$$E(\in_t \in_{t+s}) = \sigma_\in^2, \, s = 0,$$

$$= 0 \, s \neq 0 \text{ for all } t$$

we then have

$$u_t = \rho u_{t-1} + \in_t$$
$$= \rho(\rho u_{t-2} + \in_{t-1}) + \in_t$$
$$= E_t + \rho \in_{t-2} + \rho^2 \in_{t-2} + ...$$

i.e.

$$u_t = \sum_{r=0}^{\infty} \rho^r \in_{t-r}$$

$$E(u_t) = 0$$

since

$$E(\in_t) = 0 \text{ for all } t$$

Further more

$$E(u_t^2) = E(\in_t^2) + \rho^2 E(\in_t^2) + \rho^4 E(\in_{t-2}^2)$$

Since the $\in$ are serially independent, and so

$$E(u_t^2) = \left(1 + \rho^2 + \rho^4 + ...\right)\sigma_\in^2$$

Thus

$$\sigma_4^2 = \frac{\sigma_\varepsilon^2}{1 - \rho^2} \text{ for all } t$$

$$E(u_t u_{t-1}) = E(\in_t + \rho \in_t + \rho^2 \in_{t-2} + ...)$$
$$(\in_{t-1} + \rho \in_{t-2} + \rho^2 \in_{t-3} + ...)$$
$$= E\left[(\varepsilon_t + \rho(\varepsilon_{t-1} + \rho\varepsilon_{t-2} +))\right](\varepsilon_{t-1} + \rho\varepsilon_{t-2} + \rho^2\varepsilon_{t-3})$$
$$= \rho E\left[\varepsilon_{t-1} + \rho t_{t-2} + ... +)^2\right]$$
$$= \rho\sigma_u^2$$

Similarly

$$E(u\, u_{t-2}) = \rho^2 \sigma_u^2$$

In general

$$E(u_t u_{t-s}) = \rho^s \sigma_u^2 \, (s \neq 0) \qquad\qquad .....(21.33)$$

So that relation (21.29) does not satisfy the assumption of a serially independent disturbance term.

(21.33) relation may be rewritten to give

$$\frac{E(u_t u_{t-s})}{\sigma_u^2} = \rho^s \qquad\qquad .....(21.34)$$

The LH of this expression define the sth auto correlation coefficients of the u series. The auto correlation coefficient of zero order of any series is simply unity, and for a random series all coefficients of higher order will be zero.

Consequences of Auto correlated Disturbances:

1. The sampling variances of the estimates of α and β may be unduly large compared with those achievable by a slightly different method of estimation.

2. If we apply the usual least-squares formulas for the sampling variances of the regression coefficients, we are likely to obtain a serious underestimate of the variances.

3. We shall obtain in efficient prediction, i.e., predication with needlessly large sampling variances.

21.4 THE DURBIN – WATSON d STATISTIC:

Let $Z_t(t=1,2,...n)$ denote the residuals from a fitted least squares regression. We then define

$$d = \frac{\sum_{t=2}^{n}(Z_t - Z_{t-1})^2}{\sum_{t=1}^{n}Z_t^2} \qquad .....(21.35)$$

Exact significance levels for d are not available but Durbin and Watson have tabulated lower and upper bounds d_L and d_U for various values of n and k (k = no. of explanatory variables).

If $d < d_L$ reject the hypothesis of random disturbances in favor of that of positive auto correlation.

If $d > d_U$, do not reject the hypothesis

If $d_L < d < d_u$ the test is inconclusive and further observations would ideally be required.

21.5 MULTICOLLINEARITY

This arises when some or all of the explanatory variables in a relation are so highly correlated one with another that it becomes very difficult to disentangle their separate influences and obtain a reasonably precise estimate of their relative effects.

Case(i): Let a linear relation between Y, X_2 and X_3

be $\qquad Y_t = \beta_1 + \beta_2 X_{2t} + \beta_3 X_{3t} + \varepsilon_t \qquad .....(21.36)$

With n sample observations on X's we estimate the parameters in (21.36).

$$\text{Let} \quad \underline{Y} = \begin{bmatrix} y_1 \\ y_2 \\ \vdots \\ y_n \end{bmatrix}, \quad X = \begin{bmatrix} x_{21} & x_{31} \\ x_{22} & x_{32} \\ \vdots & \vdots \\ x_{2n} & x_{3n} \end{bmatrix}, \quad \hat{\underline{\beta}} = \begin{bmatrix} \hat{\beta}_2 \\ \hat{\beta}_3 \end{bmatrix}$$

Here the variables x, y are expressed in deviation from their mean.

For ex: $\qquad x_{21} = \left(x_{21} - \overline{x_2}\right), \quad y_1 = \left(y_1 - \overline{y}\right) \qquad$ etc.

The least squares estimates are

$$\hat{\underline{\beta}} = (X^1 X)^{-1} X^1 \underline{Y}$$

$$\text{Var}\ (\hat{\beta}) = \sigma^2 (X^1 X)^{-1}$$

This estimation procedure breaks down if

$$\left| X^1 X \right| = 0$$

Suppose x_2 and x_3 were connected by linear relation.

$$X_{2t} = K_2 + K_3 X_{3t} \qquad \text{for all t}$$

then $\qquad X_{2t} = K_3 X_{3t}$

and $\qquad (X'X) = \begin{bmatrix} \Sigma x_2^2 & \Sigma x_2 x_3 \\ \Sigma x_2 x_3 & \Sigma x_3^2 \end{bmatrix} = \begin{bmatrix} K_3^2 \Sigma x_{3t}^2 & K_3 \Sigma x_3^2 \\ K_3 \Sigma x_3^2 & \Sigma x_3^2 \end{bmatrix}$

$$= \Sigma x_3^2 \begin{bmatrix} K_3^2 & K_3 \\ K_3 & 1 \end{bmatrix}$$

$$\left| X^1 X \right| = 0$$

This is the case of perfect multicollinearity where the two explanatory variables are connected by exact linear relation. Therefore it is not possible to estimate the separate influences of X_2 and X_3 and our estimation method breaks down.

Case (ii): In this case the values of the explanatory variables in the sample are highly correlated but not perfectly correlated.

In the three – Variable case

$$\text{Var}\ \left(\hat{\beta}_2\right) = \frac{\sigma^2}{\Sigma x_{2.3}^2}, \quad \text{Var}\ \left(\hat{\beta}_3\right) = \frac{\sigma^2}{\Sigma x_{3.2}^2} \qquad \qquad .....(21.37)$$

where $x_{2.3} = x_2 - b_{23}x_3$, $x_{3.2} = x_3 - b_{32}x_2$

In the case of high correlation between X_2 and X_3, and $\Sigma x_{2.3}^2 \ \Sigma x_{3.2}^2$ would be very small and therefore Var $(\hat{\beta}_2)$ and Var $(\hat{\beta}_3)$ would become high. Hence the estimates of separate effects of X_2 and X_3 will be highly correlated.

Case (iii): If multicollinearity is serious, then the remedy lies in obtaining new data which will break the multicollinearity dead lock. For example for data on income and prices there will be correlation between these two variables. The use of cross-section budget data gives a wide range of income variation, which enables precise determination of the income coefficient which can be employed in time series analysis. This combined use of time – series and cross section data is now conventional practice of demand studies.

However if forecasting is a primary objective, then intercorrelation between explanatory variables may not be that serious as it may reasonably be expected to continue in future.

21.6 LAGGED VARIABLES

Sometime the relationship exists in which lagged values of the dependent variables appear on the right hand side of the relation.

For example,

$$Y_t = \alpha + \beta Y_{t-1} + \varepsilon_t \qquad\qquad .....(21.38)$$

Where ε values are assumed to be serially independent. One of the conditions for the applications of simple least squares in the case of a single explanatory variable is

$$E\left\{\varepsilon_t\left[X_{t+r} - E(X_{t+r})\right]\right\} = 0 \qquad\qquad .....(21.39)$$

$$\text{for all} \qquad t$$
$$\text{for } r = \ldots -1, 0, 1, \ldots$$

The (21.39) implies that ε_t should be independent of future values X_{t+1}, X_{t+2} etc as well as of current and past values.

In (21.38), $X_t = Y_{t-1}$ so that ε_t influences Y_t since $Y_t = X_{t+1}$. Thus ε_t is independent of X_t, X_{t-1}, … but it is not independent of X_{t+1}, X_{t+2} etc.

This causes violation of assumption of independence between disturbance term and explanatory variable. Due to this violation, least squares estimators will be biased even though disturbance term follows normal distribution. These estimators will have desirable asymptotic properties of consistency and efficiency.

Suppose we have observations Y_1, Y_2, …, Y_n generated by the relation

$$Y_t = \alpha + \beta Y_{t-1} + \varepsilon_t$$

Where ε_t is assumed to be normally and independently distributed with zero mean and constant variance σ^2.

Let us assume Y_1 is a fixed number and the remaining values depend upon the $(n-1)$ values $\in_2, \in_3, \ldots \in_n$ that come from the postulated probability distribution of ε.

Maximizing the likelihood function with respect to α and β is the same as minimizing

$$\sum_{i=2}^{n}(Y_i - \alpha - \beta Y_{i-1})^2$$

This will give the least squares estimates $\hat{\alpha}$ and $\hat{\beta}$ by solving normal equations

$$\sum_{i=2}^{n}Y_i = (n-1)\,\hat{\alpha} + \hat{\beta}\sum_{i=2}^{n}Y_{i-1}$$

$$\sum_{i=2}^{n}Y_i\,Y_{i-1} = \hat{\alpha}\sum_{i=2}^{n}Y_{i-1} + \hat{\beta}\sum_{i=2}^{n}Y_{i-1}^2$$

Thus for the special case of fixed Y_1, the least squares estimates are maximum likelihood estimates and have the large sample properties of consistency and efficiency.

We can derive comfort from large sample properties of estimators but least squares estimates may be seriously biased in small samples.

Let $\qquad Y_t = \alpha + \beta Y_{t-1} + \varepsilon_t \qquad\qquad \ldots..(21.40)$

The least squares estimate of β is

$$\hat{\beta} = \frac{\sum_{t=2}^{n}(Y_t - \bar{Y})(Y_{t-1} - \bar{\bar{Y}})}{\sum_{t=2}^{n}(Y_{t-1} - \bar{\bar{Y}})^2}$$

where $\qquad \bar{Y} = \dfrac{1}{n-1}(Y_2 + \ldots + Y_n)$

$$\bar{\bar{Y}} = \frac{1}{n-1}(Y_1 + Y_2 + \ldots + Y_{n-1})$$

From (21.40)

$$\bar{Y} = \alpha + \beta\bar{\bar{Y}} + \bar{\varepsilon}$$

$$Y_t - \bar{Y} = \beta(Y_{t-1} - \bar{\bar{Y}}) + (\varepsilon_t - \bar{\varepsilon})$$

and $\qquad \hat{\beta} = \beta + \dfrac{\sum_{t=2}^{n}(Y_{t-1} - \bar{\bar{Y}})\varepsilon_t}{\sum_{t=2}^{n}(Y_{t-1} - \bar{\bar{Y}})^2} \qquad\qquad \ldots..(21.41)$

$E(\hat{\beta})$ will be obtained by taking the expected value of the last term in the right hand side of (21.41). The evaluation of this term is difficult since the numerator and denominator depend upon the same random variable. However the sign of this term depends upon the sign of numerator since denominator is always positive.

From (21.40), we have

$$Y_1$$
$$Y_2 = \alpha + \beta Y_1 + \varepsilon_2$$
$$Y_3 = \alpha + \alpha\beta(\alpha + \beta Y_1 + \varepsilon_2) + \varepsilon_3$$
$$= \alpha + \alpha\beta + \beta^2 Y_1 + \beta\varepsilon_2 + \varepsilon_3$$
$$Y_4 = \alpha + \beta(\alpha + \alpha\beta + \beta^2 Y_1 + \beta\varepsilon_2 + \varepsilon_3) + \varepsilon_4$$
$$= \alpha + \alpha\beta + \alpha\beta^2 + \beta^3 Y_1 + \beta^2\varepsilon_2 + \beta\varepsilon_3 + \varepsilon_4$$
$$\vdots \qquad \vdots \qquad \vdots$$
$$Y_{n-1} = \alpha + \alpha\beta + \text{-----} + \alpha\beta^{n-2} + \beta^{n-2}Y_1 + \beta^{n-3}\varepsilon_2 + \ldots + \epsilon_{n-1}$$

Thus $\overline{\overline{Y}} = $ terms in α, β and Y_1 plus $B_2\varepsilon_2 + B_3\varepsilon_3 + \ldots + B_{n-1}\varepsilon_{n-1}$

where B's are averages of powers of β.

Thus if $\beta > o$, all the B's are positive.

Assuming ε's are independent.

$$E\left[\sum_{t=2}^{n}(Y_{t-1} - \overline{\overline{Y}})\varepsilon_t\right] = -\sigma^2\sum_{i=2}^{n-1}B_i$$

which gives negative bias if $\beta > 0$

Recently lagged variables have been used more often in estimating relationships. Suppose one variable is effected to movements of another variable with a lagged response. The equation for this relationship can be written as

$$Y_t = \alpha_0 X_t + \alpha_1 X_t + \alpha_2 X_{t-2} + \ldots + u_t \qquad \ldots(21.42)$$

It is reasonable to expect more remote values of X can exercise less influence than the recent X values. We can assume that α coefficients decline exponentially i.e.

$$\alpha_j = \alpha\lambda^j \qquad \text{where} \qquad \begin{array}{l} j = 1, 2\ldots \\ 0 \leq \lambda < 1 \end{array} \qquad \ldots(21.43)$$

Equation (21.42) can be written as

$$Y_{t+1} = \alpha X_{t+1} + \alpha\lambda X_t + \alpha\lambda^2 X_{t-1} + \ldots + u_{t+1}$$
$$\lambda Y_t = \alpha\lambda X_t + \alpha\lambda^2 X_{t-1} + \ldots + \lambda u_t$$

Now $\qquad Y_{t+1} = \lambda Y_t + \alpha X_{t+1} + (u_{t+1} - \lambda u_t) \qquad \ldots(21.44)$

or $\quad \Delta\ Y_t = Y_{t+1} - Y_t$

$$= \alpha X_{t+1} - (1 - \lambda)\ Y_t + (u_{t+1} - \lambda u_t) \qquad .....(21.45)$$

$$= \alpha X_{t+1} - \gamma Y_t + (u_{t+1} - \lambda u_t)$$

where $\quad \gamma = (1 - \lambda)$

The advantage of (21.44) over (21.41) is that it consists of only two explanatory variables so that multicollinearity problem can be overcome.

21.7 HETEROSCEDASTICITY

One of the assumptions of the linear regression model was

$$E(\underline{u}\underline{u}') = \sigma^2 I$$

That is the disturbances were independently distributed with a constant variance σ^2. The condition of constant variance is called 'homoscedasticity' and its opposite is called 'heteroscedasticity'.

On the usual matrix notation, the model is

$$\underline{Y} = X\underline{\beta} + \underline{u} \qquad .....(21.46)$$

$$E(\underline{u}\underline{u}') = \sigma^2 \begin{bmatrix} \dfrac{1}{\lambda_1} & 0 & \cdots & 0 \\ 0 & \dfrac{1}{\lambda_2} & \cdots & 0 \\ & & \dfrac{1}{\lambda} & \\ 0 & 0 & \cdots & \dfrac{1}{\lambda_n} \end{bmatrix} \qquad .....(21.47)$$

Let $\quad \Lambda = \begin{bmatrix} \sqrt{\lambda_1} & 0 & \cdots & 0 \\ 0 & \sqrt{\lambda_2} & \cdots & 0 \\ 0 & 0 & \cdots & \sqrt{\lambda_n} \end{bmatrix} \qquad .....(21.48)$

Premultiply Λ in the equation (21.46), we have

$$\Lambda\underline{Y} = \Lambda \times \underline{\beta} + \Lambda\underline{u} \qquad .....(21.49)$$

The variance – covariance matrix for the transformed disturbance $\Lambda\underline{u}$ is

$$E(\Lambda\underline{u}\underline{u}'\Lambda') = \sigma^2\ I \qquad .....(21.50)$$

So that we may apply simple least squares directly to relation (21.49)

$$\hat{\beta} = (X'\Lambda^2 X)^{-1} X'\Lambda\underline{Y} \qquad .....(21.51)$$

With variance – covariance matrix

$$\text{Var} \quad (\hat{\underline{\beta}}) = \sigma^2 (X^1 \Lambda^2 X)^{-1} \qquad \qquad \dots\dots(21.52)$$

For the case of single explanatory variable

$$X'\Lambda X = \begin{bmatrix} \sum\limits_{i=1}^{n} \lambda_i & \sum\limits_{i=1}^{n} \lambda_i X_i \\ \sum\limits_{i=1}^{n} \lambda_i X_i & \sum\limits_{i=1}^{n} \lambda_i X_i^2 \end{bmatrix}$$

Equation (21.50) gives two normal equations

$$\hat{\beta}_1 \Sigma\lambda_i + \hat{\beta}_2 \Sigma\lambda_i X_i = \Sigma\lambda_i Y_i$$
$$\beta_1 \Sigma\lambda_i X_i + \hat{\beta}_2 \Sigma\lambda_i X_i^2 = \Sigma\lambda_i X_i Y_i \qquad \dots\dots(21.53)$$

From (21.52), we get the variance of regression coefficient $\hat{\beta}_2$ as

$$\text{Var} \ (\hat{\beta}_2) = \frac{\sigma^2 \Sigma\lambda_i}{(\Sigma\lambda_i)(\Sigma\lambda_i X_i^2) - (\Sigma\lambda_i X_i)^2} \qquad \dots\dots(21.54)$$

If we applied least squares method directly to (21.46),

$$\underline{\beta}^* = (X^1 X)^{-1} X^1 Y \qquad \qquad \dots\dots(21.55)$$

The estimate $\underline{\beta}*$ would be unbiased. Its variance covariance matrix would be

$$E\left[(\underline{\beta}^* - \underline{\beta})(\underline{\beta}^* - \underline{\beta})^1 \right] = (X^1 X) X^1 E(\underline{\cup}\underline{\cup}') X (X^1 X)^{-1} \qquad \dots\dots(21.56)$$

In the case of single explanatory variable, the normal equations are

$$n\beta_1^* + \beta_2^* \Sigma X = \Sigma Y$$
$$\beta_1^* \Sigma X + \beta_2^* \Sigma X^2 = \Sigma XY \qquad \dots\dots(21.57)$$

$$\text{Var} \quad (\beta_2^*) = \frac{\sigma^2 \left[\left(\Sigma\dfrac{1}{\lambda_i} \right)(\Sigma X_i^2) - 2n\left(\Sigma\dfrac{1}{\lambda_i} X_i \right)(\Sigma X_i) + n^2\left(\Sigma\dfrac{1}{\lambda_i} X_i^2 \right) \right]}{\left[n\Sigma X_i^2 - (\Sigma X_i)^2 \right]^2}$$

$$\dots\dots(21.58)$$

If the elements of the matrix Λ are known, then the choice of whether to apply least squares method to the transformed data as in (21.49) or directly to the original data as in (21.46) depends upon the sampling variances given in (21.52) and (21.56).

EXAMPLE: For a simple numerical comparison let us assume X takes on the values

$$1, 2, 3, 4, 5.$$

we then find

$$\frac{\text{Var}(\hat{\beta}_2)}{\text{Var}(\beta_2^*)} = \frac{0.69}{1.24} = 0.56$$

So that the efficiency of least squares applied to original data is only about 56 percent of least squares applied to transformed data.

we may write

$$\sqrt{\lambda_i} = \frac{1}{X_i} \qquad\qquad .....(21.59)$$

If we apply (21.59) to (21.49), we obtain

$$\frac{Y_i}{X_i} = \beta_1 \frac{1}{X_1} + \beta_2 + v_i \qquad i = 1, ...n$$

where

$$v_i = {u_i}\big/{X_i}$$

Thus our estimate of β_2, the coefficients of X in the original relation is given by computing the intercept in the regression of ${Y_i}\big/{X_i}$ on $\frac{1}{X_i}$.

If

$$\lambda_i = \frac{1}{X_i}$$

For the above numerical example

$$\frac{\text{Var}(\hat{\beta}_2)}{\text{Var}(\beta_2^*)} = 0.83$$

Which shows that the application of simple least squares to the original data to be relatively more efficient that in the previous example, which is to be expected, since the second example assumes a less radical departure from constant variance than does the first. It is still better, however, to apply least squares to the transformed data as indicated by (21.49), provided the termsformation matrix Λ is known.

21.8 DUMMY VARIABLES

Econometric research in recent years provides many examples of the use of dummy variables in regression analysis. They are used to represent temporal effects such as shifts in relations between wartime and peacetime years, between different seasons, or between different political regimes. They are also used to represent qualitative variables such as sex, marital condition, and occupational or social status, and they are sometimes used to represent quantitative variables such as age, where it is thought that only broad age groupings are relevant. As an example, consider consumption expenditure is linearly related to income and in wartime there is a low and in peace time high. Our hypothesis is

$$e = \alpha_1 + \beta Y \qquad \text{war time} \qquad \qquad \dots\dots(21.60)$$

$$e = \alpha_2 + \beta Y \qquad \text{peace time} \qquad \qquad \dots\dots(21.61)$$

Where $\alpha_2 > \alpha_1$ since it is expected the income is expected to be more in peace time than wartime. Since we are making the assumption that the marginal propensity to consume, β is the same in each period, a more efficient procedure for estimating β would be to pool the data from both periods.

$$C = \alpha_1 X_1 + \alpha_2 X_2 + \beta Y \qquad \qquad \dots\dots(21.62)$$

where X's are dummy variables s.t.

$$X_1 = \begin{cases} 1 & \text{in each wartime year} \\ 0 & \text{in each peacetime year} \end{cases}$$

$$X_2 = \begin{cases} 0 & \text{in each wartime year} \\ 1 & \text{in each peacetime year} \end{cases}$$

There is no intercept term in (21.62) so estimation procedure breaks down.

In we assume 'm' wartime followed by m peacetime years, the matrix of squares and cross products is

$$\begin{bmatrix} m+n & m & n & \sum_{i=1}^{m+n} Y_i \\ m & m & 0 & \sum_{i=1}^{m} Y_i \\ n & 0 & n & \sum_{i=m+1}^{m+n} Y_i \\ \sum_{i=1}^{m+n} Y_i & \sum_{i=1}^{n} Y_i & \sum_{m+1}^{m+n} Y_i & \sum_{i=1}^{m+n} Y_i^2 \end{bmatrix}$$

The 1st column is the sum of 2nd and 3rd column in the above matrix. A conventional computing program may, however, be used simply by dropping one of the dummy variables.

$$C = \gamma_1 + \gamma_2 x_2 + \beta Y \qquad \qquad \dots\dots(21.63)$$

where $\qquad x_2 = \begin{cases} 0 \text{ in each wartime year} \\ 1 \text{ in eact peacetime year} \end{cases}$

This gives the wartime consumption function as

$$C = \gamma_1 + \beta Y$$

Peacetime function as

$$C = (\gamma_1 + \gamma_2) + \beta Y$$

Comparison with (21.60) and (21.61) shows that

$$\gamma_1 = \alpha_1 \quad \text{and} \quad \gamma_2 = \alpha_2 - \alpha_1$$

If we wish to incorporate an assumption that marginal to consume also varies with the period, then

$$C = \gamma_1 + \gamma_2 X_2 + \beta_1 Y + \beta_2 Z \qquad \qquad .....(21.64)$$

where $\qquad Z = X_2 Y$

and $\qquad X_2 = \begin{cases} 0 \text{ in each wartime year} \\ 1 \text{ in eact peacetime year} \end{cases}$

This gives wartime function as

$$C = \gamma_1 + \beta_1 Y$$

Peacetime function as

$$C = (\gamma_1 + \gamma_2) + (\beta_1 + \beta_2) Y$$

The above relations contain dummy variables and conventional quantitative variables on the right hand side.

It is also quite feasible to have nothing but dummy variables on R.H.S. For ex: we regard a dependent variable Y, the no. of hours in a year spent in reading nonfiction, as dependent on two quantitative variables, namely sex and educational attainment. Sex is represented by a dummy variable with two levels, if we postulate three levels of educational attainment, there are 6 possible combinations of dummy variables.

$$Y = \beta_1 x_1 + \beta_2 x_2 + ... + \beta_6 x_6 + u \qquad \qquad .....(21.65)$$

where X's are defined in Table (21.1).

TABLE 21.1

Edn level	Sex	X_1	X_2	X_3	X_4	X_5	X_6
1	I	1	0	0	0	0	0
1	II	1	0	0	1	0	0
II	I	1	1	0	0	0	0
II	II	1	1	0	1	1	0
III	I	1	0	1	0	0	0
III	II	1	0	1	1	0	1

from Table (21.1) and equation (21.65), we have. Here

$$E(Y/I, I) = \beta_1 \qquad\qquad X_5 = (x_2)(x_4)$$

$$E(Y/I, II) = \beta_1 + \beta_4 \qquad\qquad X_6 = (x_3)(x_4)$$

$$E(Y/II, I) = \beta_1 + \beta_2 \qquad\qquad X_4 = Sex$$

$$E(Y/II, II) = \beta_1 + \beta_2 + \beta_4 + \beta_5 \qquad X_2 = Edn.\ level\text{-}II$$

$$E(Y/III, I) = \beta_1 + \beta_3 \qquad\qquad X_3 = Edn.\ level\ III$$

$$E(Y/III, II) = \beta_1 + \beta_3 + \beta_4 + \beta_6$$

The scheme allows interaction effects. The difference between the sexes for people of Edn.level I is β_4, Edn. level II is $\beta_4 + \beta_5$, for Edn. level III is $\beta_4 + \beta_6$. Similarly for sexI, the differential effect of Edn. level II compared with I is β_2, of III compared with I is β_3, and of III with II is $\beta_3 - \beta_2$, For people of sexII, the same differential effects is $\beta_2 + \beta_5$, $\beta_3 + \beta_6$ and $\beta_3 + \beta_6 - \beta_2 - \beta_5$.

Dummy variable can also be taken on left hand side also. For ex: a person owns a car or does not, has a mortgage on his home or he does not etc. In all such cases the dependent variable takes only two values, so that we may use unity to indicate the occurrence of the event and zero to indicate its non-occurrence. If we run a multiple regression of such a dependent variable Y on several explanatory variables X, then we may interpret the calculated value of Y, for any given X, as an estimate of the 'conditional' probability of Y, given X.

In a study of mortgage – debt holdings of U.S spending units, G.H ORCUTT and Rivlin (1961) have split the problem into two parts. First they developed a prediction equation for the probability that a spending unit has mortgage debt, and then for spending units with nonzero mortgage debt. They developed a prediction equation for the amount of mortgage debt. The first equation based on data from the survey of consumer Finances, is

$$\hat{Y}_1 = -0.08 + F_{1.1} + F_{1.2} + F_{1.3} + F_{1.4}$$

where $\hat{Y}_1$ is the predicted probability that a spending unit will have mortgage debt, and the values of functions $F_{1.1}$, $F_{1.2}$ etc are obtained from the fallowing tables.

TABLE 21.2

$F_{1.1}$ *(marital status)*		$F_{1.2}$ *(age of head)*	
Marital status	*$F_{1.1}$*	*Age of head*	*$F_{1.2}$*
Un married	0	18 - 20	0
1 year	0.07	21 - 24	– 0.02
2 years	0.15	25 - 29	0.09
3 years	0.19	30 - 34	0.16
4 years	0.27	35 - 39	0.25

Table 21.2 Contd...

$F_{1.1}$ *(marital status)*		$F_{1.2}$ *(age of head)*	
Marital status	$F_{1.1}$	*Age of head*	$F_{1.2}$
5 - 9 years	0.31	40 - 44	0.22
10 - 20 years	0.25	45 - 49	0.24
over 20 years	0.14	50 - 54	0.16
		55 - 59	0.15
		60 - 64	0.12
		65 and above	0.07

Table 21.3

$F_{1.3}$ *(education)*		$F_{1.4}$ *(race)*	
Education of head	$F_{1.3}$	*Race of head*	$F_{1.4}$
None	0	White	0
Grammar	0.02	Negro	-0.10
Some high school	0.05		
High school degree	0.08	Other	-0.14
Some College	0.12		
College degree	0.13		

Source: G.H. orcutt, Martin Greenberger John Korbel and Alice M. Rivlin (1961). Microgralysis of socioeconomic systems. A simulation study, Harper × Row, New York.

Thus the estimated probability that a spending unit headed by a 32 years old Negro with some high school edn and married for 3 years will have a mortgage debt is

$$\hat{Y} = -0.08 + 0.19 + 0.16 + 0.05 - 0.10$$

$$= 0.22$$

For units with a non zero mortgage debt, the prediction equation for the amount of mortgage debt is

$$\hat{Y}_2 = 3{,}040 + F_{2.1} + F_{2.2} + F_{2.3} + F_{2.4}$$

$$= 3{,}040 + 1530 + 1560 + 1190 - 1040$$

$$= \$ 6{,}280$$

The basic feature of the second step in this analysis is the absence of any assumption about a precise functional form for the relationship. In both steps the application of least - squares regression corresponds to the estimation of cell means where cells are defined by the specific steps in the explanatory variables. For example for the marital - status variable we use seven dummy variables

$$X_1 = \begin{cases} 1 \text{ if duration is 1 year} \\ 0 \text{ if duration is not 1 year} \end{cases}$$

$$X_2 = \begin{cases} 1 \text{ if duration is 2 years} \\ 0 \text{ if duration is not 2 years} \end{cases}$$

$$X_Y = \begin{cases} 1 \text{ if duration is 20 years} \\ 0 \text{ if duration is not 20 years} \end{cases}$$

Dummy variables are also used to represent the other explanatory variables, and when a regression of Y on all dummy variables is run, the least squares coefficients of the dummy variables are simply the cell means, which are tabulated in the above functions. The above schemes assume additivity.

REFERENCES AND BIBLIOGRAPHY

1. Bal, H.k. and Bal, H.S. (1973) : Fertilizer demand for wheat crop in Punjab. *Jour . ind. Soc. Agr. Eco, 28: 68–76.*

2. Bartlett, M.S. (1947) : The use of transformation. *Biometrics,* 3:39–52.

3. Box, G.E.P (1954) Exploration and exploitation of response surfaces, some general consideration and examples, Biometrics, 10:16.

4. Cochran W.G. (1951): Testing a linear relation among variances. *Biometrics,* 17: 17–32.

5. Cochran W.G. (1953): *sampling techniques.* John Wiley & sons, Inc., New York.

6. Cochran, W.G. and Cox, G.M. (1957): *Experimental Designs.* John wiley & sons, Inc., New York.

7. Croxton and Cowden (1966): *Applied general statistics.* Prentice – Hall of India (pvt.) Ltd., New Delhi.

8. Das, M.N. (1946): A some what alternative approach for construction of symmetrical factorial designs and obtaining maximum number of factors. *Cal. Stat Atsoc. Bull.,* 13: 1–17.

9. Duncan, D.B. (1955): Multiple range and multiple F Tests. *Biometrics,* 11: 1–42

10. Duncan, D.B. (1957): Multiple range tests for correlated and heteroschedastic means. *Biometrics,* 13: 164–76.

11. Fair field smith, H. (1938) : An empirical law describing heterogeneity in the yields of Agricultural crops. *Jour. Agr . sci.,* 26: 1–29.

12. Federer, W.T. (1967): *Experimental Design.* Oxford & I.B.H publishing Co, New Delhi.

13. Finney , D.J. (1947): The principle of biological assay. *Suppl. Jour. Roy. Stat. Soc.* 9 : 46–91.

14. Finney, D.J. (1947): *probit Analysis.* Cambridge university press, Cambridge , England.

15. Finney, D.J. (1948): The Fisher-Yates test of significance in 2×2 contingency tables. *Biometrika*, 35 : 145-156.

16. Fisher R.A and Yates, F. (1948) : *Statistical Tables for Agricultural, Biological and Medical Research*. Oliver and Boyd, Edinburgh, 3rd Ed.

17. Fisher, R. A. (1947) : *The Design of Experiments*. Oliver and Boyd. Edinburgh. 4th Ed.

18. Fisher, R. A. (1950) : *Statistical Methods for Research Workers*, Oliver and Boyd, Edinburgh, 11t edn.

19. Goulden, C.H, (1939) : *Methods of Statistical Analysis* -John wiley & Sons Inc., New York.

20. Hansen, M. H. et. al., (1951) : Response errors in surveys. Jour. Amer.Stat. Assoc., 46 : 147-90.

21. Harvey, W. R. (1960) : Least-squares Analysis of data with unequal sub class number's Dept. of Dairy Science Bulletin. Ohio State University, Columbus, Ohio.

22. Henderson C.R. (1953) : Estimation of general, specific and maternal combining abilities in crosses aming inbred lines of swine : unpublished Ph. D. thesis, Iowa State College library, 199 pp.

23. Hoel, P.G. (1947) : *An Introduction of Mathematical Statistics*. John Wiley & Sons Inc., New York.

24. Johnston, J (1971) Econometric Methods Mc Grow Hill Book Co., New York.

25. Kemp, Kenneth E. (1972) : Least square analysis of variance. A procedure, A programme and Example of their use. Dept. of Statistics and the Statistical laboratory, Kansas Agricultural Experimental Station, Kansas state University Research paper 7.

26. Kempthorne, O. (1966) : The Design and Analysis of Experiments, Wiley Eastern Pvt. Ltd., New Delhi.

27. Kruskal W.H. (1957) : Historical notes on the Wilcoxon unpaired two-sample test. *Jour. Amer. Stat. Assn.*, 52 : 356-60.

28. Kruskal W.H. and Wallis, W.A. (1952) : Use of ranks in one criterion variance analysis. *Jour. Amer. Stat. Assn.,* 47 : 583.

29. Kumari, K. Ratna (1992) M.Sc. (Ag) statistics thesis submitted to APAU, Hyderabad.

30. Laxmi Devi, A (1983) Ph.D. (Extension) thesis submitted to APAU, Hyderabad.

31. Mann. H.B. and Whitney, D.R. (1947) : On a test of whether, one of two random variables is stochastically larger than the other. *Ann. Math.Stat.,* 18 : 50.

32. Massey, F.J.Jr. (1951) : The Kolmogorov-Smirnov test for goodness of fit. *J. Amer. Statis. Ass.,* 46 : 70.

33. Morrison, D.F. (1967) Multivariate Statistical Methods, Mc Graw Hill Book Co., New York.

34. Murthy, H.G. Shankara (1986). Ph.D. (Ag. Economics) thesis submitted to APAU, Hyderabad.

35. Ostle, (1966) : *Statistics in Research.* Oxford & IBH publishing Co. New Delhi.

36. Panse, V.G. and Sukhatme, P.V. (1967) : *Statistical Methods for Agricultural Workers.* Indian Council of Agril. Research , New Delhi.

37. Prasad, Y. Eswara (1987) Ph.D. (Ag. Economics) thesis submitted to UAS, Bangalore.

38. Raghavarao, D. (1971) : *Constructions and Combinational problems in Design of Experiments.* John Wiley & Sons, Inc., New York.

39. Rani, U. Swaroopa (1987) Ph.D. (Extension) thesis submitted to APAU, Hyderabad.

40. Rao, B. Narasimha (1982). Ph. D. (Entomology) thesis submitted to APAU, Hyderabad.

41. Rao, C.R. (1952) : Advanced Statistical Methods in Biometric Research. John Wiley & Sons. Inc., New York.

42. Rao, C.R. and Mitra, S.K. (1971) : *Generalized Inverse of Matrices and Its Applications.* John Wiley & Sons. Inc.,   New York.

43. Rao, G. Nageswara 1976-77 : *Application of D2-statistic in education. Journal of Research,* APAU IV (1-4) : 9-14.

44. Rao, G. Nageswara and Apte, V.D. (1972) : Tolerances in the testing of seeds. Allahabad Farmer, 55 : 21-24.

45. Reddy, A. Krishna (1980) : Studies on certain Economic traits in Nellore breed of sheep. Thesis submitted for award of M.Sc. degree A.P. Agricultural University, Hyderabad.

46. Reddy, M. Vijaya Bhaskara (1991) M.Sc. (Ag) Statistics thesis submitted to APAU, Hyderabad.

47. Reddy, S.B. Sreenivasa (1992) M.Sc. (Ag. Statistics) thesis submitted to APAU, Hyderabad.

48. Scheffe, H. (1943) : Statistical inference in the non-parametric case. *Ann. Math. Stat.,* 14 : 305-332.

49. Scheffe, H. (1953) : A method for judging all contrasts in the analysis of variance. *Biometrika,* 40 : 87-104.

50. Scheffe', H. (1967) : *The Analysis of variance.* John Wiley & Sons. Inc., New York.

51. Sekhar, P. Raja (1989) Ph.D. (Entomology) thesis submitted to APAU, Hyderabad.

52. Siegel, Sydney, (1956) : *Non-parametric Statistics for the Behavioural Sciences.* McGraw-Hill Book Co. Inc., New York.

53. Smirnov, N.V. (1948) : Table for estimating the goodness of fit of empirical distributions. Ann. Math. Stat., 19 : 279-81.

54. Snedecor, G.W. and Cochran, W.G. (1968) : Statistical Methods. Oxford & IBH Publishing Co., New Delhi.

55. Steel and Torrie (1960) : Principals and Procedures of Statistics. McGraw-Hill Book Co. Inc., New York.

56. Subba Ratnam, G.V. (1979). Ph.D. (Entomology) thesis submitted to IARI, New Delhi.

57. Sukhatme P.V 1953) : Measurement of observational errors in Surveys. Reuve de l' Institute-International du Statistique 20 : No. 2.

58. Sukhatme, P.V. (1953) : *Sampling Theory of Surveys with Applications.* Indian Council of Agricultural Research, New Delhi.

59. Tukey, J.W. (1949) : One degree of freedom for non-additivity Biometrics, 5 : 232-242.

60. Umarji, R.R. (1962) : Probability and Mathematical Statistics. Asia Publishing House, Bombay.

61. Wilcoxon F. (1947) : Probability tables for individual comparisons by ranking methods. Biometrics, 3 : 119-22.

62. Wilcoxon, F. (1945) : Individual comparisons by ranking methods. Biometrics, 1 : 80-83.

63. Yates F. and Cochran, W.G. (1938) : The analysis of groups of experiments. Jour. Agr. Sci., 28 556-580.

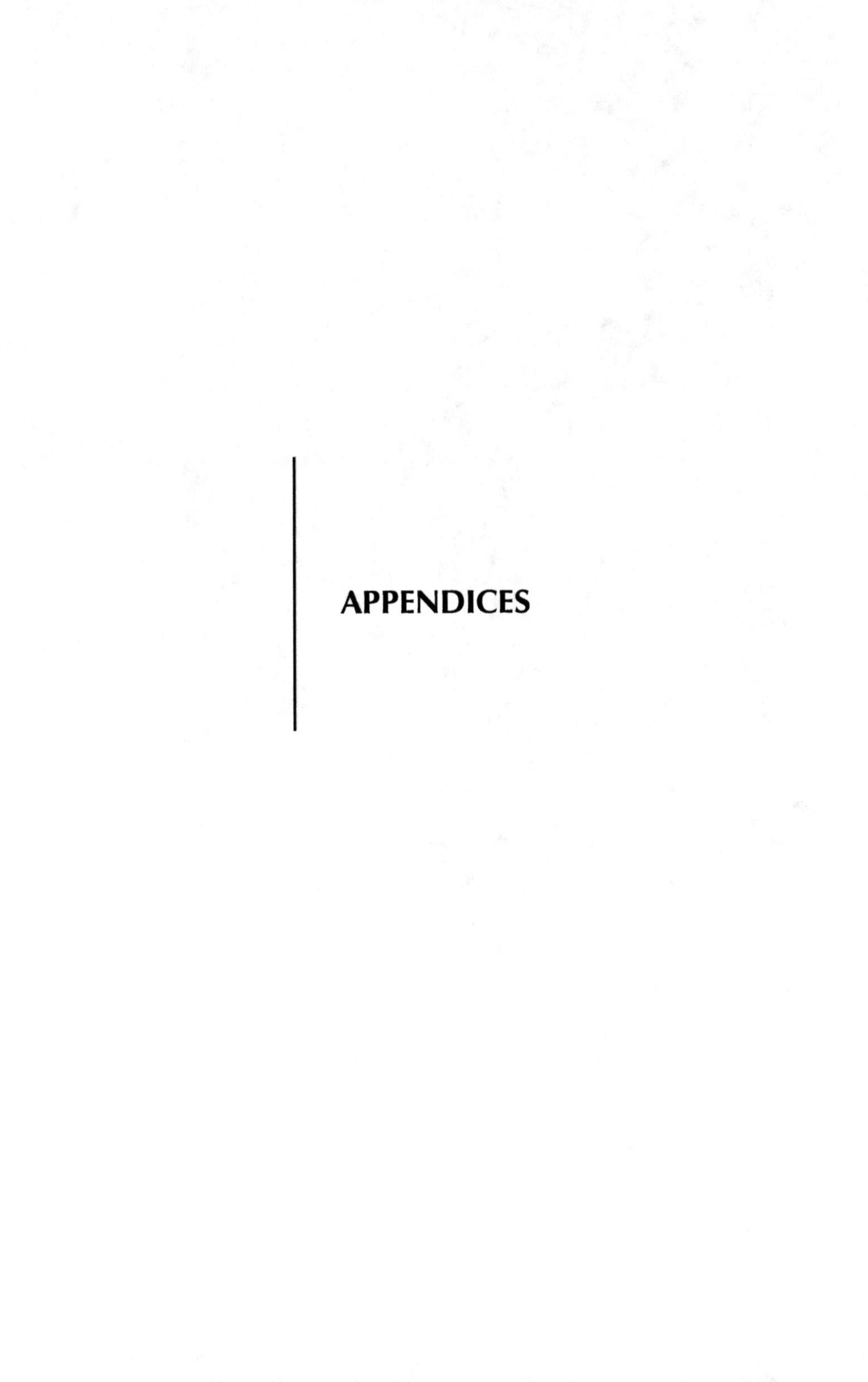

APPENDICES

TABLE I: THE NORMAL PROBABILITY INTEGRAL

x		0	1	2	3	4	5	6	7	8	9
0·0	0·	50000	49601	49202	48803	48405	48006	47608	47210	46812	46414
0·1		46017	45620	45224	44828	44433	44038	43644	43251	42858	42465
0·2		42074	41683	41294	40905	40517	40129	39743	39358	38974	38591
0·3		38209	37828	37448	37070	36693	36317	35942	35569	35197	34827
0·4		34458	34090	33724	33360	32997	32636	32276	31918	31561	31207
0·5		30854	30503	30153	29806	29460	29116	28774	28434	28096	27760
0·6		27425	27093	26763	26435	26109	25785	25463	25143	24825	24510
0·7		24196	23885	23576	23270	22965	22663	22363	22065	21770	21476
0·8		21186	20897	20611	20327	20045	19766	19489	19215	18943	18673
0·9		18406	18141	17879	17619	17361	17106	16853	16602	16354	16109
1·0		15866	15625	15386	15151	14917	14686	14457	14231	14007	13786
1·1		13567	13350	13136	12924	12714	12507	12302	12100	11900	11702
1·2		11507	11314	11123	10935	10749	10565	10383	10204	10027	98525
1·3	0·0	96800	95098	93418	91759	90123	88508	86915	85343	83793	82264
1·4		80757	79270	77804	76359	74934	73529	72145	70781	69437	68112
1·5		66807	65522	64255	63008	61780	60571	59380	58208	57053	55917
1·6		54799	53699	52616	51551	50503	49471	48457	47460	46479	45514
1·7		44565	43633	42716	41815	40930	40059	39204	38364	37538	36727
1·8		35930	35148	34380	33625	32884	32157	31443	30742	30054	29379
1·9		28717	28067	27429	26803	26190	25588	24998	24419	23852	23295
2·0		22750	22216	21692	21178	20675	20182	19699	19226	18763	18309
2·1		17864	17429	17003	16586	16177	15778	15386	15003	14629	14262
2·2		13903	13553	13209	12874	12545	12224	11911	11604	11304	11011
2·3		10724	10444	10170	99031	96419	93867	91375	88940	86563	84242
2·4	0·0²	81975	79763	77603	75494	73436	71428	69469	67557	65691	63872
2·5		62097	50366	58677	57031	55426	53861	52336	50849	49400	47988
2·6		46612	45271	43965	42692	41453	40246	39070	37926	36811	35726
2·7		34670	33642	32641	31667	30720	29798	28901	28028	27179	26354
2·8		25551	24771	24012	23274	22557	21860	21182	20524	19884	19262
2·9		18658	18071	17502	16948	16411	15889	15382	14890	14412	13949
3·0		13499	13062	12639	12228	11829	11442	11067	10703	10350	10008
3·1	0·0³	96760	93544	90426	87403	84474	81635	78885	76219	73638	71136
3·2		68714	66367	64095	61895	59765	57703	55706	53774	51904	50094

TABLE II : DISTRIBUTION OF t
PROBABILITY

d.f	.1	.05	.02	.01	.001
1	6.314	12.706	31.821	63.657	636.619
2	2.920	4.303	6.965	9.925	31.598
3	2.353	3.182	4.541	5.841	12.924
4	2.132	2.776	3.747	4.604	8.610
5	2.015	2 571	3.365	4.032	6.869
6	1.943	2.447	3.143	3.707	5.959
7	1.895	2.365	2.998	3.499	5.408
8	1.860	2.306	2.896	3.355	5.041
9	1.833	2.262	2.821	3.250	4.781
10	1.812	2.228	2.764	3.169	4.587
11	1.796	2.201	2.718	3.106	4.437
12	1.782	2.179	2.681	3.055	4.318
13	1.771	2.160	2.650	3.012	4.221
14	1.761	2.145	2.624	2.977	4.140
15	1.753	2.131	2.602	2.947	4.073
16	1.746	2.120	2.583	2.92	4.015
17	1.740	2.110	2.567	2.898	3.965
18	1.734	2.101	2.552	2.878	3.922
19	1.729	2.093	2.539	2.861	3.883
20	1.725	2.086	2.528	2.845	3.850
21	1.721	2.080	2.518	2.83	3.819
22	1.717	2.074	2.58	2.819	3.792
23	1.714	2.069	2.500	2.807	3.767
24	1.711	2.064	2.492	2.797	3.745
25	1.708	2.060	2.485	2.787	3.725
26	1.706	2.056	2.479	2.779	3.707
27	1.703	2.052	2.473	2.771	3.690
28	1.701	2.048	2.467	2.763	3.674
29	1.699	2.045	2.462	2.75	3.659
30	1.697	2.042	2.457	2.750	3.646
40	2.021	2.423	2.704	3.551	3.551
50	1.671	2.000	2.390	2.660	3.460
60	1.658	1.980	2.358	2.17	3.373
∞	1.645	1.960	2.326	2.576	3.291

Table II is taken from table Table III of Fisher and Yates : Statistical Tables for Biological, Agricultural and Medial Research, Published by Longman Group Ltd., London. (Previously published by Oliver and Boyd. Edin burgh). and by Permission of the Authors and Publishers.

TABLE III: DISTRIBUTION OF χ^2

Probability.

n	·99	·98	·95	·90	·80	·70	·50	·30	·20	·10	·05	·02	·01
1	$\cdot0^3157$	$\cdot0^3628$	·00393	·0158	·0642	·148	·455	1·074	1·642	2·706	3·841	5·412	6·635
2	·0201	·0404	·103	·211	·446	·713	1·386	2·408	3·219	4·605	5·991	7·824	9·210
3	·115	·185	·352	·584	1·005	1·424	2·366	3·665	4·642	6·251	7·815	9·837	11·345
4	·297	·429	·711	1·064	1·649	2·195	3·357	4·878	5·989	7·779	9·488	11·668	13·277
5	·554	·752	1·145	1·610	2·343	3·000	4·351	6·064	7·289	9·236	11·070	13·388	15·086
6	·872	1·134	1·635	2·204	3·070	3·828	5·348	7·231	8·558	10·645	12·592	15·033	16·812
7	1·239	1·564	2·167	2·833	3·822	4·671	6·346	8·383	9·803	12·017	14·067	16·622	18·475
8	1·646	2·032	2·733	3·490	4·594	5·527	7·344	9·524	11·030	13·362	15·507	18·168	20·090
9	2·088	2·532	3·325	4·168	5·380	6·393	8·343	10·656	12·242	14·684	16·919	19·679	21·666
10	2·558	3·059	3·940	4·865	6·179	7·267	9·342	11·781	13·442	15·987	18·307	21·161	23·209
11	3·053	3·609	4·575	5·578	6·989	8·148	10·341	12·899	14·631	17·275	19·675	22·618	24·725
12	3·571	4·178	5·226	6·304	7·807	9·034	11·340	14·011	15·812	18·549	21·026	24·054	26·217
13	4·107	4·765	5·892	7·042	8·634	9·926	12·340	15·119	16·985	19·812	22·362	25·472	27·688
14	4·660	5·368	6·571	7·790	9·467	10·821	13·339	16·222	18·151	21·064	23·685	26·873	29·141
15	5·229	5·985	7·261	8·547	10·307	11·721	14·339	17·322	19·311	22·307	24·996	28·259	30·578
16	5·812	6·614	7·962	9·312	11·152	12·624	15·338	18·418	20·465	23·542	26·296	29·633	32·000
17	6·408	7·255	8·672	10·085	12·002	13·531	16·338	19·511	21·615	24·769	27·587	30·995	33·409
18	7·015	7·906	9·390	10·865	12·857	14·440	17·338	20·601	22·760	25·989	28·869	32·346	34·805
19	7·633	8·567	10·117	11·651	13·716	15·352	18·338	21·689	23·900	27·204	30·144	33·687	36·191
20	8·260	9·237	10·851	12·443	14·578	16·266	19·337	22·775	25·038	28·412	31·410	35·020	37·566
21	8·897	9·915	11·591	13·240	15·445	17·182	20·337	23·858	26·171	29·615	32·671	36·343	38·932
22	9·542	10·600	12·338	14·041	16·314	18·101	21·337	24·939	27·301	30·813	33·924	37·659	40·289
23	10·196	11·293	13·091	14·848	17·187	19·021	22·337	26·018	28·429	32·007	35·172	38·968	41·638
24	10·856	11·992	13·848	15·659	18·062	19·943	23·337	27·096	29·553	33·196	36·415	40·270	42·980
25	11·524	12·697	14·611	16·473	18·940	20·867	24·337	28·172	30·675	34·382	37·652	41·566	44·314
26	12·198	13·409	15·379	17·292	19·820	21·792	25·336	29·246	31·795	35·563	38·885	42·856	45·642
27	12·879	14·125	16·151	18·114	20·703	22·719	26·336	30·319	32·912	36·741	40·113	44·140	46·963
28	13·565	14·847	16·928	18·939	21·588	23·647	27·336	31·391	34·027	37·916	41·337	45·419	48·278
29	14·256	15·574	17·708	19·768	22·475	24·577	28·336	32·461	35·139	39·087	42·557	46·693	49·588
30	14·953	16·306	18·493	20·599	23·364	25·508	29·336	33·530	36·250	40·256	43·773	47·962	50·892

Table III is taken from Table IV of Fisher and Yates: *Statistical Tables for Biological, Agricultural and Medical Research*, Published by Longman Group Ltd., London. (Previously Published by Oliver and Boyd, Edinburgh), and by Permission of the Authors and Publishers.

Table IV (a) F-distribution

5 per cent points

df_2 \ df_1	1	2	3	4	5	6	7	8	9	10	12	15	20	24	30	40	60	120	∞
5	6.61	5.79	5.41	5.19	5.05	4.95	4.88	4.82	4.77	4.74	4.68	4.62	4.56	4.53	4.50	4.46	4.43	4.40	4.36
6	5.99	5.14	4.76	4.53	4.39	4.28	4.21	4.15	4.10	4.06	4.00	3.94	3.87	3.84	3.81	3.77	3.74	3.70	3.67
7	5.59	4.74	4.35	4.12	3.97	3.87	3.79	3.73	3.68	3.64	3.57	3.51	3.44	3.41	3.38	3.34	3.30	3.27	3.23
8	5.32	4.46	4.07	3.84	3.69	3.58	3.50	3.44	3.39	3.35	3.28	3.22	3.15	3.12	3.08	3.04	3.01	2.97	2.93
9	5.12	4.26	3.86	3.63	3.48	3.37	3.29	3.23	3.18	3.14	3.07	3.01	2.94	2.90	2.86	2.83	2.79	2.75	2.71
10	4.96	4.10	3.71	3.48	3.33	3.22	3.14	3.07	3.02	2.98	2.91	2.84	2.77	2.74	2.70	2.66	2.62	2.58	2.54
11	4.84	3.98	3.59	3.36	3.20	3.09	3.01	2.95	2.90	2.85	2.79	2.72	2.65	2.61	2.57	2.53	2.49	2.45	2.40
12	4.75	3.89	3.49	3.26	3.11	3.00	2.91	2.85	2.80	2.75	2.69	2.62	2.54	2.51	2.47	2.43	2.38	2.34	2.30
13	4.67	3.81	3.41	3.18	3.03	2.92	2.83	2.77	2.71	2.67	2.60	2.53	2.46	2.42	2.38	2.34	2.30	2.25	2.21
14	4.60	3.74	3.34	3.11	2.96	2.85	2.76	2.70	2.65	2.60	2.53	2.46	2.39	2.35	2.31	2.27	2.22	2.18	2.13
15	4.54	3.68	3.29	3.06	2.90	2.79	2.73	2.64	2.59	2.54	2.48	2.40	2.33	2.29	2.25	2.20	2.16	2.11	2.07
16	4.49	3.63	3.24	3.01	2.85	2.74	2.66	2.59	2.54	2.49	2.42	2.35	2.28	2.24	2.19	2.15	2.11	2.06	2.01
17	4.45	3.59	3.20	2.96	2.81	2.70	2.61	2.55	2.49	2.45	2.38	2.31	2.23	2.19	2.15	2.10	2.06	2.01	1.96
18	4.41	3.55	3.16	2.93	2.77	2.66	2.58	2.51	2.46	2.41	2.34	2.27	2.19	2.15	2.11	2.06	2.02	1.97	1.92
19	4.38	3.52	3.13	2.90	2.74	2.63	2.54	2.48	2.42	2.38	2.31	2.23	2.16	2.11	2.07	2.03	1.98	1.93	1.88
20	4.35	3.49	3.10	2.87	2.71	2.60	2.51	2.45	2.39	2.35	2.28	2.20	2.12	2.08	2.04	1.99	1.95	1.90	1.84
21	4.32	3.47	3.07	2.84	2.68	2.57	2.49	2.42	2.37	2.32	2.25	2.18	2.10	2.05	2.01	1.96	1.92	1.87	1.81
22	4.30	3.44	3.05	2.82	2.66	2.55	2.46	2.40	2.34	2.30	2.23	2.15	2.07	2.03	1.98	1.94	1.89	1.84	1.78
23	4.28	3.42	3.03	2.80	2.64	2.53	2.44	2.37	2.32	2.27	2.20	2.13	2.05	2.00	1.96	1.91	1.86	1.81	1.76
24	4.26	3.40	3.01	2.78	2.62	2.51	2.42	2.36	2.30	2.25	2.18	2.11	2.03	1.98	1.94	1.89	1.84	1.79	1.73
25	4.24	3.39	2.99	2.76	2.60	2.49	2.40	2.34	2.28	2.24	2.16	2.09	2.01	1.96	1.92	1.87	1.82	1.77	1.71
26	4.23	3.37	2.98	2.74	2.59	2.47	2.39	2.32	2.27	2.22	2.15	2.07	1.99	1.95	1.90	1.85	1.80	1.75	1.69
27	4.21	3.35	2.96	2.73	2.57	2.46	2.37	2.31	2.25	2.20	2.13	2.06	1.97	1.93	1.88	1.84	1.79	1.73	1.67
28	4.20	3.34	2.95	2.71	2.56	2.45	2.36	2.29	2.24	2.19	2.12	2.04	1.96	1.91	1.87	1.82	1.77	1.71	1.65
29	4.18	3.33	2.93	2.70	2.55	2.43	2.35	2.28	2.22	2.18	2.10	2.03	1.94	1.90	1.85	1.81	1.75	1.70	1.64
30	4.17	3.32	2.92	2.69	2.53	2.42	2.33	2.27	2.21	2.16	2.09	2.01	1.93	1.89	1.84	1.79	1.74	1.68	1.62
40	4.08	3.23	2.84	2.61	2.45	2.34	2.25	2.18	2.12	2.08	2.00	1.92	1.84	1.79	1.74	1.69	1.64	1.58	1.51
60	4.00	3.15	2.76	2.53	2.37	2.25	2.17	2.10	2.04	1.99	1.92	1.84	1.75	1.70	1.65	1.59	1.53	1.47	1.39
120	3.92	3.07	2.68	2.45	2.29	2.18	2.09	2.02	1.96	1.91	1.83	1.75	1.66	1.61	1.55	1.50	1.43	1.35	1.25
∞	3.84	3.00	2.60	2.37	2.21	2.10	2.01	1.94	1.88	1.83	1.75	1.67	1.57	1.52	1.46	1.39	1.32	1.22	1.00

Adopted from Table 18, Biometrika Tables for Statisticians, Vol. 1, 3rd edition (1966) with the kind permission of Publishers of Biometrika.

Table IV (b) F-distribution

1 per cent points

df_2 \ df_1	1	2	3	4	5	6	7	8	9	10	12	15	20	24	30	40	60	120	∞
5	16.26	13.27	12.06	11.39	10.97	10.67	10.46	10.29	10.16	10.05	9.89	9.72	9.55	9.47	9.38	9.29	9.20	9.11	9.02
6	13.74	10.92	9.78	9.15	8.75	8.47	8.26	8.10	7.98	7.87	7.72	7.56	7.40	7.31	7.23	7.14	7.06	6.97	6.88
7	12.25	9.55	8.45	7.85	7.46	7.19	6.99	6.84	6.72	6.62	6.47	6.31	6.16	6.07	5.99	5.91	5.82	5.74	5.65
8	11.26	8.65	7.59	7.01	6.63	6.37	6.18	6.03	5.91	5.81	5.67	5.52	5.36	5.28	5.20	5.12	5.03	4.95	4.86
9	10.56	8.02	6.99	6.42	6.06	5.80	5.61	5.47	5.35	5.26	5.11	4.96	4.81	4.73	4.65	4.57	4.48	4.40	4.31
10	10.04	7.56	6.55	5.99	5.64	5.39	5.20	5.06	4.94	4.85	4.71	4.56	4.41	4.33	4.25	4.17	4.08	4.00	3.91
11	9.65	7.21	6.22	5.67	5.32	5.07	4.89	4.74	4.63	4.54	4.40	4.25	4.10	4.02	3.94	3.86	3.78	3.69	3.60
12	9.33	6.93	5.95	5.41	5.06	4.82	4.64	4.50	4.39	4.30	4.16	4.01	3.86	3.78	3.70	3.62	3.54	3.45	3.36
13	9.07	6.70	5.74	5.21	4.86	4.62	4.44	4.30	4.19	4.10	3.96	3.82	3.66	3.59	3.51	3.43	3.34	3.25	3.17
14	8.86	6.51	5.56	5.04	4.70	4.46	4.28	4.14	4.03	3.94	3.80	3.66	3.51	3.43	3.35	3.27	3.18	3.09	3.00
15	8.68	6.36	5.42	4.89	4.56	4.32	4.14	4.00	3.89	3.80	3.67	3.52	3.37	3.29	3.21	3.13	3.05	2.96	2.87
16	8.53	6.23	5.29	4.77	4.44	4.20	4.03	3.89	3.78	3.69	3.55	3.41	3.26	3.18	3.10	3.02	2.93	2.84	2.75
17	8.40	6.11	5.18	4.67	4.34	4.10	3.93	3.79	3.68	3.59	3.46	3.31	3.16	3.08	3.00	2.92	2.83	2.75	2.65
18	4.29	3.01	3.09	2.58	2.25	2.01	2.84	2.71	2.60	2.51	2.37	2.23	2.08	3.00	2.92	2.84	2.75	1.66	1.92
19	8.18	5.93	5.01	4.50	4.17	3.94	3.77	3.63	3.52	3.43	3.30	3.15	3.00	2.92	2.84	2.76	2.67	2.58	2.49
20	8.10	5.85	4.94	4.43	4.10	3.87	3.70	3.56	3.46	3.37	3.23	3.09	2.94	2.86	2.78	2.69	2.61	2.52	2.42
21	8.02	5.78	4.87	4.37	4.04	3.81	3.64	3.51	3.40	3.31	3.17	3.03	2.88	2.80	2.72	2.64	2.55	2.46	2.36
22	7.95	5.72	4.82	4.31	3.99	3.76	3.59	3.45	3.35	3.26	3.12	2.98	2.83	2.75	2.67	2.58	2.50	2.40	2.31
23	7.88	5.66	4.76	4.26	3.94	3.71	3.54	3.41	3.30	3.21	3.07	2.93	2.78	2.70	2.62	2.54	2.45	2.35	2.26
24	7.82	5.61	4.72	4.22	3.90	3.67	3.50	3.36	3.26	3.17	3.03	2.89	2.74	2.66	2.58	2.49	2.40	2.31	2.21
25	7.77	5.57	4.68	4.18	3.86	3.63	3.46	3.32	3.22	3.13	2.99	2.85	2.70	2.62	2.54	2.45	2.36	2.27	2.17
26	7.72	5.53	4.64	4.14	3.82	3.59	3.42	3.29	3.18	3.09	2.96	2.82	2.66	2.58	2.50	2.42	2.33	2.23	2.13
27	7.68	5.49	4.60	4.11	3.78	3.56	3.39	3.26	3.15	3.06	2.93	2.78	2.63	2.55	2.47	2.38	2.29	2.20	2.10
28	7.64	5.45	4.57	4.07	3.75	3.53	3.36	3.23	3.12	3.03	2.90	2.75	2.60	2.52	2.44	2.35	2.26	2.17	2.06
29	7.60	5.42	4.54	4.04	3.73	3.50	3.33	3.20	3.09	3.00	2.87	2.73	2.57	2.49	2.41	2.33	2.23	2.14	2.03
30	7.56	5.39	4.51	4.02	3.70	3.47	3.30	3.17	3.07	2.98	2.84	2.70	2.55	2.47	2.39	2.30	2.21	2.11	2.01
40	7.31	5.18	4.31	3.83	3.51	3.29	3.12	2.99	2.89	2.80	2.66	2.52	2.37	2.29	2.20	2.11	2.02	1.92	1.80
60	7.08	4.98	4.13	3.65	3.34	3.12	2.95	2.82	2.72	2.63	2.50	2.35	2.20	2.12	2.03	1.94	1.84	1.73	1.60
120	6.85	4.79	3.95	3.48	3.17	2.96	2.79	2.66	2.56	2.47	2.34	2.19	2.03	1.95	1.86	1.76	1.66	1.53	1.38
∞	6.63	4.61	3.78	3.32	3.02	2.80	2.64	2.51	2.41	2.32	2.18	2.04	1.88	1.79	1.70	1.59	1.47	1.32	1.00

THE CORRELATION COEFFICIENT

TABLE V: Values of the Correlation Coefficient for Different Levels of Significance

n	$\cdot 1$	$\cdot 05$	$\cdot 02$	$\cdot 01$	$\cdot 001$	n	$\cdot 1$	$\cdot 05$	$\cdot 02$	$\cdot 01$	$\cdot 001$
1	·98769	·99692	·999507	·999877	·9999988	16	·4000	·4683	·5425	·5897	·7084
2	·90000	·95000	·98000	·990000	·99900	17	·3887	·4555	·5285	·5751	·6932
3	·8054	·8783	·93433	·95873	·99116	18	·3783	·4438	·5155	·5614	·6787
4	·7293	·8114	·8822	·91720	·97406	19	·3687	·4329	·5034	·5487	·6652
5	·6694	·7545	·8329	·8745	·95074	20	·3598	·4227	·4921	·5368	·6524
6	·6215	·7067	·7887	·8343	·92493	25	·3233	·3809	·4451	·4869	·5974
7	·5822	·6664	·7498	·7977	·8982	30	·2960	·3494	·4093	·4487	·5541
8	·5494	·6319	·7155	·7646	·8721	35	·2746	·3246	·3810	·4182	·5189
9	·5214	·6021	·6851	·7348	·8471	40	·2573	·3044	·3578	·3932	·4896
10	·4973	·5760	·6581	·7079	·8233	45	·2428	·2875	·3384	·3721	·4648
11	·4762	·5529	·6339	·6835	·8010	50	·2306	·2732	·3218	·3541	·4433
12	·4575	·5324	·6120	·6614	·7800	60	·2108	·2500	·2948	·3248	·4078
13	·4409	·5139	·5923	·6411	·7603	70	·1954	·2319	·2737	·3017	·3799
14	·4259	·4973	·5742	·6226	·7420	80	·1829	·2172	·2565	·2830	·3568
15	·4124	·4821	·5577	·6055	·7246	90	·1726	·2050	·2422	·2673	·3375
						100	·1638	·1946	·2301	·2540	·3211

Table V is reproduced from Table VII, of Fisher and Yates: *Statistical Tables for Biological, Agricultural and Medical Research*. Published by Longman Group Ltd., London. (Previously Published by Oliver and Boyd, Edinburgh), and by Permission of the Authors and Publishers.

TABLE VI: PROBITS

Transformation of the Sigmoid Dosage Mortality Curve to a Straight Line. (C. I. Bliss.)

	0·0	0·1	0·2	0·3	0·4	0·5	0·6	0·7	0·8	0·9	1	2	3	4	5
0	...	1·9098	2·1218	2·2522	2·3479	2·4242	2·4879	2·5427	2·5911	2·6344					
1	2·6737	2·7096	2·7429	2·7738	2·8027	2·8299	2·8556	2·8799	2·9031	2·9251					
2	2·9463	2·9665	2·9859	3·0046	3·0226	3·0400	3·0569	3·0732	3·0890	3·1043		For more detail see			
3	3·1192	3·1337	3·1478	3·1616	3·1750	3·1881	3·2009	3·2134	3·2256	3·2376		values for 95-100.			
4	3·2493	3·2608	3·2721	3·2831	3·2940	3·3046	3·3151	3·3253	3·3354	3·3454					
5	3·3551	3·3648	3·3742	3·3836	3·3928	3·4018	3·4107	3·4195	3·4282	3·4368	9	18	27	36	45
6	3·4452	3·4536	3·4618	3·4699	3·4780	3·4859	3·4937	3·5015	3·5091	3·5167	8	16	24	32	40
7	3·5242	3·5316	3·5389	3·5462	3·5534	3·5605	3·5675	3·5745	3·5813	3·5882	7	14	21	28	36
8	3·5949	3·6016	3·6083	3·6148	3·6213	3·6278	3·6342	3·6405	3·6468	3·6531	6	13	19	26	32
9	3·6592	3·6654	3·6715	3·6775	3·6835	3·6894	3·6953	3·7012	3·7070	3·7127	6	12	18	24	30
10	3·7184	3·7241	3·7298	3·7354	3·7409	3·7464	3·7519	3·7574	3·7628	3·7681	6	11	17	22	28
11	3·7735	3·7788	3·7840	3·7893	3·7945	3·7996	3·8048	3·8099	3·8150	3·8200	5	10	16	21	26
12	3·8250	3·8300	3·8350	3·8399	3·8448	3·8497	3·8545	3·8593	3·8641	3·8689	5	10	15	20	24
13	3·8736	3·8783	3·8830	3·8877	3·8923	3·8969	3·9015	3·9061	3·9107	3·9152	5	9	14	18	23
14	3·9197	3·9242	3·9286	3·9331	3·9375	3·9419	3·9463	3·9506	3·9550	3·9593	4	9	13	18	22
15	3·9636	3·9678	3·9721	3·9763	3·9806	3·9848	3·9890	3·9931	3·9973	4·0014	4	8	13	17	21
16	4·0055	4·0096	4·0137	4·0178	4·0218	4·0259	4·0299	4·0339	4·0379	4·0419	4	8	12	16	20
17	4·0458	4·0498	4·0537	4·0576	4·0615	4·0654	4·0693	4·0731	4·0770	4·0808	4	8	12	16	19
18	4·0846	4·0884	4·0922	4·0960	4·0998	4·1035	4·1073	4·1110	4·1147	4·1184	4	8	11	15	19
19	4·1221	4·1258	4·1295	4·1331	4·1367	4·1404	4·1440	4·1476	4·1512	4·1548	4	7	11	15	18
20	4·1584	4·1619	4·1655	4·1690	4·1726	4·1761	4·1796	4·1831	4·1866	4·1901	4	7	11	14	18
21	4·1936	4·1970	4·2005	4·2039	4·2074	4·2108	4·2142	4·2176	4·2210	4·2244	3	7	10	14	17
22	4·2278	4·2312	4·2345	4·2379	4·2412	4·2446	4·2479	4·2512	4·2546	4·2579	3	7	10	13	17
23	4·2612	4·2644	4·2677	4·2710	4·2743	4·2775	4·2808	4·2840	4·2872	4·2905	3	7	10	13	16
24	4·2937	4·2969	4·3001	4·3033	4·3065	4·3097	4·3129	4·3160	4·3192	4·3224	3	6	10	13	16

	0	1	2	3	4	5	6	7	8	9					
25	4·3255	4·3287	4·3318	4·3349	4·3380	4·3412	4·3443	4·3474	4·3505	4·3536	3	6	9	12	16
26	4·3567	4·3597	4·3628	4·3659	4·3689	4·3720	4·3750	4·3781	4·3811	4·3842	3	6	9	12	15
27	4·3872	4·3902	4·3932	4·3962	4·3992	4·4022	4·4052	4·4082	4·4112	4·4142	3	6	9	12	15
28	4·4172	4·4201	4·4231	4·4260	4·4290	4·4319	4·4349	4·4378	4·4408	4·4437	3	6	9	12	15
29	4·4466	4·4495	4·4524	4·4554	4·4583	4·4612	4·4641	4·4670	4·4698	4·4727	3	6	9	12	14
30	4·4756	4·4785	4·4813	4·4842	4·4871	4·4899	4·4928	4·4956	4·4985	4·5013	3	6	9	11	14
31	4·5041	4·5070	4·5098	4·5126	4·5155	4·5183	4·5211	4·5239	4·5267	4·5295	3	6	8	11	14
32	4·5323	4·5351	4·5379	4·5407	4·5435	4·5462	4·5490	4·5518	4·5546	4·5573	3	6	8	11	14
33	4·5601	4·5628	4·5656	4·5684	4·5711	4·5739	4·5766	4·5793	4·5821	4·5848	3	5	8	11	14
34	4·5875	4·5903	4·5930	4·5957	4·5984	4·6011	4·6039	4·6066	4·6093	4·6120	3	5	8	11	14
35	4·6147	4·6174	4·6201	4·6228	4·6255	4·6281	4·6308	4·6335	4·6362	4·6389	3	5	8	11	13
36	4·6415	4·6442	4·6469	4·6495	4·6522	4·6549	4·6575	4·6602	4·6628	4·6655	3	5	8	11	13
37	4·6681	4·6708	4·6734	4·6761	4·6787	4·6814	4·6840	4·6866	4·6893	4·6919	3	5	8	11	13
38	4·6945	4·6971	4·6998	4·7024	4·7050	4·7076	4·7102	4·7129	4·7155	4·7181	3	5	8	10	13
39	4·7207	4·7233	4·7259	4·7285	4·7311	4·7337	4·7363	4·7389	4·7415	4·7441	3	5	8	10	13
40	4·7467	4·7492	4·7518	4·7544	4·7570	4·7596	4·7622	4·7647	4·7673	4·7699	3	5	8	10	13
41	4·7725	4·7750	4·7776	4·7802	4·7827	4·7853	4·7879	4·7904	4·7930	4·7955	3	5	8	10	13
42	4·7981	4·8007	4·8032	4·8058	4·8083	4·8109	4·8134	4·8160	4·8185	4·8211	3	5	8	10	13
43	4·8236	4·8262	4·8287	4·8313	4·8338	4·8363	4·8389	4·8414	4·8440	4·8465	3	5	8	10	13
44	4·8490	4·8516	4·8541	4·8566	4·8592	4·8617	4·8642	4·8668	4·8693	4·8718	3	5	8	10	13
45	4·8743	4·8769	4·8794	4·8819	4·8844	4·8870	4·8895	4·8920	4·8945	4·8970	3	5	8	10	13
46	4·8996	4·9021	4·9046	4·9071	4·9096	4·9122	4·9147	4·9172	4·9197	4·9222	3	5	8	10	13
47	4·9247	4·9272	4·9298	4·9323	4·9348	4·9373	4·9398	4·9423	4·9448	4·9473	3	5	8	10	13
48	4·9498	4·9524	4·9549	4·9574	4·9599	4·9624	4·9649	4·9674	4·9699	4·9724	3	5	8	10	13
49	4·9749	4·9774	4·9799	4·9825	4·9850	4·9875	4·9900	4·9925	4·9950	4·9975	3	5	8	10	13

Table VI is reproduced from Table IX of Fisher and Yates: *Statistical Tables for Biological, Agricultural and Medical Research.* Published by Longman Group Ltd., London. (Previously Published by Oliver and Boyd, Edinburgh), and by Permission of the Authors and Publishers.

TABLE VI: PROBITS—*continued*

	0·0	0·1	0·2	0·3	0·4	0·5	0·6	0·7	0·8	0·9	1	2	3	4	5
50	5·0000	5·0025	5·0050	5·0075	5·0100	5·0125	5·0150	5·0175	5·0201	5·0226	3	5	8	10	13
51	5·0251	5·0276	5·0301	5·0326	5·0351	5·0376	5·0401	5·0426	5·0451	5·0476	3	5	8	10	13
52	5·0502	5·0527	5·0552	5·0577	5·0602	5·0627	5·0652	5·0677	5·0702	5·0728	3	5	8	10	13
53	5·0753	5·0778	5·0803	5·0828	5·0853	5·0878	5·0904	5·0929	5·0954	5·0979	3	5	8	10	13
54	5·1004	5·1030	5·1055	5·1080	5·1105	5·1130	5·1156	5·1181	5·1206	5·1231	3	5	8	10	13
55	5·1257	5·1282	5·1307	5·1332	5·1358	5·1383	5·1408	5·1434	5·1459	5·1484	3	5	8	10	13
56	5·1510	5·1535	5·1560	5·1586	5·1611	5·1637	5·1662	5·1687	5·1713	5·1738	3	5	8	10	13
57	5·1764	5·1789	5·1815	5·1840	5·1866	5·1891	5·1917	5·1942	5·1968	5·1993	3	5	8	10	13
58	5·2019	5·2045	5·2070	5·2096	5·2121	5·2147	5·2173	5·2198	5·2224	5·2250	3	5	8	10	13
59	5·2275	5·2301	5·2327	5·2353	5·2378	5·2404	5·2430	5·2456	5·2482	5·2508	3	5	8	10	13
60	5·2533	5·2559	5·2585	5·2611	5·2637	5·2663	5·2689	5·2715	5·2741	5·2767	3	5	8	10	13
61	5·2793	5·2819	5·2845	5·2871	5·2898	5·2924	5·2950	5·2976	5·3002	5·3029	3	5	8	10	13
62	5·3055	5·3081	5·3107	5·3134	5·3160	5·3186	5·3213	5·3239	5·3266	5·3292	3	5	8	11	13
63	5·3319	5·3345	5·3372	5·3398	5·3425	5·3451	5·3478	5·3505	5·3531	5·3558	3	5	8	11	13
64	5·3585	5·3611	5·3638	5·3665	5·3692	5·3719	5·3745	5·3772	5·3799	5·3826	3	5	8	11	13
65	5·3853	5·3880	5·3907	5·3934	5·3961	5·3989	5·4016	5·4043	5·4070	5·4097	3	5	8	11	14
66	5·4125	5·4152	5·4179	5·4207	5·4234	5·4261	5·4289	5·4316	5·4344	5·4372	3	5	8	11	14
67	5·4399	5·4427	5·4454	5·4482	5·4510	5·4538	5·4565	5·4593	5·4621	5·4649	3	6	8	11	14
68	5·4677	5·4705	5·4733	5·4761	5·4789	5·4817	5·4845	5·4874	5·4902	5·4930	3	6	8	11	14
69	5·4959	5·4987	5·5015	5·5044	5·5072	5·5101	5·5129	5·5158	5·5187	5·5215	3	6	9	11	14
70	5·5244	5·5273	5·5302	5·5330	5·5359	5·5388	5·5417	5·5446	5·5476	5·5505	3	6	9	12	14
71	5·5534	5·5563	5·5592	5·5622	5·5651	5·5681	5·5710	5·5740	5·5769	5·5799	3	6	9	12	15
72	5·5828	5·5858	5·5888	5·5918	5·5948	5·5978	5·6008	5·6038	5·6068	5·6098	3	6	9	12	15
73	5·6128	5·6158	5·6189	5·6219	5·6250	5·6280	5·6311	5·6341	5·6372	5·6403	3	6	9	12	15
74	5·6433	5·6464	5·6495	5·6526	5·6557	5·6588	5·6620	5·6651	5·6682	5·6713	3	6	9	12	16

75	5·6745	5·6776	5·6808	5·6840	5·6871	5·6903	5·6935	5·6967	5·6999	5·7031	3	6	10	13	16
76	5·7063	5·7095	5·7128	5·7160	5·7192	5·7225	5·7257	5·7290	5·7323	5·7356	3	7	10	13	16
77	5·7388	5·7421	5·7454	5·7488	5·7521	5·7554	5·7588	5·7621	5·7655	5·7688	3	7	10	13	17
78	5·7722	5·7756	5·7790	5·7824	5·7858	5·7892	5·7926	5·7961	5·7995	5·8030	3	7	10	14	17
79	5·8064	5·8099	5·8134	5·8169	5·8204	5·8239	5·8274	5·8310	5·8345	5·8381	4	7	11	14	18
80	5·8416	5·8452	5·8488	5·8524	5·8560	5·8596	5·8633	5·8669	5·8705	5·8742	4	7	11	14	18
81	5·8779	5·8816	5·8853	5·8890	5·8927	5·8965	5·9002	5·9040	5·9078	5·9116	4	7	11	15	19
82	5·9154	5·9192	5·9230	5·9269	5·9307	5·9346	5·9385	5·9424	5·9463	5·9502	4	8	12	15	19
83	5·9542	5·9581	5·9621	5·9661	5·9701	5·9741	5·9782	5·9822	5·9863	5·9904	4	8	12	16	20
84	5·9945	5·9986	6·0027	6·0069	6·0110	6·0152	6·0194	6·0237	6·0279	6·0322	4	8	13	17	21
85	6·0364	6·0407	6·0450	6·0494	6·0537	6·0581	6·0625	6·0669	6·0714	6·0758	4	9	13	18	22
86	6·0803	6·0848	6·0893	6·0939	6·0985	6·1031	6·1077	6·1123	6·1170	6·1217	5	9	14	18	23
87	6·1264	6·1311	6·1359	6·1407	6·1455	6·1503	6·1552	6·1601	6·1650	6·1700	5	10	15	19	24
88	6·1750	6·1800	6·1850	6·1901	6·1952	6·2004	6·2055	6·2107	6·2160	6·2212	5	10	15	21	26
89	6·2265	6·2319	6·2372	6·2426	6·2481	6·2536	6·2591	6·2646	6·2702	6·2759	5	11	16	22	27
90	6·2816	6·2873	6·2930	6·2988	6·3047	6·3106	6·3165	6·3225	6·3285	6·3346	6	12	18	24	29
91	6·3408	6·3469	6·3532	6·3595	6·3658	6·3722	6·3787	6·3852	6·3917	6·3984	6	13	19	26	32
92	6·4051	6·4118	6·4187	6·4255	6·4325	6·4395	6·4466	6·4538	6·4611	6·4684	7	14	21	28	35
93	6·4758	6·4833	6·4909	6·4985	6·5063	6·5141	6·5220	6·5301	6·5382	6·5464	8	16	24	31	39
94	6·5548	6·5632	6·5718	6·5805	6·5893	6·5982	6·6072	6·6164	6·6258	6·6352	9	18	27	36	45
95	6·6449	6·6546	6·6646	6·6747	6·6849	6·6954	6·7060	6·7169	6·7279	6·7392					
	97	100	101	102	105	106	109	110	113	115					
96	6·7507	6·7624	6·7744	6·7866	6·7991	6·8119	6·8250	6·8384	6·8522	6·8663					
	117	120	122	125	128	131	134	138	141	145					
97	6·8808	6·8957	6·9110	6·9268	6·9431	6·9600	6·9774	6·9954	7·0141	7·0335					
	149	153	158	163	169	174	180	187	194	202					

Continued on next page.

Table VI is reproduced from Table IX of Fisher and Yates: *Statistical Tables for Biological, Agricultural and Medical Research*. Published by Longman Group Ltd., London (Previously Published by Oliver and Boyd, Edinburgh), and by Permission of the Authors and Publishers.

TABLE VI: PROBITS—*continued*

	0·00	0·01	0·02	0·03	0·04	0·05	0·06	0·07	0·08	0·09	1	2	3	4	5
98·0	7·0537	7·0558	7·0579	7·0600	7·0621	7·0642	7·0663	7·0684	7·0706	7·0727	2	4	6	8	11
98·1	7·0749	7·0770	7·0792	7·0814	7·0836	7·0858	7·0880	7·0902	7·0924	7·0947	2	4	7	9	11
98·2	7·0969	7·0992	7·1015	7·1038	7·1061	7·1084	7·1107	7·1130	7·1154	7·1177	2	5	7	9	12
98·3	7·1201	7·1224	7·1248	7·1272	7·1297	7·1321	7·1345	7·1370	7·1394	7·1419	2	5	7	10	12
98·4	7·1444	7·1469	7·1494	7·1520	7·1545	7·1571	7·1596	7·1622	7·1648	7·1675	3	5	8	10	13
98·5	7·1701	7·1727	7·1754	7·1781	7·1808	7·1835	7·1862	7·1890	7·1917	7·1945	3	5	8	11	14
98·6	7·1973	7·2001	7·2029	7·2058	7·2086	7·2115	7·2144	7·2173	7·2203	7·2232	3	6	9	12	14
98·7	7·2262	7·2292	7·2322	7·2353	7·2383	7·2414	7·2445	7·2476	7·2508	7·2539	3	6	9	12	15
98·8	7·2571	7·2603	7·2636	7·2668	7·2701	7·2734	7·2768	7·2801	7·2835	7·2869	3	7	10	13	17
98·9	7·2904	7·2938	7·2973	7·3009	7·3044	7·3080	7·3116	7·3152	7·3189	7·3226	4	7	11	14	18
99·0	7·3263	7·3301	7·3339	7·3378	7·3416	7·3455	7·3495	7·3535	7·3575	7·3615	4	8	12	16	20
99·1	7·3656	7·3698	7·3739	7·3781	7·3824	7·3867	7·3911	7·3954	7·3999	7·4044	4	9	13	17	22
99·2	7·4089	7·4135	7·4181	7·4228	7·4276	7·4324	7·4372	7·4422	7·4471	7·4522	5	10	14	19	24
99·3	7·4573	7·4624	7·4677	7·4730	7·4783	7·4838	7·4893	7·4949	7·5006	7·5063	5	11	16	22	27
99·4	7·5121	7·5181	7·5241	7·5302	7·5364	7·5427	7·5491	7·5556	7·5622	7·5690	6	13	19	25	32
99·5	7·5758	7·5828	7·5899	7·5972	7·6045	7·6121	7·6197	7·6276	7·6356	7·6437					
99·6	7·6521	7·6606	7·6693	7·6783	7·6874	7·6968	7·7065	7·7164	7·7266	7·7370					
99·7	7·7478	7·7589	7·7703	7·7822	7·7944	7·8070	7·8202	7·8338	7·8480	7·8627					
99·8	7·8782	7·8943	7·9112	7·9290	7·9478	7·9677	7·9889	8·0115	8·0357	8·0618					
99·9	8·0902	8·1214	8·1559	8·1947	8·2389	8·2905	8·3528	8·4316	8·5401	8·7190					

Table VI is reproduced from Table IX of Fisher and Yates: *Statistical Tables for Biological, Agricultural and Medical Research*, Published by Longman Group Ltd., London. (Previously Published by Oliver and Boyd, Edinburgh), and by Permission of the Authors and Publishers.

TABLE VII: PROBITS

Weighting Coefficients and Probit Values to be used for Final Adjustments
(Adapted from Bliss, 1935)

Expected Probit Y	Minimum Working Probit $Y-P/Z$	Range $1/Z$	Maximum working Probit $Y+Q/Z$	Weighting Coefficient Z^2/PQ	Expected Probit Y	Minimum Working Probit $Y-P/Z$	Range $1/Z$	Maximum working Probit $Y+Q/Z$	Weighting Coefficient Z^2/PQ
1·1	0·8579	5034	5035	·00082	5·0	3·7467	2·5066	6·2533	·63662
1·2	0·9522	3425	3426	·00118	5·1	3·7401	2·5192	6·2593	·63431
1·3	1·0462	2354	2355	·00167	5·2	3·7186	2·5573	6·2759	·62742
1·4	1·1400	1634	1635	·00235	5·3	3·6798	2·6220	6·3018	·61609
1·5	1·2335	1146	1147	·00327	5·4	3·6203	2·7154	6·3357	·60052
1·6	1·3266	811·5	812·8	·00451	5·5	3·5360	2·8404	6·3764	·58099
1·7	1·4194	580·5	581·9	·00614	5·6	3·4220	3·0010	6·4230	·55788
1·8	1·5118	419·4	420·9	·00828	5·7	3·2724	3·2025	6·4749	·53159
1·9	1·6038	306·1	307·7	·01104	5·8	3·0794	3·4519	6·5313	·50260
2·0	1·6954	225·6	227·3	·01457	5·9	2·8335	3·7582	6·5917	·47144
2·1	1·7866	168·00	169·79	·01903	6·0	2·5230	4·1327	6·6557	·43863
2·2	1·8772	126·34	128·22	·02459	6·1	2·1324	4·5903	6·7227	·40474
2·3	1·9673	95·96	97·93	·03143	6·2	1·6429	5·1497	6·7926	·37031
2·4	2·0568	73·62	75·68	·03977	6·3	1·0295	5·8354	6·8649	·33589
2·5	2·1457	57·05	59·20	·04979	6·4	0·2606	6·6788	6·9394	·30199
2·6	2·2340	44·654	46·888	·06169	6·5	−0·795	7·721	7·0158	·26907
2·7	2·3214	35·302	37·623	·07563	6·6	−1·921	9·015	7·0940	·23753
2·8	2·4081	28·189	30·597	·09179	6·7	−3·459	10·633	7·1739	·20774
2·9	2·4938	22·736	25·230	·11026	6·8	−5·411	12·666	7·2551	·17994
3·0	2·5786	18·522	21·101	·13112	6·9	−7·902	15·240	7·3376	·15436
3·1	2·6624	15·240	17·902	·15436	7·0	−11·101	18·522	7·4214	·13112
3·2	2·7449	12·666	15·411	·17994	7·1	−15·230	22·736	7·5062	·11026
3·3	2·8261	10·633	13·459	·20774	7·2	−20·597	28·189	7·5919	·09179
3·4	2·9060	9·015	11·921	·23753	7·3	−27·623	35·302	7·6786	·07564
3·5	2·9842	7·721	10·705	·26907	7·4	−36·888	44·654	7·7661	·06168
3·6	3·0606	6·6788	9·7394	·30199	7·5	−49·20	57·05	7·8543	·04979
3·7	3·1351	5·8354	8·9705	·33589	7·6	−65·68	73·62	7·9432	·03977
3·8	3·2074	5·1497	8·3571	·37031	7·7	−87·93	95·96	8·0327	·03143
3·9	3·2773	4·5903	7·8676	·40474	7·8	−118·22	126·34	8·1228	·02458
4·0	3·3443	4·1327	7·4770	·43863	7·9	−159·79	168·00	8·2134	·01903
4·1	3·4083	3·7582	7·1665	·47144	8·0	−217·3	225·6	8·3046	·01457
4·2	3·4687	3·4519	6·9206	·50260	8·1	−297·7	306·1	8·3962	·01104
4·3	3·5251	3·2025	6·7276	·53159	8·2	−410·9	419·4	8·4882	·00828
4·4	3·5770	3·0010	6·5780	·55788	8·3	−571·9	580·5	8·5806	·00614
4·5	3·6236	2·8404	6·4640	·58099	8·4	−802·8	811·5	8·6734	·00451
4·6	3·6643	2·7154	6·3797	·60052	8·5	−1137	1146	8·7666	·00327
4·7	3·6982	2·6220	6·3202	·61609	8·6	−1625	1634	8·8600	·00235
4·8	3·7241	2·5573	6·2814	·62741	8·7	−2345	2354	8·9538	·00167
4·9	3·7407	2·5192	6·2599	·63431	8·8	−3416	3425	9·0478	·00118
5·0	3·7467	2·5066	6·2533	·63662	8·9	−5025	5034	9·1421	·00082

Table VII is reproduced from Table IX₂, of Fisher and Yates: *Statistical Tables for Biological, Agricultural and Medical Research*, Published by Longman Group Ltd., London. (Previously Published by Oliver and Boyd, Edinburgh), and by Permission of the Authors and Publishers.

TABLE VIII: TABLE OF CRITICAL VALUES OF r IN THE RUNS TEST*

Given in the bodies of Table VIII(a) and Table VIII(b) are various critical values of r for various values of n_1 and n_2. For the one-sample runs test, any values of r which is equal to or smaller than that shown in Table VIII(a) or equal to or larger than that shown in Table VIII(b) is significant at the .05 level.

Table VIII(a)

n_1 \ n_2	2	3	4	5	6	7	8	9	10	11	12	13	14	15	16	17	18	19	20
2											2	2	2	2	2	2	2	2	2
3					2	2	2	2	2	2	2	2	2	3	3	3	3	3	3
4			2	2	2	3	3	3	3	3	3	3	3	3	4	4	4	4	4
5			2	2	3	3	3	3	3	4	4	4	4	4	4	4	5	5	5
6		2	2	3	3	3	3	4	4	4	4	5	5	5	5	5	5	6	6
7		2	2	3	3	3	4	4	5	5	5	5	5	6	6	6	6	6	6
8		2	3	3	3	4	4	5	5	5	6	6	6	6	6	7	7	7	7
9		2	3	3	4	4	5	5	5	6	6	6	7	7	7	7	8	8	8
10		2	3	3	4	5	5	5	6	6	7	7	7	7	8	8	8	8	9
11		2	3	4	4	5	5	6	6	7	7	7	8	8	8	9	9	9	9
12	2	2	3	4	4	5	6	6	7	7	7	8	8	8	9	9	9	10	10
13	2	2	3	4	5	5	6	6	7	7	8	8	9	9	9	10	10	10	10
14	2	2	3	4	5	5	6	7	7	8	8	9	9	9	10	10	10	11	11
15	2	3	3	4	5	6	6	7	7	8	8	9	9	10	10	11	11	11	12
16	2	3	4	4	5	6	6	7	8	8	9	9	10	10	11	11	11	12	12
17	2	3	4	4	5	6	7	7	8	9	9	10	10	11	11	11	12	12	13
18	2	3	4	5	5	6	7	8	8	9	9	10	10	11	11	12	12	13	13
19	2	3	4	5	6	6	7	8	8	9	10	10	11	11	12	12	13	13	13
20	2	3	4	5	6	6	7	8	9	9	10	10	11	12	12	13	13	13	14

* Adapted from Swed, Frieda S., and Eisenhart, C. 1943. Tables for testing randomness of grouping in a sequence of alternatives. *Ann. Math. Statist.*, 14, 83–86, with the kind permission of the authors and publisher.

TABLE VIII: TABLE OF CRITICAL VALUES OF r IN THE RUNS TEST*
(Continued)

Table VIII(b)

n_1 \ n_2	2	3	4	5	6	7	8	9	10	11	12	13	14	15	16	17	18	19	20
2																			
3																			
4				9	9														
5			9	10	10	11	11												
6			9	10	11	12	12	13	13	13	13								
7				11	12	13	13	14	14	14	14	15	15	15					
8				11	12	13	14	14	15	15	16	16	16	16	17	17	17	17	17
9					13	14	14	15	16	16	16	17	17	18	18	18	18	18	18
10					13	14	15	16	16	17	17	18	18	18	19	19	19	20	20
11					13	14	15	16	17	17	18	19	19	19	20	20	20	21	21
12					13	14	16	16	17	18	19	19	20	20	21	21	21	22	22
13						15	16	17	18	19	19	20	20	21	21	22	22	23	23
14						15	16	17	18	19	20	20	21	22	22	23	23	23	24
15						15	16	18	18	19	20	21	22	22	23	23	24	24	25
16							17	18	19	20	21	21	22	23	23	24	25	25	25
17							17	18	19	20	21	22	23	23	24	25	25	26	26
18							17	18	19	20	21	22	23	24	25	25	26	26	27
19							17	18	20	21	22	23	23	24	25	26	26	27	27
20							17	18	20	21	22	23	24	25	25	26	27	27	28

* Adapted from Swed, Frieda S., and Eisenhart, C. 1943. Tables for testing randomness of grouping in a sequence of alternatives. *Ann. Math. Statist.*, **14**, 83–86, with the kind permission of the authors and publisher.

TABLE IX: TABLE OF CRITICAL VALUES OF d (OR c) IN THE FISHER TEST*,†

Totals in right margin		b (or a)†	Level of significance			
			.05	.025	.01	.005
$R_1 = 3$	$R_2 = 3$	3	0	—	—	—
$R_1 = 4$	$R_2 = 4$	4	0	0	—	—
	$R_2 = 3$	4	0	—	—	—
$R_1 = 5$	$R_2 = 5$	5	1	1	0	0
		4	0	0	—	—
	$R_2 = 4$	5	1	0	0	—
		4	0	—	—	—
	$R_2 = 3$	5	0	0	—	—
	$= 2$	5	0	—	—	—
$R_1 = 6$	$R_2 = 6$	6	2	1	1	0
		5	1	0	0	—
		4	0	—	—	—
	$R_2 = 5$	6	1	0	0	0
		5	0	0	—	—
		4	0	—	—	—
	$R_2 = 4$	6	1	0	0	0
		5	0	0	—	—
	$R_2 = 3$	6	0	0	—	—
		5	0	—	—	—
	$R_2 = 2$	6	0	—	—	—
$R_1 = 7$	$R_2 = 7$	7	3	2	1	1
		6	1	1	0	0
		5	0	0	—	—
		4	0	—	—	—
	$R_2 = 6$	7	2	2	1	1
		6	1	0	0	0
		5	0	0	—	—
		4	0	—	—	—
	$R_2 = 5$	7	2	1	0	0
		6	1	0	0	—
		5	0	—	—	—
	$R_2 = 4$	7	1	1	0	0
		6	0	0	—	—
		5	0	—	—	—
	$R_2 = 3$	7	0	0	0	—
		6	0	—	—	—
	$R_2 = 2$	7	0	—	—	—

* Adapted from Finney, D. J. 1948. The Fisher-Yates test of significance in 2×2 contingency tables. *Biometrika*, **35**, 149–154, with the kind permission of the author and the publisher.

TABLE IX: TABLE OF CRITICAL VALUES OF d (OR c) IN THE FISHER TEST*,† (*Continued*)

Totals in right margin		b (or a)†	Level of significance			
			.05	.025	.01	.005
$R_1 = 8$	$R_2 = 8$	8	4	3	2	2
		7	2	2	1	0
		6	1	1	0	0
		5	0	0	—	—
		4	0	—	—	—
	$R_2 = 7$	8	3	2	2	1
		7	2	1	1	0
		6	1	0	0	—
		5	0	0	—	—
	$R_2 = 6$	8	2	2	1	1
		7	1	1	0	0
		6	0	0	0	—
		5	0	—	—	—
	$R_2 = 5$	8	2	1	1	0
		7	1	0	0	0
		6	0	0	—	—
		5	0	—	—	—
	$R_2 = 4$	8	1	1	0	0
		7	0	0	—	—
		6	0	—	—	—
	$R_2 = 3$	8	0	0	0	—
		7	0	0	—	—
	$R_2 = 2$	8	0	0	—	—
$R_1 = 9$	$R_2 = 9$	9	5	4	3	3
		8	3	3	2	1
		7	2	1	1	0
		6	1	1	0	0
		5	0	0	—	—
		4	0	—	—	—
	$R_2 = 8$	9	4	3	3	2
		8	3	2	1	1
		7	2	1	0	0
		6	1	0	0	—
		5	0	0	—	—
	$R_2 = 7$	9	3	3	2	2
		8	2	2	1	0
		7	1	1	0	0
		6	0	0	—	—
		5	0	—	—	—

†When b is entered in the middle column, the significance levels are for d. When a is used in place of b, the significance levels are for c.

TABLE IX: TABLE OF CRITICAL VALUES OF d (OR c) IN THE FISHER TEST*,† (*Continued*)

Totals in right margin		b (or a)†	Level of significance			
			.05	.025	.01	.005
$R_1 = 9$	$R_2 = 6$	9	3	2	1	1
		8	2	1	0	0
		7	1	0	0	—
		6	0	0	—	—
		5	0	—	—	—
	$R_2 = 5$	9	2	1	1	1
		8	1	1	0	0
		7	0	0	—	—
		6	0	—	—	—
	$R_2 = 4$	9	1	1	0	0
		8	0	0	0	—
		7	0	0	—	—
		6	0	—	—	—
	$R_2 = 3$	9	1	0	0	0
		8	0	0	—	—
		7	0	—	—	—
	$R_2 = 2$	9	0	0	—	—
$R_1 = 10$	$R_2 = 10$	10	6	5	4	3
		9	4	3	3	2
		8	3	2	1	1
		7	2	1	1	0
		6	1	0	0	—
		5	0	0	—	—
		4	0	—	—	—
	$R_2 = 9$	10	5	4	3	3
		9	4	3	2	2
		8	2	2	1	1
		7	1	1	0	0
		6	1	0	0	—
		5	0	0	—	—
	$R_2 = 8$	10	4	4	3	2
		9	3	2	2	1
		8	2	1	1	0
		7	1	1	0	0
		6	0	0	—	—
		5	0	—	—	—
	$R_2 = 7$	10	3	3	2	2
		9	2	2	1	1
		8	1	1	0	0
		7	1	0	0	—
		6	0	0	—	—
		5	0	—	—	—

* Adapted from Finney, D. J. 1948. The Fisher-Yates test of significance in < 2 contingency tables. *Biometrika*, **35**, 149–154, with the kind permission of the author and the publisher.

TABLE IX: TABLE OF CRITICAL VALUES OF d (OR c) IN THE FISHER TEST*,† (*Continued*)

Totals in right margin		b (or a)†	Level of significance			
			.05	.025	.01	.005
$R_1 = 10$	$R_2 = 6$	10	3	2	2	1
		9	2	1	1	0
		8	1	1	0	0
		7	0	0	—	—
		6	0	—	—	—
	$R_2 = 5$	10	2	2	1	1
		9	1	1	0	0
		8	1	0	0	—
		7	0	0	—	—
		6	0	—	—	—
	$R_2 = 4$	10	1	1	0	0
		9	1	0	0	0
		8	0	0	—	—
		7	0	—	—	—
	$R_2 = 3$	10	1	0	0	0
		9	0	0	—	—
		8	0	—	—	—
	$R_2 = 2$	10	0	0	—	—
		9	0	—	—	—
$R_1 = 11$	$R_2 = 11$	11	7	6	5	4
		10	5	4	3	3
		9	4	3	2	2
		8	3	2	1	1
		7	2	1	0	0
		6	1	0	0	—
		5	0	0	—	—
		4	0	—	—	—
	$R_2 = 10$	11	6	5	4	4
		10	4	4	3	2
		9	3	3	2	1
		8	2	2	1	0
		7	1	1	0	0
		6	1	0	0	—
		5	0	—	—	—
	$R_2 = 9$	11	5	4	4	3
		10	4	3	2	2
		9	3	2	1	1
		8	2	1	1	0
		7	1	1	0	0
		6	0	0	—	—
		5	0	—	—	—

† When b is entered in the middle column, the significance levels are for d. When a is used in place of b, the significance levels are for c.

TABLE IX: TABLE OF CRITICAL VALUES OF d (OR c) IN THE FISHER TEST*,† (*Continued*)

Totals in right margin		b (or a)†	Level of significance			
			.05	.025	.01	.005
$R_1 = 11$	$R_2 = 8$	11	4	4	3	3
		10	3	3	2	1
		9	2	2	1	1
		8	1	1	0	0
		7	1	0	0	—
		6	0	0	—	—
		5	0	—	—	—
	$R_2 = 7$	11	4	3	2	2
		10	3	2	1	1
		9	2	1	1	0
		8	1	1	0	0
		7	0	0	—	—
		6	0	0	—	—
	$R_2 = 6$	11	3	2	2	1
		10	2	1	1	0
		9	1	1	0	0
		8	1	0	0	—
		7	0	0	—	—
		6	0	—	—	—
	$R_2 = 5$	11	2	2	1	1
		10	1	1	0	0
		9	1	0	0	0
		8	0	0	—	—
		7	0	—	—	—
	$R_2 = 4$	11	1	1	1	0
		10	1	0	0	0
		9	0	0	—	—
		8	0	—	—	—
	$R_2 = 3$	11	1	0	0	0
		10	0	0	—	—
		9	0	—	—	—
	$R_2 = 2$	11	0	0	—	—
		10	0	—	—	—
$R_1 = 12$	$R_2 = 12$	12	8	7	6	5
		11	6	5	4	4
		10	5	4	3	2
		9	4	3	2	1
		8	3	2	1	1
		7	2	1	0	0
		6	1	0	0	—
		5	0	0	—	—
		4	0	—	—	—

* Adapted from Finney, D. J. 1948. The Fisher-Yates test of significance in 2 × 2 contingency tables. *Biometrika*, **35**, 149–154, with the kind permission of the author and the publisher.

TABLE IX: TABLE OF CRITICAL VALUES OF d (OR c) IN THE FISHER TEST*,† (*Continued*)

Totals in right margin		b (or a)†	Level of significance			
			.05	.025	.01	.005
$R_1 = 12$	$R_2 = 11$	12	7	6	5	5
		11	5	5	4	3
		10	4	3	2	2
		9	3	2	2	1
		8	2	1	1	0
		7	1	1	0	0
		6	1	0	0	—
		5	0	0	—	—
	$R_2 = 10$	12	6	5	5	4
		11	5	4	3	3
		10	4	3	2	2
		9	3	2	1	1
		8	2	1	0	0
		7	1	0	0	0
		3	0	0	—	—
		5	0	—	—	—
	$R_2 = 9$	12	5	5	4	3
		11	4	3	3	2
		10	3	2	2	1
		9	2	2	1	0
		8	1	1	0	0
		7	1	0	0	—
		6	0	0	—	—
		5	0	—	—	—
	$R_2 = 8$	12	5	4	3	3
		11	3	3	2	2
		10	2	2	1	1
		9	2	1	1	0
		8	1	1	0	0
		7	0	0	—	—
		6	0	0	—	—
	$R_2 = 7$	12	4	3	3	2
		11	3	2	2	1
		10	2	1	1	0
		9	1	1	0	0
		8	1	0	0	—
		7	0	0	—	—
		6	0	—	—	—

When b is entered in the middle column, the significance levels are for d. When a is used in place of b, the significance levels are for c.

TABLE IX: TABLE OF CRITICAL VALUES OF d (OR c) IN THE
FISHER TEST*,† (*Continued*)

Totals in right margin		b (or a)†	Level of significance			
			.05	.025	.01	.005
$R_1 = 12$	$R_2 = 6$	12	3	3	2	2
		11	2	2	1	1
		10	1	1	0	0
		9	1	0	0	0
		8	0	0	—	—
		7	0	0	—	—
		6	0	—	—	—
	$R_2 = 5$	12	2	2	1	1
		11	1	1	1	0
		10	1	0	0	0
		9	0	0	0	—
		8	0	0	—	—
		7	0	—	—	—
	$R_2 = 4$	12	2	1	1	0
		11	1	0	0	0
		10	0	0	0	—
		9	0	0	—	—
		8	0	—	—	—
	$R_2 = 3$	12	1	0	0	0
		11	0	0	0	—
		10	0	0	—	—
		9	0	—	—	—
	$R_2 = 2$	12	0	0	—	—
		11	0	—	—	—
$R_1 = 13$	$R_2 = 13$	13	9	8	7	6
		12	7	6	5	4
		11	6	5	4	3
		10	4	4	3	2
		9	3	3	2	1
		8	2	2	1	0
		7	2	1	0	0
		6	1	0	0	—
		5	0	0	—	—
		4	0	—	—	—
	$R_2 = 12$	13	8	7	6	5
		12	6	5	5	4
		11	5	4	3	3
		10	4	3	2	2
		9	3	2	1	1
		8	2	1	1	0
		7	1	1	0	0
		6	1	0	0	—
		5	0	0	—	—

* Adapted from Finney, D. J. 1948. The Fisher-Yates test of significance in 2×2 contingency tables. *Biometrika*, **35**, 149–154, with the kind permission of the author and the publisher.

TABLE IX: TABLE OF CRITICAL VALUES OF d (OR c) IN THE FISHER TEST*,† (*Continued*)

Totals in right margin		b (or a)†	Level of significance			
			.05	.025	.01	.005
$R_1 = 13$	$R_2 = 11$	13	7	6	5	5
		12	6	5	4	3
		11	4	4	3	2
		10	3	3	2	1
		9	3	2	1	1
		8	2	1	0	0
		7	1	0	0	0
		6	0	0	—	—
		5	0	—	—	—
	$R_2 = 10$	13	6	6	5	4
		12	5	4	3	3
		11	4	3	2	2
		10	3	2	1	1
		9	2	1	1	0
		8	1	1	0	0
		7	1	0	0	—
		6	0	0	—	—
		5	0	—	—	—
	$R_2 = 9$	13	5	5	4	4
		12	4	4	3	2
		11	3	3	2	1
		10	2	2	1	1
		9	2	1	0	0
		8	1	1	0	0
		7	0	0	—	—
		6	0	0	—	—
		5	0	—	—	—
	$R_2 = 8$	13	5	4	3	3
		12	4	3	2	2
		11	3	2	1	1
		10	2	1	1	0
		9	1	1	0	0
		8	1	0	0	—
		7	0	0	—	—
		6	0	—	—	—
	$R_2 = 7$	13	4	3	3	2
		12	3	2	2	1
		11	2	2	1	1
		10	1	1	0	0
		9	1	0	0	0
		8	0	0	—	—
		7	0	0	—	—
		6	0	—	—	—

† When b is entered in the middle column, the significance levels are for d. When a is used in place of b, the significance levels are for c.

Table IX: Table of Critical Values of d (or c) in the Fisher Test*,† (*Continued*)

Totals in right margin		b (or a)†	Level of significance			
			.05	.025	.01	.005
$R_1 = 13$	$R_2 = 6$	13	3	3	2	2
		12	2	2	1	1
		11	2	1	1	0
		10	1	1	0	0
		9	1	0	0	—
		8	0	0	—	—
		7	0	—	—	—
	$R_2 = 5$	13	2	2	1	1
		12	2	1	1	0
		11	1	1	0	0
		10	1	0	0	—
		9	0	0	—	—
		8	0	—	—	—
	$R_2 = 4$	13	2	1	1	0
		12	1	1	0	0
		11	0	0	0	—
		10	0	0	—	—
		9	0	—	—	—
	$R_2 = 3$	13	1	1	0	0
		12	0	0	0	—
		11	0	0	—	—
		10	0	—	—	—
	$R_2 = 2$	13	0	0	0	—
		12	0	—	—	—
$R_1 = 14$	$R_2 = 14$	14	10	9	8	7
		13	8	7	6	5
		12	6	6	5	4
		11	5	4	3	3
		10	4	3	2	2
		9	3	2	2	1
		8	2	2	1	0
		7	1	1	0	0
		6	1	0	0	—
		5	0	0	—	—
		4	0	—	—	—

* Adapted from Finney, D. J. 1948. The Fisher-Yates test of significance in 2×2 contingency tables. *Biometrika*, **35**, 149–154, with the kind permission of the author and the publisher.

TABLE IX: TABLE OF CRITICAL VALUES OF d (OR c) IN THE FISHER TEST*,† (Continued)

Totals in right margin		b (or a)†	Level of significance			
			.05	.025	.01	.005
$R_1 = 14$	$R_2 = 13$	14	9	8	7	6
		13	7	6	5	5
		12	6	5	4	3
		11	5	4	3	2
		10	4	3	2	2
		9	3	2	1	1
		8	2	1	1	0
		7	1	1	0	0
		6	1	0	—	—
		5	0	0	—	—
	$R_2 = 12$	14	8	7	6	6
		13	6	6	5	4
		12	5	4	4	3
		11	4	3	3	2
		10	3	3	2	1
		9	2	2	1	1
		8	2	1	0	0
		7	1	0	0	—
		6	0	0	—	—
		5	0	—	—	—
	$R_2 = 11$	14	7	6	6	5
		13	6	5	4	4
		12	5	4	3	3
		11	4	3	2	2
		10	3	2	1	1
		9	2	1	1	0
		8	1	1	0	0
		7	1	0	0	—
		6	0	0	—	—
		5	0	—	—	—
	$R_2 = 10$	14	6	6	5	4
		13	5	4	4	3
		12	4	3	3	2
		11	3	3	2	1
		10	2	2	1	1
		9	2	1	0	0
		8	1	1	0	0
		7	0	0	0	—
		6	0	0	—	—
		5	0	—	—	—

† When b is entered in the middle column, the significance levels are for d. When a is used in place of b, the significance levels are for c.

TABLE IX: TABLE OF CRITICAL VALUES OF d (OR c) IN THE FISHER TEST*,† (Continued)

Totals in right margin		b (or a)†	Level of significance			
			.05	.025	.01	.005
$R_1 = 14$	$R_2 = 9$	14	6	5	4	4
		13	4	4	3	3
		12	3	3	2	2
		11	3	2	1	1
		10	2	1	1	0
		9	1	1	0	0
		8	1	0	0	—
		7	0	0	—	—
		6	0	—	—	—
	$R_2 = 8$	14	5	4	4	3
		13	4	3	2	2
		12	3	2	2	1
		11	2	2	1	1
		10	2	1	0	0
		9	1	0	0	0
		8	0	0	0	—
		7	0	0	—	—
		6	0	—	—	—
	$R_2 = 7$	14	4	3	3	2
		13	3	2	2	1
		12	2	2	1	1
		11	2	1	1	0
		10	1	1	0	0
		9	1	0	0	—
		8	0	0	—	—
		7	0	—	—	—
	$R_2 = 6$	14	3	3	2	2
		13	2	2	1	1
		12	2	1	1	0
		11	1	1	0	0
		10	1	0	0	—
		9	0	0	—	—
		8	0	0	—	—
		7	0	—	—	—
	$R_2 = 5$	14	2	2	1	1
		13	2	1	1	0
		12	1	1	0	0
		11	1	0	0	0
		10	0	0	—	—
		9	0	0	—	—
		8	0	—	—	—

* Adapted from Finney, D. J. 1948. The Fisher-Yates test of significance in 2×2 contingency tables. *Biometrika*, **35**, 149–154, with the kind permission of the author and the publisher.

TABLE IX: TABLE OF CRITICAL VALUES OF d (OR c) IN THE FISHER TEST*,† (*Continued*)

Totals in right margin		b (or a)†	Level of significance			
			.05	.025	.01	.005
$R_1 = 14$	$R_2 = 4$	14	2	1	1	1
		13	1	1	0	0
		12	1	0	0	0
		11	0	0	—	—
		10	0	0	—	—
		9	0	—	—	—
	$R_2 = 3$	14	1	1	0	0
		13	0	0	0	—
		12	0	0	—	—
		11	0	—	—	—
	$R_2 = 2$	14	0	0	0	—
		13	0	0	—	—
		12	0	—	—	—
$R_1 = 15$	$R_2 = 15$	15	11	10	9	8
		14	9	8	7	6
		13	7	6	5	5
		12	6	5	4	4
		11	5	4	3	3
		10	4	3	2	2
		9	3	2	1	1
		8	2	1	1	0
		7	1	1	0	0
		6	1	0	0	—
		5	0	0	—	—
		4	0	—	—	—
	$R_2 = 14$	15	10	9	8	7
		14	8	7	6	6
		13	7	6	5	4
		12	6	5	4	3
		11	5	4	3	2
		10	4	3	2	1
		9	3	2	1	1
		8	2	1	1	0
		7	1	1	0	0
		6	1	0	—	—
		5	0	—	—	—

† When b is entered in the middle column, the significance levels are for d.   When A is used in place of b, the significance levels are for c.

TABLE IX: TABLE OF CRITICAL VALUES OF d (OR c) IN THE FISHER TEST*,† (*Continued*)

Totals in right margin		b (or a)†	Level of significance			
			.05	.025	.01	.005
$R_1 = 15$	$R_2 = 13$	15	9	8	7	7
		14	7	7	6	5
		13	6	5	4	4
		12	5	4	3	3
		11	4	3	2	2
		10	3	2	2	1
		9	2	2	1	0
		8	2	1	0	0
		7	1	0	0	—
		6	0	0	—	—
		5	0	—	—	—
	$R_2 = 12$	15	8	7	7	6
		14	7	6	5	4
		13	6	5	4	3
		12	5	4	3	2
		11	4	3	2	2
		10	3	2	1	1
		9	2	1	1	0
		8	1	1	0	0
		7	1	0	0	—
		6	0	0	—	—
		5	0	—	—	—
	$R_2 = 11$	15	7	7	6	5
		14	6	5	4	4
		13	5	4	3	3
		12	4	3	2	2
		11	3	2	2	1
		10	2	2	1	1
		9	2	1	0	0
		8	1	1	0	0
		7	1	0	0	—
		6	0	0	—	—
		5	0	—	—	—
	$R_2 = 10$	15	6	6	5	5
		14	5	5	4	3
		13	4	4	3	2
		12	3	3	2	2
		11	3	2	1	1
		10	2	1	1	0
		9	1	1	0	0
		8	1	0	0	—
		7	0	0	—	—
		6	0	—	—	—

* Adapted from Finney, D. J. 1948. The Fisher-Yates test of significance in 2 × 2 contingency tables. *Biometrika*, **35**, 149–154, with the kind permission of the author and the publisher.

TABLE IX: TABLE OF CRITICAL VALUES OF d (OR c) IN THE
FISHER TEST*,† (*Continued*)

Totals in right margin		b (or a)†	Level of significance			
			.05	.025	.01	.005
$R_1 = 15$	$R_2 = 9$	15	6	5	4	4
		14	5	4	3	3
		13	4	3	2	2
		12	3	2	2	1
		11	2	2	1	1
		10	2	1	0	0
		9	1	1	0	0
		8	1	0	0	—
		7	0	0	—	—
		6	0	—	—	—
	$R_2 = 8$	15	5	4	4	3
		14	4	3	3	2
		13	3	2	2	1
		12	2	2	1	1
		11	2	1	1	0
		10	1	1	0	0
		9	1	0	0	—
		8	0	0	—	—
		7	0	—	—	—
		6	0	—	—	—
	$R_2 = 7$	15	4	4	3	3
		14	3	3	2	2
		13	2	2	1	1
		12	2	1	1	0
		11	1	1	0	0
		10	1	0	0	0
		9	0	0	—	—
		8	0	0	—	—
		7	0	—	—	—
	$R_2 = 6$	15	3	3	2	2
		14	2	2	1	1
		13	2	1	1	0
		12	1	1	0	0
		11	1	0	0	0
		10	0	0	0	—
		9	0	0	—	—
		8	0	—	—	—
	$R_2 = 5$	15	2	2	2	1
		14	2	1	1	1
		13	1	1	0	0
		12	1	0	0	0
		11	0	0	0	—
		10	0	0	—	—
		9	0	—	—	—

† When b is entered in the middle column, the significance levels are for d. When
A is used in place of b, the significance levels are for c.

TABLE IX: TABLE OF CRITICAL VALUES OF d (OR c) IN THE
FISHER TEST*,† (*Continued*)

Totals in right margin		b (or a)†	Level of significance			
			.05	.025	.01	.005
$R_1 = 15$	$R_2 = 4$	15	2	1	1	1
		14	1	1	0	0
		13	1	0	0	0
		12	0	0	0	—
		11	0	0	—	—
		10	0	—	—	—
	$R_2 = 3$	15	1	1	0	0
		14	0	0	0	0
		13	0	0	—	—
		12	0	0	—	—
		11	0	—	—	—
	$R_2 = 2$	15	0	0	0	—
		14	0	0	—	—
		13	0	—	—	—

* Adapted from Finney, D. J.   1948.   The Fisher-Yates test of significance in 2×2 contingency tables. *Biometrika*, **35**, 149–154, with the kind permission of the author and the publisher.

† When b is entered in the middle column, the significance levels are for d.   When a is used in place of b, the significance levels are for

TABLE X. TABLE OF PROBABILITIES ASSOCIATED WITH VALUES AS SMALL AS OBSERVED VALUES OF W IN THE MANN-WHITNEY TEST[*]

$n_2 = 3$

W \ n_1	1	2	3
0	.250	.100	.050
1	.500	.200	.100
2	.750	.400	.200
3		.600	.350
4			.500
5			.650

$n_2 = 4$

W \ n_1	1	2	3	4
0	.200	.067	.028	.014
1	.400	.133	.057	.029
2	.600	.267	.114	.057
3		.400	.200	.100
4		.600	.314	.171
5			.429	.243
6			.571	.343
7				.443
8				.557

$n_2 = 5$

W \ n_1	1	2	3	4	5
0	.167	.047	.018	.008	.004
1	.333	.095	.036	.016	.008
2	.500	.190	.071	.032	.016
3	.667	.286	.125	.056	.028
4		.429	.196	.095	.048
5		.571	.286	.143	.075
6			.393	.206	.111
7			.500	.278	.155
8			.607	.365	.210
9				.452	.274
10				.548	.345
11					.421
12					.500
13					.579

$n_2 = 6$

W \ n_1	1	2	3	4	5	6
0	.143	.036	.012	.005	.002	.001
1	.286	.071	.024	.010	.004	.002
2	.428	.143	.048	.019	.009	.004
3	.571	.214	.083	.033	.015	.008
4		.321	.131	.057	.026	.013
5		.429	.190	.086	.041	.021
6		.571	.274	.129	.063	.032
7			.357	.176	.089	.047
8			.452	.238	.123	.066
9			.548	.305	.165	.090
10				.381	.214	.120
11				.457	.268	.155
12				.545	.331	.197
13					.396	.242
14					.465	.294
15					.535	.350
16						.409
17						.469
18						.531

[*] Reproduced from Mann, H. B., and Whitney, D. R. 1947. On a test of whether one of two random variables is stochastically larger than the other. *Ann. Math. Statist.*, **18**, 52–54, with the kind permission of the authors and the publisher.

TABLE X. TABLE OF PROBABILITIES ASSOCIATED WITH VALUES AS SMALL AS OBSERVED VALUES OF W IN THE MANN-WHITNEY TEST* (*Continued*)

$$n_2 = 7$$

n_1	1	2	3	4	5	6	7
0	.125	.028	.008	.003	.001	.001	.000
1	.250	.056	.017	.006	.003	.001	.001
2	.375	.111	.033	.012	.005	.002	.001
3	.500	.167	.058	.021	.009	.004	.002
4	.625	.250	.092	.036	.015	.007	.003
5		.333	.133	.055	.024	.011	.006
6		.444	.192	.082	.037	.017	.009
7		.556	.258	.115	.053	.026	.013
8			.333	.158	.074	.037	.019
9			.417	.206	.101	051	.027
10			.500	.264	.134	069	.036
11			.583	.324	.172	.090	.049
12				.394	.216	.117	.064
13				.464	.265	.147	.082
14				.538	.319	.183	.104
15					.378	.223	.130
16					.438	.267	.159
17					.500	.314	.191
18					.562	.365	.228
19						.418	.267
20						.473	.310
21						.527	.355
22							.402
23							.451
24							.500
25							.549

* Reproduced from Mann, H. B., and Whitney, D. R.   1947.   On a test of whether one of two random variables is stochastically larger than the other. *Ann. Math. Statist.*, **18**, 52–54, with the kind permission of the authors and the publisher.

INDEX

Q